Territory, Identity and Spatial Planning

Planning is essential to delivering environmental sustainability, social inclusion, improved and integrated public transport, economic development and urban regeneration, and must be analysed in relation to other aspects of New Labour's regional project, not least the government's concern with business competitiveness. At another level, the current reforms which privilege regional scale policy interventions will inevitably require changes in the divisions of powers and responsibilities at local and national levels. In other words, devolution involves a major rescaling of both spatial planning and development, which is unfolding rapidly and unevenly across Britain.

This book addresses these issues, highlighting important case studies and drawing on the experience of a team of eminent contributors each with a different perspective on territory, identity and space in a devolved UK.

Mark Tewdwr-Jones is Professor of Spatial Planning and Governance and Director of Research at University College London's Bartlett School of Planning. His research interests are related to planning, politics, community planning, and urban and regional development. He is currently involved in a number of research projects, including work across the European Union relating to European spatial planning, the scope and relevance of local planning, and the relationship between architecture, cities, and identity.

Philip Allmendinger is Professor of Planning and Director of the Centre of Planning Studies at the University of Reading. He is a chartered planner and surveyor with interests in theory, politics, regional planning and development. He is currently undertaking a number of research projects including work on resourcing planning and an ESRC funded project on integrated spatial planning, multi-level governance and state rescaling.

Territory, Identity and Spatial Planning

Spatial governance in a
fragmented nation

Edited by Mark Tewdwr-Jones
and Philip Allmendinger

Routledge
Taylor & Francis Group

LONDON AND NEW YORK

First published 2006 by Routledge
2 Park Square, Milton Park, Abingdon, Oxon OX14 4RN

Simultaneously published in the USA and Canada
by Routledge
270 Madison Avenue, New York, NY 10016

Routledge is an imprint of the Taylor & Francis Group, an informa business

© 2006 Mark Tewdwr-Jones and Philip Allmendinger for selection and
editorial material; individual chapters, the contributors

Typeset in Goudy by
GreenGate Publishing Services, Tonbridge
Printed and bound in Great Britain by
The Cromwell Press, Trowbridge, Wiltshire

British Library Cataloguing in Publication Data
A catalogue record for this book is available from the British Library

Library of Congress Cataloging in Publication Data
Tewdwr-Jones, Mark.
Territory identity and spatial planning : spatial governance in a fragmented nation / by Mark Tewdwr-Jones and Philip Allmendinger–1st ed.
        p.cm.
    Includes index.
    ISBN 0-415-36034-X (hb : alk. paper)–ISBN 0-415-36035-8
(pb : alk. paper) 1. Regional planning–Great Britain.
    2. Decentralization in government–Great Britain. 3. Space in
economics. I. Allmendinger, Philip, 1968 - II. Title.

HT395.G7T456 2006
307.1'20941–dc22

                                                                2005026791

ISBN10: 0-415-36034-X (hbk)
ISBN10: 0-415-36035-8 (pbk)
ISBN10: 0-203-00800-6 (ebk)

ISBN 13: 978-0-415-36034-0 (hbk)
ISBN 13: 978-0-415-36035-7 (pbk)
ISBN 13: 978-0-203-00800-3 (ebk)

Contents

Illustrations

Boxes

Contributors

Philip Allmendinger is Professor of Planning and Director of the Centre of Planning Studies at the University of Reading. He is currently working on an ESRC funded project with Graham Haughton and Dave Counsell (Hull) and Geoff Vigar (Newcastle) that explores how spatial planning systems are evolving in different ways across the UK.

Mark Baker is Senior Lecturer in Planning Policy & Practice within the School of Environment and Development at Manchester University. He is a chartered town planner with previous professional planning experience in UK local and central government. His teaching and research interests focus on the operation of the UK planning system and, especially, regional and strategic planning.

Nick Clifton is Senior Research Associate at the Centre for Advanced Studies (CASS), Cardiff University. He has published in the fields of regional economics, networks and innovation. Nick has recently carried out research for the Economic and Social Research Council (ESRC), the WDA, and the National Assembly for Wales.

Philip Cooke is University Research Professor and founding Director (1993) of the Centre for Advanced Studies, University of Wales, Cardiff. His research interests lie in studies of Economics of Biotechnology (partner in ESRC CESAGen Research Centre), Regional Innovation Systems, Knowledge Economies, and Policy Actions for Business Clusters and Networks. He co-edited a book entitled *Regional Innovation Systems* in 1998, a fully revised 2nd edition of which was published in 2004. He co-authored a book on network governance, called *The Associational Economy*, also published in 1998, and is co-author of *The Governance of Innovation in Europe* published in 2000. His authored book *Knowledge Economies: Clusters, Learning and Cooperative Advantage* was published by Routledge in 2002. In 2004 he co-edited *Regional Economies as Knowledge Laboratories* (Edward Elgar). Prof. Cooke was adviser to Lord Sainsbury's Biotechnology Clusters mission in 1999 and subsequently UK government Cluster Policy and Innovation Review adviser.

Dave Counsell is Research Fellow in the Department of Geography at the University of Hull.

Iain Deas is Senior Lecturer in Planning in the School of Environment and Development at the University of Manchester. His research interests focus on: the dynamics of regionalism, regional policy, regional institutional change and the politics of geographical scale; evaluation of the impact of area-based urban regeneration initiatives; and the measurement of social and economic circumstances in cities.

Geraint Ellis is Senior Lecturer and Director of Undergraduate Planning Studies in the School of Planning, Architecture and Civil Engineering at Queen's University, Belfast. He also holds voluntary Directorships with Belfast Healthy Cities, Community Technical Aid (NI) and Sustainable NI.

David Goode is former Head of Environment at the Great London Authority and is now Professor of Geography at University College London's Department of Geography.

Mark Goodwin is Professor within the Department of Geography at the University of Exeter. Mark's research has centred on an analysis of the structures and processes of sub-national government. He is currently a member of the ESRC's Strategic Research Board and sits on the Strategic Advisory Committee of the Rural Economy and Land Use Research Programme, a £23 million cross-research council initiative funded by the BBSRC, ESRC and NERC.

Sir Peter Hall is Bartlett Professor of Planning and Regeneration at University College London, Chair of ReBlackpool (the Blackpool Urban Regeneration Company) and President of the Town and Country Planning Association. Author or editor of nearly forty books on urban development and planning, he received the Gold Medal of the Royal Town Planning Institute in 2003 and the Balzan International Prize in 2005. At the Urban Summit in 2005 he received the Deputy Prime Minister's award for lifetime achievement in planning.

Neil Harris is Lecturer in the School of City and Regional Planning at Cardiff University. His research interests focus on the statutory planning system and include more recently the production of spatial planning frameworks. Neil is a Chartered Town Planner and a member of various practice-based organisations in Wales.

Graham Haughton is Professor of Human Geography at the University of Hull.

Patsy Healey is Professor Emeritus at Newcastle University. She is a specialist in planning theory, planning systems and practices. Her recent work has focused on developing an institutionalist approach to policy analysis and planning, as applied to issues of urban governance, urban regeneration policies and strategic approaches to urban region spatial planning.

Alan Hooper is Professor of Housing and Planning in the School of City and Regional Planning, Cardiff University. His research and teaching interests relate to the interface between housing and planning policy, especially with regard to land supply for housing provision and the structure and operation of the housebuilding industry.

Ole B. Jensen is Professor of Urban Theory at the Department of Architecture and Design, Aalborg University, Denmark. His main research areas are: urban theory and power, the cultural sociology of space, the sociology of mobility, and European Union spatial policies.

Peter John is the Hallsworth Chair of Governance at the University of Manchester, where he is co-director of the Institute for Political and Economic Governance (IPEG). He is known for his work on public policy, urban politics, and citizen involvement. He is author of *Analyzing Public Policy* (1998) and *Local Governance in Western Europe* (2001).

Martin Jones is Professor of Human Geography and Director of the Institute of Geography and Earth Sciences, at the University of Wales Aberystwyth. His research focuses on the geographies of state restructuring and rise of regional and local economic development. Martin has recently co-directed a research project on devolution and economic governance for the ESRC as part of the Devolution and Constitutional Change programme and is author or co-author of *New Institutional Spaces* (Routledge, 1999), *State-Space: A Reader* (Blackwell, 2003), and *An Introduction to Political Geography* (2004).

Rhys A. Jones is Senior Lecturer in Human Geography at the University of Wales, Aberystwyth, with specific research interests in political geography. In recent years, his research has focused on the social and cultural constitution of state forms in the UK, in historic and contemporary contexts.

Michael Keating is Professor of Regional Studies at the European University Institute in Florence, and Professor of Scottish Politics at the University of Aberdeen. His most recent book is *The Government of Scotland. Public Policy Making after Devolution* (Edinburgh University Press, 2005).

Greg Lloyd is Head of the School of Town and Regional Planning at the University of Dundee. He currently serves on the Tayside Economic Forum. He has undertaken numerous funded research studies, and is currently involved in research into modernising planning practice, its relations with community planning, and local vulnerability.

Roger MacGinty is Lecturer in Politics at the Department of Politics, University of York. He has written extensively on intra-state conflict and its management. His last book (edited with John Darby) was *Contemporary Peacemaking: Conflict, Violence and Peace Processes* (Palgrave).

Gordon MacLeod is Reader in Urban and Regional Studies in the Department of Geography, University of Durham. He has written numerous articles on regional governance and urban politics in a range of academic journals, and is currently researching on the 'new urban politics' and the privatisation of the urban realm.

Janice Morphet has been a planner for over thirty-five years and during that time has been engaged in both academic and practice roles. She was Chief Executive of Rutland from 1996 to 2000 and Professorial Head of the School of Planning at the former Birmingham Polytechnic between 1986 and 1990. She has also been a member of a number of RTPI boards and committees and served as an external examiner continuously for over 20 years. Janice was recently appointed a Visiting Professor at the Bartlett School of Planning at UCL and to the RAE 2008 Planning Panel. Janice is currently a Senior Adviser at ODPM.

Richard Munton is Professor of Human Geography and Director of the Environment Institute at University College London. He has written extensively on planning, environmental and agricultural issues, including urban development processes, green belt policy, land ownership and urban governance. He has acted as adviser to numerous bodies including DEFRA, English Nature, Countryside Agency, House of Lords, RICS, NERC and ESRC.

Steven Musson is Lecturer in Human Geography at the University of Reading. He was previously Lecturer in Human Geography at Birkbeck, University of London. His research focuses on the evolution of regional government in England and the geography of public investment.

William J. V. Neill (Bill Neill) teaches urban planning at Queen's University Belfast. He previously taught at the University of Manchester and worked for many years for the Michigan Department of Commerce. His latest books are *Urban Planning and Cultural Identity* (Routledge 2004) and *Cultural Inclusion in the European City* (co-edited with Hanns-Uve Schwedler) Palgrave, 2006. He is a member of the International Committee of the Royal Town Planning Institute.

Deborah Peel is Lecturer at the School of Town and Regional Planning at the University of Dundee. She is an executive member of the Royal Town Planning Institute in Scotland, and sits on the UK Task Force on Marine Spatial Planning. She is a member of the Scottish Executive E-planning Group, and sits on the Planning Summer School Committee. She has undertaken funded research for the Calouste Gulbenkian Foundation, the Carnegie Trust for the Universities of Scotland, and the Scottish Executive. She is currently involved in research into modernising planning practice, and university real estate.

Mike Raco is Lecturer in Human Geography in the Department of Geography, King's College London. His research has focused on the themes of governance, local economic development, urban regeneration, devolution and sustainable communities. He has published widely in the fields of urban studies, geography and planning.

Tim Richardson is Reader in Town and Regional Planning at the University of Sheffield and Visiting Professor at Aalborg University, Denmark. His research explores how dominant ideas emerge in spatial planning, and how the embedding of these ideas in institutions frames the possibilities of thought, communication and action in planning practice. This work seeks to contribute to current planning debates by showing the value of critical analysis of the contested shaping of ideas.

Mark Tewdwr-Jones is Professor of Spatial Planning and Governance, and Director of Research, at Bartlett School of Planning, UCL. His main research interests are urban planning and politics, regionalism and devolution, spatial policy, and film space and identity. His most recent books include *The Planning Polity* (2002), the co-edited *Planning Futures* (2002), and the co-authored *Second Homes* (2005).

Adam Tickell is Professor of Geography and the Research Director in Social Sciences and Law at the University of Bristol. His research explores English regionalism, the political geography of finance and the role of think tanks in developing the idea of the free market.

Geoff Vigar is Senior lecturer in the School of Architecture Planning and Landscape at Newcastle University and a member of the School's Global Urban Research Unit.

Cecilia Wong is Professor of Spatial Planning at the University of Manchester and an editor of *Town Planning Review*. Her research interests include quantitative measurement, spatial analysis of socio-economic change, policy and strategy monitoring and evaluation, planning methodology and spatial planning issues.

Preface

Spatial planning is situated at the juncture of the integration of official state policy making and a requirement for joined-up government, with a determination to embed strategy making within specific distinctive territories that can be sponsored or owned by either government or governance actors. Inevitably, there is room here for distinctiveness and experimentation that will illustrate not only new commitments and tasks for spatial planning as a coordinating and integrating activity, but also scope for disagreements, contention and intra-territorial conflict between those agencies and actors searching to utilise spatial strategy making for their particular needs.

As a result of these tensions, the varied ebbs and flows in policy making, and unequal territorial claims on spatial agendas, spatial planning is deeply implicated in the re-territorialisation of the British state and a number of related significant issues and areas of interest arise from such changes. These include:

- The need and opportunity to integrate spatial development through new regional strategies including activities such as economic development, transport, planning, sustainable development, energy, water and in some cases biodiversity;
- The scope for and implications of policy divergence and intra- and inter-regional rivalry between various actors and institutions;
- The tension between regional autonomy, a reawakening of identity, and the national interest, including wider questions concerning the balance between the two and ongoing debate about the role of the centre;
- The relationship between the evolving new regionalist agenda and existing institutions, processes and stakeholders, and whether this supports or inhibits new distinctive forms of strategy making, and on whose terms; and
- The role and extent of 'region building' and the ways in which the region is discursively constructed by differing actors, being politically, socially and symbolically shaped and contested in various ways.

This book explores these and other issues by providing a multi-disciplinary study of territory, identity and spatial planning, and the ongoing processes of devolution and decentralisation in the UK, through the lens of governance, identity and distinctiveness in policy development. Contributors to the book are drawn from a wide array of social science disciplines, including political and economic geography, planning, urban studies, political studies and environmental studies, and one of the intentions for this was to foster further inter-disciplinary studies of these important and ongoing themes for research and conceptual development.

This chapter provides an introduction to some of these issues and an overview of the book. Clearly an understanding of the underpinning ideas behind Labour's policies and approach is an important starting point in any assessment of the changing nature of spatial planning and the re-territorialisation of the UK state. Following this, the book sets out briefly an overview of the themes of territory, identity and spatial planning that will serve as a necessary conceptual context for the chapters that follow. The main part of the chapter summarises the structure and contents of the book, chapter by chapter, together with the reasons for the ordering of parts and chapters in this particular way and the themes that start to emerge.

Part 1: Theoretical contexts for territory, identity and spatial planning

In Part 1 of the book, we aim to provide a political and conceptual context for many of the themes that emerge in later chapters. Chapter 1 has explored the nature of New Labour and tries to identify a number of conceptual themes or ideas that help understand the re-territorialisation of the UK state, the driving forces behind it and the relationship between government policy and changes to spatial planning. In Chapter 2, Michael Keating outlines the historical and political context of the current forms of devolution and decentralisation. He argues that the UK is a multinational state and it is this that has historically impelled the demand for devolution. Nationality claims imply an element of self-determination, rather than a mere decentralisation of the state, and they are asymmetrical. Changing conditions in Europe have altered the meaning of independence and self-government, so that many nationalist movements have abandoned the search for their own state in favour of building systems of action within the new transitional networks. Functional change and the new regionalism mean that it is possible to construct a territorial system of action without necessarily building a state, and nationalities and nationalist movements have responded to this. Nationality may also provide motifs and resources for collective action and social mobilisation. Political parties in the UK, however, have been slow to recognise these possibilities, so that the debate about devolution and independence is still largely conducted in the old language.

In Chapter 3, Mark Goodwin, Martin Jones and Rhys A. Jones provide a robust theoretical context by discussing the economic and political geographies of devolution and state restructuring in the United Kingdom. The chapter examines current theoretical debates on the remaking of governance, before offering a conceptual framework through which to explore the shifting politico-economic geographies of the UK. They extend Bob Jessop's strategic-relational approach to the state by arguing that it is no longer enough to simply refer to a multivariate 'hollowing out' of the nation state in an era of economic and political restructuring. They suggest that devolution represents a geographically uneven 'filling-in' of the state's institutional and scalar matrix, which is leading to an increasingly complex spatial division of the state.

Within Chapter 4, attention turns towards Europe and how European Union spatial policies are shaping new policy narratives at multiple spatial scales and within new complex governance settings. Ole B. Jensen and Tim Richardson's chapter discusses the identification of a Europeanisation of spatial narratives, where notions of a seamless flow space of Europe are reproduced, as part of the reproduction of a 'Europe of Monotopia'. Within EU territories, this process of Europeanisation is manifested in the shaping of policies for mobility, infrastructure and urban and regional development, as well as in the general expression of becoming more European (alongside other 'scales' of territorial identity). Within their discussion, these 'hard' and 'soft' policy issues are linked, to show how territories within the UK are reorienting their policy narratives of territory, identity and space through a complex engagement with Europe.

Patsy Healey's contribution in Chapter 5 focuses discussion on forms of planning, spatial strategy making, and the task of integrating strategies and different forms of spatial regulation and policy development. In contemporary discussions of the purpose and organisation of planning systems and practices in the UK and elsewhere in Europe, a recurring theme is the search for integration. Arguments for a more strategic focus to the management of development and land use change are often justified in terms of the need to overcome fragmented initiatives and competing projects. An 'integrated' approach is put forward as a way of linking diverse policy objectives (the search for a beneficial economic, social and environmental 'balance'); as a way of connecting issues as they play out spatially (for example, housing and economic development, or land use and transport); or as a way of linking different types of government intervention (especially regulatory power and investment power). An 'integrated' approach may also be used to refer to increasing the connections between levels of government, or to linking multiple stakeholders in pursuit of an agreed framework or strategy. In the UK, these debates arise in several contexts, from the government's regionalisation agenda, its modernisation agenda, and its search for sustainable approaches to development. This chapter explores the contemporary expectations and challenges of place-focused 'integration' in the context of expectations and experiences of 'strategic spatial planning'. The discussion will centre on four questions: What underlies the search for a new 'integrative' focus in UK public policy and governance? What does a focus on space, place and territory have to offer as an 'integrative focus'? What are the potentials and limits of such a focus in the developing UK context? And how justified are the various claims being made for the 'integrative' role of the planning system and its practices?

Part 2: Studies of territorial and spatial planning

In Part 2, we offer some case studies of spatial planning strategy making and governance and spatial policy development as they are starting to emerge in different parts of the UK; these case studies are intended to provide a snapshot of the forms of spatial governance and new forms of planning that are being promoted by devolved administrations and regional governance bodies. The intention here is to record the reasons for recent developments and an analysis of the spatial policy changes occurring with a particular quest to identify whether this leads to difference and distinctiveness in the form of spatial planning across different regions of the UK.

Iain Deas' contribution in Chapter 6 discusses a history of region-building efforts through an analysis of the North West region of England, rehearsing the tensions that have cross-cut regional identity between Liverpool and Manchester, urban–suburban–rural, and historical fragmentation, contrasting it with the influence of proximity to Scotland and Wales. A brief outline is provided of some of the efforts in the 1960s such as Regional Economic Planning Councils and their demise, before tracing the roots of the re-territorialisation of spatial planning post-1990. A brief outline of national policy changes and their impact on the North West follows through the advent of Government Offices, the rise of Regional Planning Guidance, and the establishment of the North West Development Agency and the North West Regional Assembly from 1999. The final section of the chapter debates scalar conflict and spatial planning and the importance of alternative territorial configurations (sub-regions, city-regions etc.), informed by neo-liberal arguments about competitiveness, global inter-urban and inter-regional competition, and the relationship to regional identity, and territorial relations in and around spatial planning.

The situation in the North West is contrasted to developments in another English region: in Chapter 7, Dave Counsell and Graham Haughton address the changing institutional landscape of regional governance in Yorkshire and the Humber. After a brief historical contextualisation, the chapter examines contemporary regional strategy making processes with their strong emphasis on integration between sectoral strategies, more particularly between spatial planning and economic development. In the case of Yorkshire and Humber, the key linking document is 'Advancing Together', produced by the Regional Assembly. Linking to Patsy Healey's discussion, they focus on the use of integrative techniques, such as sustainability appraisal and the Regional Sustainability Framework, to examine issues surrounding barriers to integration such as the different timescales of the strategies and different regulatory requirements and stakeholder arrangements. Finally, and linking to Ole B. Jensen and Tim Richardson's chapter, they use the analysis of the region to look at the interaction between EU, national, regional and local scales of policy making.

The following three chapters deal with the Celtic nations and developments in devolution and spatial planning. In Chapter 8, Geraint Ellis and William J.V. Neill discuss the recent problems of strategic spatial planning in Northern Ireland where a continuing conflict over the re-territorialisation of the polity within an Irish state provides the main underlying discourse within which environmental possibilities are framed. The intense spatiality of cultural identity conflict where symbolic landscapes take precedence over ecological ones precludes mature debate over a sustainable development spatial agenda. 'Visioning' as a plan-making approach to bolstering commonplace identity has been of but limited success here in creating through an alternative language a new enabling horizon of territorial possibility. In this context, their research shows that despite opportunities for North–South cooperation on the theme of sustainable development, realisation remains underdeveloped. It is not helped by a tendency amongst planners to remain passive in the midst of cultural conflict endorsing a language of fudge rather than endeavouring, to some degree, to talk truth to power. They argue that only in confronting rather than avoiding spatial and environmental issues, impregnated as they are with cultural resonance, can a true sense of shared territory and citizenship possibly emerge.

In the case of Wales, Neil Harris and Alan Hooper address the development of the Wales Spatial Plan in Chapter 9. They identify the preparation of a national spatial planning framework for Wales as an early commitment of the devolved National Assembly for Wales. The chapter focuses on the development of the framework from its initial conception through to the recent publication of the plan. In doing so, it focuses on both the practical and political potential of such a spatial planning instrument. The case study highlights a series of methodological and practical issues involved in spatial planning, ranging from the availability of data through to the political sensitivities of spatial planning and representation of spatial information.

In Chapter 10, Philip Allmendinger contrasts the asymmetrical nature of the devolution settlement with Scotland's historic legal, educational and governmental distinctiveness that has led to high expectations of divergence and close scrutiny of progress since Scottish devolution. The multi-scalar nature of spatial planning has provided early indications of the tensions involved in managing a federalising UK with regional, national and pan-national policy coordination and integration. This chapter explores the nature of these tensions and expectations as well as the emerging outcomes for Scottish and UK spatial planning. Specifically, the chapter covers the devolution settlement and planning including the history and development of planning in Scotland, the role of the Scottish Parliament and Executive in planning including impacts of committees and ministerial portfolios, forces pushing for and against policy divergence in planning and expectations of change, and recent initiatives and their implications. The chapter concludes with some observations on the likely trajectory of change and relations with UK planning and governance.

Peter Hall's discussion of devolution to London, in Chapter 11, focuses on another of the first stages of the Blair government's devolution project: the elections for the London Mayor and the Greater London Assembly. This was a novel form of devolution, since as well as devolving power to an executive Mayor (the Assembly being given only weak powers of scrutiny), it required the Mayor to develop a series of strategic plans – for economic development, for transport, and for the environment – which would then be integrated into a spatial development strategy. The Mayor's London Plan, which was published in draft in 2002 and in final form in 2004, is the first English example of a regional spatial strategy. The strategy essentially seeks to steer development from the more prosperous and economically overheated western side of London towards the eastern side, which suffers from multiple deprivation but also offers greater opportunities for physical development and regeneration. It thus accords with central government's Thames Gateway strategy, continued by the Blair government from the previous Major administration and further developed in the Sustainable Communities strategy of 2003. Attention is focused on the tensions and uncertainties caused by planning for London's growth between the inner metropolitan area and Greater London as a whole.

Part 3: Institutions of governance and substantive policy roles

In Part 3, building on the conceptual framework provided in Part 1 and the scalar examinations provided in Part 2, we focus on substantive policy areas and perceptions, with the aim of illustrating in what ways the various devolved and decentralised bodies have attempted to formulate and practise distinctive policies to meet the particular needs of their regions. The intention here is to examine whether identity and difference in institutional and governmental structures is giving rise to differentiation in policy sectors and the reasons for this.

The first chapter that deals with substantive policy areas, Chapter 12, considers housing. Mark Baker and Cecilia Wong analyse the Government's Sustainable Communities Plan and the Barker Review of housing supply, to consider the current institutional arrangements, planning processes and policies for regional housing. They provide an analysis of regional housing requirements and compare this with national housing projections and completions. The chapter discusses issues such as regional targets for brownfield development, affordable housing and the 'monitorium' on housing recently imposed by several local planning authorities in the North West region and elsewhere. The chapter discusses the extent to which the regional housing policy framework adds up to a coherent national picture for England as a whole and assesses the implications of the Barker Review for planning and housing in the English regions.

Geoff Vigar takes up the themes of differentiation and distinctiveness in Chapter 13. In seeking to manage contemporary mobility trends, governments often look to innovate in terms of policy mechanisms and/or associated institutions. In the UK, a new emphasis has emerged in the last decade on the regional scale for much governance activity both in relation to transportation and other policy areas. The chapter examines progress toward using the regional scale as an effective level through which to deliver transport policy goals, focusing particularly on a new policy mechanism (the Regional Transport Strategy). Vigar notes the difficulties in the UK that principally relate to the institutions for strategy making as currently configured at this scale, a continuing centralisation of governance activity and precipitately the difficulties of constructing and operating policy processes given this institutional context.

In Chapter 14, Philip Cooke and Nick Clifton explore another important devolved policy field, economic development, for which territories like Scotland, Wales and Northern Ireland can be characterised as being in more or less permanent policy crisis as the UK government from 1997 onwards pursued a strong sterling policy outside the eurozone that has devastated in particular manufacturing employment. To this are added new pressures on regional competitiveness caused by global economic recession from March 2000. The chapter explores these dynamics and associated policy responses thereafter indicating areas of convergence and divergence across the devolved territories, and the authors pose some pertinent questions about the role of the central state in controlling devolved economic policy, and the ability of the institutions to create distinctive and successful economic policy.

Following Peter Hall's contribution in Part 2 of the book, David Goode and Richard Munton refocus attention on London for a discussion of the pursuit of sustainable development in Chapter 15. The opportunity to put in place an overall framework for sustainable development was widely recognised and proposals were made jointly by public sector organisations on how that might best be achieved after London devolution in 2000. The Greater London Authority Act 1999 specified in considerable detail the range of functions to be addressed by the new authority, and was high on expectation but provided limited new powers. The three core functions of the GLA, to promote economic and social development and to improve London's environment, provided a basis for promoting sustainable development, but the Act went further by specifically requiring the GLA to contribute to sustainable development in the UK through its decision making, particularly in preparing strategies and policies for London's future. This chapter examines how far this has been achieved, by examining the notion of sustainable cities, how cities might be made more sustainable, and asks how far progress in London can be attributed to the GLA's particular governance arrangements.

The final contribution of Part 3 concerns a slightly different substantive area for devolved governance, that of democratic accountability and public perceptions. Roger MacGinty in Chapter 16 provides a discussion of identity and public attitudes towards the Northern Ireland Assembly. This chapter discusses time series survey evidence to track continuity and change in identity in post-devolution Northern Ireland. Proponents of devolution argue that the granting of autonomy to sub-state groups will satisfy identity aspirations while maintaining the integrity of the meta-state. His research shows continued attachment to exclusive identity blocs despite the (albeit troubled) institution of devolution. Indeed, the continued salience of exclusive nationalist/unionist political projects has thus far rendered devolution unsustainable and complicated the new patterns of governance it attempted to introduce. Key questions to emerge from this debate concern whether new devolved institutions within territories are owned by communities and on whose and what sort of terms.

Part 4: Complexities and interdependencies in spatial governance

Part 4 of the book brings together many of the themes emerging in the previous sections, with a return to more conceptual issues surrounding complexities and intergovernmental and inter-scalar relations, and those concerning territory, identity and spatial planning.

The first chapter, Chapter 17, considers government, governance and decentralisation. Steven Musson, Adam Tickell and Peter John discuss how England is historically highly centralised. Civil service decentralisation has focused on routine administrative activities and whilst elected local government has carved out some autonomy, power and control remained at the centre. Devolution and other elements of the regional policy agenda seek to challenge this structure of government. However, the process of transformation is geographically variegated and institutionally complex. This chapter draws on research in London and the South East of England, and argues that despite the expansion of regional government under New Labour, the central state maintains tight control at most scales of government in the two regions. While an increasing range of policy administration is being decentralised, much continuity remains, with the central state retaining control over the core policy instruments.

Greg Lloyd and Deborah Peel address another scalar relationship in Chapter 18, that of city-regionalism. Their chapter explains the current vogue for city-regionalism as a means of modern territorial management that seeks to reconcile a host of economic, social and environmental ambitions. The current political paradigms of reconciling markets and interventions encourage this spatial redefinition. Their work identifies the contemporary articulation of city-regionalism in terms of three converging ideas, each of which separately advocates city-regions as the appropriate re-territorialisation of national space. This occurs by teasing out city-regionalism in terms of 'new' regionalism, 'new' urbanism, and 'new' (spatial) planning and governance. These ideas have resulted in an apparent convergence of thinking about city-regionalism, but a divergence in its spatial articulation in terms of scale and form. The chapter explores the implications of this particular construction of space in terms of the realities of political, economic, social and environmental capacities, but cautions against an overly simplistic interpretation of city-regionalism which obscures the need to critically consider issues such as changing relationships between the state and the market, the state and the citizen, and the ecology of a city-region.

Janice Morphet's contribution in Chapter 19 focuses on the local state level and its distinctiveness across the UK under devolution and its relationship to other tiers of government. Her work provides an overview of the origins and current debates on the principles of new localism as they are being developed in the UK by both government ministers and think tanks. The discussion is also located within an international context and situates the issue within the broader considerations of democratic engagement, space and place control, and local self-management. A particular theme developed in the chapter is the implications and consequences of new localism for the objective of democratic re-engagement, one of the core foundations of decentralisation.

In Chapter 20, Mike Raco discusses the relationships between new identities, the restructuring of scalar politics, and the devolution settlement. He discusses the links between forces of globalisation, the hollowing out of the nation state and the (re)emergence of regional identities. New regionalist thinking and the regional common enterprise are discussed in relation to regional–national identities of the UK and the spatial structures of devolution. He then considers the processes through which specific types of regional identities are defined and mobilised as part of the devolution programme, and whether this gives rise to the building of new regional–national identities. The chapter concludes by offering debate on territorial statecraft and the political and institutional processes involved in post-devolution regional politics, including place–space tensions and the evolution of new regional governmentalities.

Some of these themes are taken further by Gordon MacLeod and Martin Jones in their discussion of UK devolution, institutional legacies, territorial fixes and network topologies in Chapter 21. Following a theme developed by many contributors to the book, they describe the asymmetrical nature of devolution and re-institutionalisation of the UK. They view the process of devolution and decentralisation through an interpretation provided by scales, networks, and territories. They argue that there are now two alternative approaches to the analysis of space: through the politics of scalar structuration, including the rescaling of policy and planning; and a topological perspective to scalar politics which considers the practices of performing devolution through flows, mobility and exchange, thereby transcending traditional governmental boundaries and institutional territories. Taking two examples by way of illustration, the contested boundaries of governing Cornwall in South West England and the enactment of 'The Northern Way', a multi-nodal networked region in North West England, they argue that the problems of spatial politics can only be understood both through an analysis of scalar structuration and topologically in order to avoid mere 'plastic achievements' of territory and governance.

Mark Tewdwr-Jones provides the final contribution in Chapter 22. Drawing together the themes emerging from each of the preceding chapters, the discussion considers conceptual developments in regional and spatial governance and poses questions on the emerging form of regional governance and spatial strategy making, relating to the permeability and durability of devolved arrangements, the relationship between governance and government, the complex nature of governance processes, difficulties in the expectations on new governance processes between the eyes of different regional stakeholders and institutions of government, and the relationship between intention and practice. He contends that aspects of the ongoing and emerging relationships between pre-existing and new flows and institutions of policy making of the state with particular responsibility for spatial strategy making, form a framework to understand sub-national governance. Within this understanding, different levels, actors and audiences of government and new forms of governance 'flex their muscles' and claim ownership of responsibility for spatial plan making and strategic coordination. The overall determinant in the bedding down of power and responsibility between actors, and between government and governance, is the desire for integration and new strategic development.

Acknowledgements

The editors are grateful to all the contributors for the provision of well-written and grounded contributions within not only a tight conceptual framework but also a tight time frame, and the fact that they were asked to be prepared to progress three sets of revisions each following editorial scrutiny, indicates their enthusiasm to engage and participate in this project. We wish to thank the staff at Routledge for this constant support for the book, including Caroline Mallinder and Helen Ibbotson, and we are grateful to our copy editor Jane Fieldsend for progressing the manuscript so smoothly from first draft to final product. Any errors, if they remain, are our own.

Mark Tewdwr-Jones
London

Philip Allmendinger
Reading
August 2005

Part 1

Theoretical contexts for territory, identity and spatial planning

When people say England, they sometimes mean Great Britain, sometimes the United Kingdom, sometimes the British Isles – but never England.

(George Mikes)

The building of states, in whatever level, is intrinsically a ruling-class operation.

(Raymond Williams)

1 Territory, identity and spatial planning

Philip Allmendinger and Mark Tewdwr-Jones

Introduction

Although forms of decentralisation and devolution in the UK predate the Labour government – particularly the introduction of Government Offices in 1994 by John Major's administration – it was not until the election of 1997 and Tony Blair's victory that the devolution and decentralisation agenda took off in earnest. Although some commentators argue that decentralisation and devolution are part of a move towards a greater centralisation of power the majority view is that such changes are part of what Iain Deas, in this book, refers to as the 'internationalisation of capital, the re-expressed pattern of state territoriality and in some accounts, the changing centrality of the nation state'. As such, the re-territorialisation of the UK state is driven by a policy reaction to a mix of global economic logic and reinvigorated regional–national territorial identities.

The changing nature of planning and the emphasis upon spatial planning has inevitably been part of this trend. As the contributors to this volume point out, spatial planning is a contributor to and a reflection of a more fundamental reform of territorial management that aims to, inter alia, improve integration of different forms of spatial development activity, not least economic development. So, at one level, devolution and decentralisation and its implications for spatial planning must be analysed in respect of other aspects of Labour's regional project, not least the government's concern with business competitiveness. At another level, the current reforms which privilege regional scale policy interventions will inevitably require changes in the divisions of powers and responsibilities at local and national levels. In other words, devolution and decentralisation involve a major rescaling of both spatial planning and development, which is unfolding rapidly and unevenly across Britain. Much of this rescaling is occurring within terms and boundaries set by the state, since these new activities are sponsored or supported by various tiers of government. But other recent activities within spatial development are choreographed by non-state actors and possibly organised territorially and culturally.

We commence our review with a detailed overview of the Blair project, its meanings and manifestations, in order to assist in understanding the origin of change and modernisation within the British state. We then go on to review the three key building blocks of this project: territory; identity; and spatial planning. These discussions form the context for the remaining parts and chapters of the book.

Labour: towards an understanding of the political context

Defining Labour (the pre-fix 'new' seems now largely redundant) and the Labour project under Blair is problematic as any stance will undoubtedly be allied to a normative position. The result is a diversity of perspectives with little agreement. Freeden (1999), for example, has characterised Labour as brimming with complex ideas, values and practices that draw on all major ideological traditions. This view of Labour as an eclectic mix of theories and values has been rejected by some (see Buckler and Dolowitz, 2000) while debates over the 'newness' of Labour persuasively argue for both continuities *and* radical breaks with Labour's past (Hay, 1997; Vincent, 1998; Rubinstein, 2000). Different analyses also point to divergent positions regarding the core values of Labour (see Kenny and Smith, 1997 and Hay, 1999 for two very different interpretations). Similarly, those who focus on ideology (e.g. White, 1999; Finlayson, 1999) and others who concentrate on policy change and outcomes (e.g. Savage and Atkinson, 2001; Brivati and Bale, 1997) come to different conclusions on its 'newness' and significance. Such diversity is not helped by the claim of Driver and Martell (1999) who state that there are different emphases within Labour depending on the policy area. Similarly, this is further complicated by the assertion of Marquand (1999) that the government contains inherent and unresolved tensions over ideology and policy.

There are obviously many different views – described as 'the bewildering variety of often seemingly incommensurate perspectives' by Hay (1999: 19). Where consensus does exist, it is at a level of abstraction and generality that is of little use in helping to pin down any distinctiveness. Kenny and Smith (1997) are not alone in claiming that the problem of definition is compounded by the *sui generis* nature of Labour in defying established categories and axes of Left and Right. In transcending such categories, there is also the need to recognise, as Finlayson (1999) points out, that Labour, like the New Right before it, is not a hermetically sealed application of theory to society but an open-ended and strategic production of collective political identities. 'Defining' Labour is likely to be little more than an historical snapshot as the project itself moves on and develops.

Given these issues, we posit a framework for better understanding the nature of Labour, its distinctiveness and significance rather than a definitive position per se. Such a framework locates Labour within a wide social, economic and political context that acknowledges both exogenous and endogenous influences in seeking to address a number of the issues identified above. The first relates to the need to acknowledge the role of structure and agency in any investigation of Labour (Hay, 1999). Emphases on individuals or upon economic or social factors tend to be simplistic and/or reductionist. Thus, Tony Blair, like Margaret Thatcher before him, is a central figure in the direction of his party though his influence needs to be tempered by other considerations. Second, the framework allows for a variety of influences upon Labour and thereby does not preclude different values or approaches existing simultaneously in the minds of its proponents or in different policy or institutional arenas. While identifying the influences and their relationship to each other, it does not, a priori, rank them in terms of influence. Third, the framework allows for variation through time and space and this is a critical element given the natural evolution of political movements and their variance in different places

(Gamble and Wright, 1999). Again, as will become clear further on, sensitivity to the changing nature of Labour is significant in the field of spatial planning and the re-territorialisation of the UK state.

One could identify a variety of exogenous and endogenous influences upon Labour though given the restrictions of space we will focus upon three main categories: the relationship between Labour and social democracy; the influence and significance of The Third Way theory; and the legacy of the New Right, statecraft and the trajectory of global social democracy.

Labour and social democracy

The focus of much academic attention has been the relationship between (New) Labour and social democracy, or Old Labour. This whole area of debate is littered with disagreement though a broad cleavage exists between those who argue that Labour is a distinct break from the past and those who highlight continuities with social democratic or liberal thought updated for 'new times' (see below). Notwithstanding these views, assessments of Labour fall into four broad schools of thought.

- 'Mainstream continuity'. Labour is continuing with the core concerns of social democracy (e.g. Rubinstein, 2000; Gamble and Wright, 1999).
- 'Social democracy in new times'. There are both continuities and significant differences in Labour (e.g., Kenny and Smith, 1997; Vincent, 1998; Shaw, 1994; Jayasuriya, 2000; Buckler and Dolowitz, 2000; White, 1999; Harris, 1999).
- 'Distinct break with the past'. Labour has accepted the neo-liberal hegemony of Thatcherism and left social democracy behind (e.g. Driver and Martell, 2000; Gray, 1996; Callinicos, 2001; Hay, 1997).
- 'The Labour project is *sui generis*'. Labour amounts to a form of pragmatism that has 'picked and mixed' different philosophies and ideological positions (e.g., Freeden, 1998; Le Grand, 1998).

Intellectually, there has been a great deal of heat though little light to help illuminate the question of what is Labour. Blair himself has commented that 'New Labour' is 'a modernised social democracy, passionate in its commitment to social justice and the goals of the centre-left, but flexible, innovative and forward-looking in the means to achieve them' (Blair, 1998: 1). More specifically, he has identified Labour as being a fusion of democratic socialism and liberalism.

In what ways does the debate around Labour and social democracy help us better understand what Labour is or is supposed to be? Overall, the various debates are inconclusive and, naturally, based upon deeply normative positions. The nature of this theoretical debate, therefore, naturally points towards the need to add evidence to enable comparison against various claims.

Labour and The Third Way

The second area where we might look to help identify and comprehend Labour is the debate concerning 'The Third Way'. Even though it is now generally used as a pejorative term 'the third way' encompasses a range of ideas that have had an influence upon the thinking of Blair and the Labour project. A number of authors have been at pains to stress that the third way is not an ideology (White, 2001; Plant, 2001; Jayasuriya, 2000) – although, to a large degree, this depends on what is meant by ideology. Without straying too much into tangential semantics, we shall follow the majority view that the third way is a collection of political values developed from analysis of contemporary social relations (Finlayson, 1999), a perspective that has a long tradition in social democratic thinking (Plant, 2001: 557). Such values are legitimised not from ethical principles but from social 'facts'. The two principal sources of these facts are the 'new times' debate and the sociology of Anthony Giddens. The basic premise of new times (and its close relative, globalisation) concerns changes in the regime of accumulation and the wider implications for society throughout the 1970s and 1980s.

The differences include class-dealignment and more fluid political identities requiring a rethinking of socialist strategy to appeal beyond class based alliances. There was also a change in the attitude of the individual vis-à-vis the role of the state that emphasises a 'new individualism' and autonomy based on the principle that state interference should be minimised (Finlayson, 1999: 274). Against the backdrop of the 'end of ideology' debate (Fukuyama, 1989) and the neo-liberal hegemony of the New Right the issue facing social democrats was less a political one and more a managerial one – how best to shape growth in the interests of all.

Giddens (1998, 2000) justifies his third way theory on the basis of both new times and globalisation: 'modernising social democrats have to come to grips with the transitions taking place in world society. Since the advance of globalisation is so much stronger than it has been before, it makes sense to suppose that the global order is different from the past' (1998a: 122). Politics needs to be rethought to meet the requirements of new times. The high degree of fluidity in the economy, for example, requires the state to facilitate social capital to encourage an entrepreneurial culture and risk taking.

According to its main academic proponent, the third way as a political project has six main features (Giddens, 2000: 50–54):

1   Society and the issues that politics needs to address are no longer divided along exclusively left–right political axes.
2   A balance needs to be struck between government, the economy and the communities of civil society in the interests of social solidarity and social justice.
3   Those who profit from social goods should both use them responsibly and give something back to the wider social community. This is the basis of the much discussed 'rights and responsibilities'.
4   In the economic sphere government should seek to develop 'human social capital' to enable people to participate better in the risk society of late-capitalism.
5   Society should be orientated towards egalitarian principles and an equality of opportunity and a concern with inequality of outcomes.
6   Globalisation should be taken seriously through both encouraging the development of new global institutions and reaping the social and economic outcomes.

What this amounts to, depending upon your viewpoint, is a position that either 'splits the difference' between the traditional left–right axes in British politics or stands in a complex relation between the two (see Tables 1.1 and 1.2).

Table 1.1 Traditional British political axes

Social democracy (The old left)	Neo-liberalism (The New Right)	Third Way (The centre-left)
Class politics of the left	Class politics of the right	Modernising movement of the centre
Old mixed economy	Market fundamentalism	New mixed economy
Corporatism: state dominates over civil society	Minimal state	New democratic state
Internationalism	Conservative nation	Cosmopolitan nation
Strong welfare state, protecting 'from cradle to grave'.	Welfare safety net	Social investment state

Source: Giddens, 1998: 18

Table 1.2 Modern British political axes

Dimension	Old Left	Third Way	New Right
Approach	Leveller	Investor	Deregulator
Outcome	Equality	Inclusion	Inequality
Citizenship	Rights	Both rights and responsibilities	Responsibilities
Mixed economy of welfare	State	Public/private; civil society	Private
Mode	Command and control	Cooperation/ Partnership	Competition
Expenditure	High	Pragmatic	Low
Benefits	High	Low?	Low
Accountability	Central state/ upwards	Both?	Market/downwards
Politics	Left	Left of centre?/ Post-ideological	Right

Source: Powell, 1999: 14

Despite the many claims to the contrary, Plant (2001) argues that the third way is not valueless or simply pragmatic. According to Blair (1998), the third way project is built upon four values (see also Le Grand, 1998):

- **Equal worth** This derives from the idea that social justice requires that each individual be considered to be of equal worth. It involves a proactive role for government in tackling discrimination and prejudice and promoting a multicultural and multi-ethnic society.
- **Opportunity for all** Social democracy in the Labour Party has traditionally emphasised equality of outcome while the third way promotes an equality of opportunity. Again, this implies a proactive role for the state in levelling the playing field.
- **Responsibility** The phrase 'rights and responsibilities' is hackneyed through overuse by Labour but it does neatly sum up the view that citizens have the

right to expect the state to help when necessary; in return there are also responsibilities revolving around the family and civil society.

- **Community** As opposed to the atomistic New Right view of individuals the third way sees people as social. People and communities should, in partnership with the state, seek to become more involved with issues that affect them.

The problem for some is not a lack of values but their vagueness. As White (1999, 2001) and Driver and Martell (2000) point out, the ambiguity of terms such as 'equal worth' leaves a large interpretative space and the consequence that there are many third ways. Giddens (2000: 31) identifies four within Europe alone. More significantly for us, White (1999: 11–13) highlights two cleavages within third way thinking. The first is between 'leftists' and 'centrists' with regard to the issue of opportunity – the former have a more egalitarian view of opportunity while the latter are more concerned with meritocratic opportunity. Different policy prescriptions flow from these two positions. Leftists argue for greater income redistribution while centrists would pursue equality of opportunity and, as meritocrats, let those who do well be rewarded.

The second cleavage concerns the differences between liberals and communitarians on the issue of civic responsibility. The 'rights and responsibilities' theme of Labour outlined earlier leaves open the extent to which individuals are responsible. Communitarians take a broad view of such responsibilities while liberals have a much more circumscribed and limited perspective based upon the concept of personal freedom. Again, there are policy implications. Communitarians would be more ready to punish an individual's use of drugs while a liberal would argue that the implications of drug use – such as crime – should be the aim of punishment. This leads to 'significant differences of opinion on values and public policy and these differences in turn define distinct, potentially opposing, political projects' (White, 1999: 13). It is possible to perceive a 'leftist liberal' third way or a 'centrist communitarian' third way.

This variation in the themes and values of the third way raises the issue of the extent to which third way principles have influenced policy and thereby Labour. Putative third way evaluations are emerging and back up the idea of the third way as a broad framework of ideas within which distinctive and opposing policies can co-exist (Seldon, 2001; Savage and Atkinson, 2001; Coates and Lawler, 2000; Powell, 1999). This has led some to define a number of third ways in respect of distinctive policy areas (see Rouse and Smith, 1999; Hewitt, 1999 and Powell, 1999: 287) to conclude that, in relation to the National Health Service, the third way exhibits tensions if not inconsistencies.

From experience, one suspects that the heterogeneous and 'looser' frameworks of White (1999) which point towards tensions and varying emphases with the third way are closer to the truth. Superficially at least, such an approach provides a convincing understanding of the apparent inconsistencies in Labour's environmental and urban planning policy, as discussed briefly earlier in the chapter. But both the legacy of social democracy and the putative third way require a final coalition of influences – the legacy of the New Right.

The legacy of the New Right and statecraft

The third set of influences upon the Labour project is a broad group or collection that was significant as both justification for and influence upon the outcome of

Labour modernisation in the 1980s and 1990s (Hay, 1999). Modernisation originated from the recognition by some in the party that the political landscape had changed not only due to 'new times' but also in the political reaction to it in the UK and overseas. It has now become hackneyed to claim, as Labour sympathisers sometimes do, that Margaret Thatcher helped save the Labour Party. The broad view is summed up by Harris (1999: 45): 'New Labour is born of the recognition, at long last, of how distanced the Labour Party, and the left generally, was from the mainstream political, social and economic perspectives of the mass of the public'. That Thatcher and the New Right played midwife in this renaissance is not in doubt. The extent of the overlap is a highly contested territory. Two broad views coexist (Kenny and Smith, 1997). The first view argues that Labour has deliberately adopted a neo-liberal stance to make itself electable in the post-Thatcherite political era. The second, is that Labour has adopted other significant elements of Conservatism, specifically, the commitment to traditional morality based on Christian ethics and the importance of the family.

Both classifications of Labour are backed up by evidence of policy change – economic policy designed to control inflation and the commitment to law and order, for example. But critics of these positions also argue that there have been policy changes that point away from a Thatcherite inheritance – for example, the adoption of the Social Chapter of the Maastricht Treaty, the minimum wage, devolution, and a Freedom of Information Act. According to Hall (1998), such contradictions arise because Labour cannot decide if it is a radical centre or centre-left party. Kenny and Smith (1997) counter that such interpretations based upon left–right axes are misleading. Labour is more eclectic in its lineage and separated from Thatcherism by 'incontestable liberalism' (Beer, 2001). Part of the problem, as Hay (1999: 20) and many others have pointed out, is that it is far from agreed what Thatcherism or the New Right actually is. The majority and, in our view, most accurate portrayal has been advanced by Kerr and Marsh (1999: 187): 'Thatcherism was an evolving process which changed significantly over the eighteen years the Conservatives were in office, in the process becoming more coherent and more radical, ultimately culminating in the reforms which characterised the party's third and fourth terms'.

Perhaps, as Heffernan (2001: vii) points out, what we can say is that Labour changed a great deal in the 1980s and 1990s. Statecraft, or the art of winning elections, guided the party's modernisation. This meant that in some areas of policy Labour did shift towards capturing various themes that had been electorally popular for the Conservatives or, at least, had been used as an unfavourable contrast to erstwhile Labour. However, this is not to imply that any shift was uni-directional or that it was purely pragmatic and unprincipled. As the debates earlier in the paper highlighted, policies and positions vary greatly within the social democratic tradition. Blair's personal beliefs are undoubtedly centre-right. But, following Kenny and Smith (1997), other changes brought about by modernisation were less easily identifiable on the left–right axes. Just as Thatcherism was a heady mix of traditional Conservative values and liberal economic orthodoxy so Labour is as much eclectic as unified in terms of philosophy and policies. Like Thatcherism, a narrative has developed around Labour that seems to emphasise its closeness to the ideology of the New Right and ignores the complexity of contradictions to this. This is not to say that Labour has not changed, but that any change is more complex than the simplistic notion of a shift to the ideological right.

Neither does a simplistic analysis fully explain the extent to which any shift was an on-going phenomenon, to what extent it affected the New Right and the degree to which neo-liberalism was institutionalised by the Thatcher and Major governments and presented to any party that sought power in the UK as a fait accompli. Overall, the trajectory of social democratic parties across the world would seem to suggest so.

So what can we conclude about the existence and significance of a phenomenon called 'Blairism' and its effect on the form of re-institutionalisation and modernisation of government, and on decentralisation and devolution? There is clearly substantial disagreement over both existence and significance. Those who point towards continuity with the New Right legacy tend to focus on the economic and, to a lesser degree, what they perceive as the authoritarian, centralising characteristics of the Blair governments. On the other hand, those who argue that Labour is new or represents discontinuity with both the Thatcher and Major administrations and the varied legacy of social democracy tend to highlight the social dimensions of policy and its fusion with the economic 'inevitability' of neo-liberalism. Perhaps this is as noticeable at the regional level as it is at a national scale.

Of course, the fusion of what were previously disparate strands of thought is what defined the New Right as new. Labour in the 1990s based itself on a fusion of ideas but it also changed its attitude towards issues such as the economy. The main area of change in attitude has been on the economy. Labour has been achingly aware of globalisation. According to some it has been and is the most significant factor in shaping party thinking since the early 1990s: 'the significance of globalisation and claims made about globalisation to the political economy of Labour can scarcely be overstated' (Hay, 1999: 31), and 'The perception that we live in a globalised world is the key to Labour's loss of faith in Keynesian macro-economics' (Driver and Martell, 1998: 42). The underpinnings of third way theory are derived from an appreciation of economic globalisation that acts as a point of departure in social democratic thinking generally and provides, at the very least, the justification for the 'newness' in Labour. But such 'newness' is tempered by the party's insistence on being both new and old – old principles in changed times. If the philosophy is the same and the means have changed, can we say that Blairism is distinct? In policy terms there are both continuities and discontinuities with the immediate past governments though this tells us as much about the nature of the British political system as it does about the radicalism of Labour. This may mean that any attempt to foster a clean break from the past, institutionally and through devolved power, is tempered by a determination for the central state, and its ideological foundations, to drive clear policy directions forward and with certainty. Before going on to debate these issues and how they manifest themselves across the UK through spatial planning and core substantive policy areas, we first provide a brief overview of some definitions that form the core concerns of this book.

Exploring the themes of spatial governance

Territory

A territory is a given area of land under the jurisdiction of the state, or an organised division of a country that has a particular set of powers and jurisdiction. Our use of the word 'territory' as one of the themes of spatial governance and of this book reflects its close relationship to the role of planning strategy making which, in itself, is a function of government and governance within defined boundaries. We are more concerned here with the territory as a geographical and political form/vessel, rather than with the processes of 'territorialization', 'de-territorialisation' and 're-territorialization', concepts that are currently and have been permeating academic debate political and economic geography over the past ten years (Brenner, 2004). Having said that, these processes do nevertheless manifest themselves within the processes of devolution, decentralisation, spatial strategy making and the role of the state now unfolding across the UK. And these processes do involve 'spaces of flows' (Castells, 1996) through global processes of capital between subnational territories. These new spaces focus on learning regions, global cities, industrial districts, and new economic spaces (Scott, 1996), described as a rescaling of activities from the national to multiple spatial scales (Boyer and Hollingsworth, 1997). This, in itself, is continuing to fascinate geographers since they transcend the very study of geography and territory as we have traditionally seen them, those 'static entitles frozen permanently into geographical space' (Brenner, 2004: 7). Neil Smith sums up these changes succinctly:

> Geographical scale is traditionally treated as a neutral metric of physical space: specific scales of social activity are assumed to be largely given as in the distinction between urban, regional, national and global events and processes ... There is now, however, a considerable literature arguing that the geographical scales of human activity are not neutral 'givens', not fixed universals of social experience ... geographical scales are the product of economic, political and social activities and relationships; as such they are as changeable as those relationships themselves.
>
> (Smith, 1995: 60–1)

Scale may be defined as 'the level of geographical resolution at which a given phenomenon is thought of, acted on or studied' (Agnew, 1997: 100). There are various ways of studying scales, including: the scaling and rescaling of processes; the relationality of scales between geographical scales upwards and downwards; the amalgam of various scalar organisations creating a vast institutional landscape; scalar fixes, involving the shifting sands of institutions and organisations that are fixed temporarily that create benefits for modern capitalism; and scalar transformations, where one scale transcends into new scales and creates its own legacies and forms (Brenner, 2004).

Within the literature, and indeed many of the contributions to this book, geographers, planners, political scientists and others are utilising many of these concepts to describe the changing economic, political, social and institutional events unfolding across the UK and other capitalist countries. Within the planning literature, these concepts have taken some time to take hold, principally, we would argue, because planners are overtly concerned with strategies, their politics, power, coercion and implementation, and delivery, within fixed territorial limits. In other words, planning – as a political creation and function of the state – is legitimised, operationalised and realised only by the will of the state and, of course, its relationship to increasingly global economic forces – a point that has long been a criticism of planning over many decades (see Cooke, 1983; Reade, 1987, for example).

We have focused on territory since it relates directly to the geographically organised jurisdiction of the state and, therefore, analytically, can encompass planning as a strategy making, integrating and participatory activity that is worthy of study, with all its associated political, economic and social externalities. That is not to say that we necessarily approve of planning within its current state manifestation with any form of alacrity; rather, we have argued over many years that planning as a state function is constantly misused politically and economically, to the detriment socially and environmentally, of non-national and multinational interests (see, for example, Allmendinger, 2001; Tewdwr-Jones, 2002). But we live in new times politically, economically and socially, and the fact that planning has survived as a state activity in the UK, indicates the political, economic and social potential that is inherent within the discipline.

Planning, conceptually, is currently undergoing a metamorphosis as it is stretched, tugged, expanded and loaded, with increasing expectations of it delivering at various spatial scales, within and outside existing territories, that suggest both forces of continuity with and radical breaks from the past. The political and economic geography literature is now serving a useful purpose within planning in assisting in the analysis of this multifarious change within territories, an attempt – in essence – to 'make sense' of it all. But in many ways, planning as an activity is within a whirlwind politically and geographically, as the chapter goes on to explain. The biggest change currently being enacted in practice is the extension of planning from its narrow regulatory base within various territories and various scales simultaneously, to a broader integrating and spatial governing activity, particularly at regional and sub-regional levels (Tewdwr-Jones, 2004; Allmendinger and Haughton, 2006). This is being rolled out with trepidation, politically, and uncertainty, professionally, but within existing political and socio-geographical territories to ensure legal authority. For many planners, it is a vastly different type of planning from the certain, legal and political activity they have been used to.

The second phase of the planning overhaul is to transcend those existing geographical, territorial and institutional boundaries, if planning is to serve a more forward purpose in the decades ahead. There is general acceptance that historical boundaries, administrative delineations, and professional silos will not deliver the type of spatial planning and governance in the future that is, politically, being expected. The more difficult part of reform, which could benefit the social and environmental as a counterpoint to the economic and political, is to encourage strategy makers to 'think outside the box' of their own predetermined territory. For strategists, this is the equivalent of asking them to strip naked; but change is in the air professionally and managerially, and indeed politically in current times, there is little option.

There are bound to be difficulties with planning within all territories in the next decade. Spatial strategy making is expected to become a collaborative tool of public services and policy development, alongside planning's traditional role as a land use regulatory activity, while being stretched across several tiers of government and owned across state and non-state agents of governance. There are broader issues that fall out of the discussion that individual territories, in their contention over spatial governance, will have to contend with:

- ownership, and the rights and responsibilities of elected government versus stakeholder interests at the sub-national scale;

- legitimacy, and the rights of a professional elite in this day and age to make supposedly still rational decisions for the public benefit; and
- control, and the attempts of the central state to retain responsibility for broader development, economic and infrastructure issues while ideologically proposing more neighbourhood, local, sub-regional and regional policy making processes, both within established territorial boundaries, and transcending them.

All these issues are raised in one form or another in the contributions to this book. The hope is that by providing a snapshot of the reforms and changes currently underway across the UK, we will be in a better position to determine whether these changes to planning at the start of the twenty-first century are either growing pains, or perhaps the onset of arthritis.

Identity

We do not intend to set out in any detail the meaning of identity and its origins. This has already been covered within the planning literature very recently (see Neill, 2004; Hague 2005; and Reeves, 2005, for example) from a broad social science perspective and its associations to places. But we are interested in examining the relationship – if any – between identity and policy making as it may be manifesting itself in different parts of the UK. Identity remains an ongoing problem for politicians and policy makers. To some extent, the UK has been a fusion of diverse identities, stemming from various political and state acts, three thousand years of history, invasion, and war, and migrants arriving from foreign shores to mix with indigenous peoples in successive generations, all of which create their stories – both fact and fiction – about nationhood and identity (Duffy, 2001). State and political changes tend to be manufactured politically, such as the 1536 union between England and Wales, the 1707 Act of Union between England and Scotland, and the 1801 Act of Union between England and Ireland, that latter effectively creating the United Kingdom of Great Britain and Northern Ireland just two hundred years ago (O'Neill, 2004). There are those who prefer to take a protectionist stance towards a particular existing or perceived identity becoming tainted over time, and this identity may be territorially distinctive, or not. Others view nationhood and identity through the very political and cultural mechanisms recreated by past generations to give rise to the notion of a distinct identity, as the Victorians did to good effect throughout the nineteenth century towards a 'British Empire', while others believe that – for example – the very essence of Britain can be summed up with reference to such modern political labels or middle class creations as the Last Night of the Proms, a green and pleasant land, Radio 4, or the Church of England.

In reality, the notions of identity that are popularised today through the media and through cultural entertainment can themselves be based on particular conceptions of a way of living, of behaving, of the protection of a territory, of possession of an historical common past, of the notion psychologically of a community, and of a claim to political rule (Easthope, 1999). Identity is also sometimes defined more by what it is not; Scots, Irish and Welsh people, for example, often claim offence if labelled mistakenly as English, and assert themselves because of a neglect, unintentional or deliberate, on the part of an English observer to recognise their 'distinctiveness'. The call, or push, towards devolution within the UK is bound up with political and cultural claims, related to territory, for a desire for home rule, self government, or simply greater recognition from the central state. It may also relate to social and economic differentiation too, in economic disparities, health problems and deprivation, with maps of differentials corresponding geographically to the north, as opposed to the more prosperous south, of the nation, with a popular notion that there is inequity or unfairness in the application of state policies or funding to ameliorate the differences, or even a perception of territorial advantage of some regions over others.

The policies and actions of the UK government over decades can be viewed from this territorial and identity perspective, which in itself creates political opposition and calls for change and protest in support of greater needs that have a unique identity badge. The UK has witnessed these changes most markedly during the twentieth century through relationships between the UK government, based in London, and each of the Celtic countries, where the degree of opposition from nationalist quarters and calls for power, recognition and autonomy have varied between a vocal minority, as in the case of Wales, to outright hostility and terrorism in the case of Northern Ireland. The devolution process for Scotland and Wales, and the peace process in Northern Ireland, since 1997 have shifted this relationship markedly, to the point that in the most extreme oppositional case, the IRA announced in July 2005 that the 'war' with the UK government was, in effect, at an end.

Conversely, the regionalisation attempts in England by the UK government through the creation of Regional Assemblies has stalled, following the failed referendum in the North East region in November 2004. The North East, widely regarded as one of the English regions with one of the most clear regional identities, voted against the government plans, not perhaps because of a lack of proactive campaign for place-based governance, but rather because the form of the decentralisation package on offer from the central state did not meet the expectations of the region or was not regarded, perhaps, as genuine. This result may have halted the government's regionalisation plans, but it has not meant that regional voices, regional culture and strategy making in distinctive ways remain unaddressed across the constituent parts of England, and the vote has created more of a problem for the central state in defining its new role and relationship with the partly empowered regions than it has for the most appropriate model for sub-national government.

Identities are not configured merely through existing or predetermined governmental boundaries. People possess multiple identities and senses of attachment and loyalty and, in a country that politically has promoted multiculturalism, there are now other forms of opposition emerging within different territorial and scalar forms. These trends mirror what has been occurring in other parts of Europe, amid constant concerns about immigration and asylum seekers. But extremism and a preparation to utilise violence and terrorism to push cultural, religious or political identity within nation states is not new; today it has merely taken on a different form. The Notting Hill riots in London in the 1950s, and the inner city riots in the UK on occasions through the 1980s, occurred within deprived communities and possessed a clear race hallmark and anti-government stamp. The 7 July 2005 bombing attacks on London take us into new and more extreme territory.

The 2005 attacks led to calls politically and in the media for strength and determination – 'Britain can take it' – similar to the sentiments expressed throughout the Second World War against foreign invasion, and a desire to reawaken national identity against the perpetrators. Far right groups, such as the extremist British National Party, have orchestrated attacks against communities, businesses and mosques, thus galvanising extreme support from other sections of British society. When it was reported that the suicide bombers had either been born in the UK or else had become naturalised citizens, there was a feeling of bewilderment on the part of the public. Keith Best, the head of the UK Immigration Service, remarked in a news television interview at the time that he was 'appalled' that an individual who had sworn allegiance to the Queen upon becoming a British citizen could carry out such an atrocity (BBC News 24, 15 July 2005). Representatives of the minority groups called for more attention from the government towards the needs of minority groups in society and in urban areas where young people feel isolated, disillusioned or agitated by the actions of the government, particularly over its foreign policy.

Britain is now undergoing a period of quiet contemplation on the state's actions in embracing and including minority groups into its society and issues of identity and government are at the surface. The UK is becoming increasingly multicultural and that means the way its citizens perceive, react to, and oppose government and the state will take on different forms in different places. But attitudes within Britain are already changing markedly. A BBC opinion poll of 1004 adults on the subject of multiculturalism in Britain, conducted by MORI in August 2005, revealed that 62 per cent of respondents believed that multiculturalism made Britain a better place to live, while 38 per cent believed that it threatened it. Furthermore, 68 per cent disagreed with the view that the policy of multiculturalism should be abandoned as some right wing political parties are advocating, although 54 per cent surveyed thought that 'parts of the country don't feel like Britain any more because of immigration' (BBC, 2005). Most citizens will engage with those that are elected to govern on their behalf, through democratic means, but some – regrettably – will prefer more reactionary responses. The largest of the UK cities will notice the change more markedly; London is already home to over seven million people who possess 200 different nationalities and speak 300 different languages. Birmingham is the first city in the UK to have a non-white citizen majority. Other urban areas, such as Leeds and Bradford, also possess large non-white populations. Within rural areas, the change will be less marked, and possess a cross section of the UK population that will become more unusual for its homogeneity compared to the rest of the UK.

Many of the attempts to redesign the structures of government and policy making have occurred over the decades with a view to placating perceived inequities or inadequate representations, either through the designation of new forms of institutions with new powers, or else the decentralisation of central state activities to new scales; these may be 'new' in the sense that they are geographical levels of policy making that may not have been touched so pointedly by national government in the past. Invariably, such processes and re-institutionalisation are made from the state's perspective, in what they are permitted to see happen or promote at sub-national levels. They are therefore designed and implemented within processes aimed to further cement the institutions of the central state, and may be viewed as patriarchal in nature. These top-down changes occur within rigid and often predetermined state, government, territorial and funding boundaries that miss the stories and meaning of how places act and interact (Sandercock, 2004; Massey, 2005). The challenge for those in government and involved in strategy making is how to retain the apparatus of administrative government, with all its rules, codes and structures, while designing new meaningful processes and policies that genuinely address the needs and concerns of multiple identities, in different ways, in different cities and regions. Spatial planning should be central to this new envisioning.

Spatial planning

As academic contributions have attempted to make sense of ongoing drivers and processes of change, the political and planning practice worlds have continued to evolve at a rapid pace. We are currently entering one of the most turbulent periods for spatial strategy making in the UK that finds form in three ways – legislative, devolved, and coordinative. The legislative and policy framework of planning has undergone yet another period of reform, intended to modernise planning as a public service for the twenty-first century (Allmendinger *et al.*, 2003). In contrast to the legislative reforms to planning during the Thatcher and Major years, when planning was reduced to nothing more than a regulatory rump (Thornley, 1993; Allmendinger and Tewdwr-Jones, 1997), reforms under the Blair government have – perhaps somewhat surprisingly – potentially reinvigorated planning into a proactive coordinating activity, intended to assist in delivering development as part of continued economic growth. The Planning and Compulsory Purchase Act 2004 completely reforms the planning policy and strategy making function at national, regional, sub-regional, local and community levels: Regional Planning Guidance Notes are replaced with 'Regional Spatial Strategies'; 'Sub-Regional Strategies' are introduced for the first time; Structure Plans, Local Plans and Unitary Development Plans are replaced with 'Local Development Frameworks', 'Action Area Plans' and 'Masterplans'; 'Community Strategies' are to find implementation at neighbourhood level within Local Development Frameworks; and 'public consultation' is to be replaced by 'public participation'.

The second form of spatial strategy reform concerns scale and uniqueness. For far too long, the UK government expected planning to be devised and implemented uniformly across all parts of Britain: one country, one system. But such an ethos lies counter to processes of devolution and regionalisation. It also contributed to the notion that planning was failing to deliver development in the right locations, to cater for the desires and expectations of local and regional actors, and produced standardized plans and policies that did not deliver. The intention is now for planning to become increasingly differentiated in different parts of the UK as a consequence of devolution, decentralisation and regionalisation (Tewdwr-Jones, 2002; Haughton and Counsell, 2004), and is being facilitated – albeit nervously – by a central state that realizes that a national planning process (originally devised in the aftermath of the 1939–45 war years) is incompatible to government policies intended to foster regional economic competitiveness (Tewdwr-Jones, 1999). Scotland, Wales and Northern Ireland are developing their own spatial strategies separate from England with a desire for differentiation (Allmendinger, 2002; Berry *et al.*, 2001; Harris *et al.*, 2002). Of course these processes are nevertheless occurring within the confines of pre-established boundaries; but the ethos marks a distinct break away from past practices.

The third aspect of reform concerns integration and the new role for spatial strategy making in achieving coordination between disparate actors and strategies. New forms of planning are being explored as tools to help resolve community, sub-regional and regional problems (Counsell *et al.*, 2003), alongside a renaissance for planning as the means of achieving policy integration and coordination and the promotion of sustainable development (Tewdwr-Jones, 2004). The emphasis here is to look at planning not as a delivery process per se, in the style of planning under the welfare state, but rather as a strategic capacity and political integration mechanism intended to cement the increasingly fragmented agents of the state, all of whom possess their own agendas, political objectives, strategies, and resources, but who need to cooperate in order to deliver projects and developments. Planning is being looked at to ensure compatible working and strategic coordination when desired. This process, dubbed 'spatial planning' rather than town and country planning, extends the remit of planning beyond mere land use development and the 1980s/1990s regulatory rump into a coordinative process where professional planners' duties concern management of agency integration (Healey, 1997; Lloyd and Illsley, 1999). Such a responsibility has already formed a significant element to new government policy on the role of and expectations on planning post-2004 (ODPM, 2004), and influenced the professional planning body's agenda on the role it expects its members to perform in the twenty-first century (RTPI, 2003). These three drivers of change within spatial plan making are transforming the activity and scope of planning, across scales and across territories in varied ways and at varied times.

Conclusions: the rebirth of civics and planning

In conclusion, what makes the subject of territory, identity and spatial planning an interesting and essential one at the present time for us, is the potential that now exists to identify and build more meaningful democratic processes across the UK. These new political spaces should be distinctive and embedded in particular places, where citizens feel able to interact, voice their opinions, and utilise the forum of participation, policy making and governance to create places and futures that they feel are their own. In essence, we believe it is time that sub-national territories are allowed to re-create a new sense of civics. Spatial planning can become a central mechanism in this pursuit if used as an interactive, integrative and participatory mechanism, but with a realistic assessment of the handicap that institutional government and state power can frequently hold over such opportunities. As such, we should talk of and expect a range of spatial plannings to emerge and flourish.

We will still require institutional spaces of government and policy making, and professionals and officials who are able to form a nexus between citizens and the state. But the re-engagement of planners and civil servants into the objective of civics, at the neighbourhood, local and sub-regional levels, will require immense changes within the state, not least in liberalising attitudes within the centre towards sub-national territories. This will necessitate not only the central state being prepared to decentralise institutions to other tiers, it means being prepared to relinquish power and control. It will also require greater capacity and a desire for governing on the part of those in neighbourhoods, cities and regions, to challenge the centre and be prepared to take on greater responsibilities.

The changes we have witnessed in the UK over the past decade demonstrate a marked shift in the decentralising ethos on the part of the central state, and degrees of enthusiasm towards engendering localisation and regionalisation, but not necessarily with the aim of creating genuine localism or regionalism. The reforms since 1997 – that are described and analysed in detail within this edited collection – mark out what might be viewed as phase one of the decentralising approach; for us, it has not happened in all places as swiftly or as successfully as we would have preferred. But it has nevertheless commenced.

One of the biggest barriers the contributors to this book identify in the decentralising process is the degree to which the central state is prepared to 'let go' in fostering alternative forms of governance, and a (natural) inability on the part of all those in government to look at places and spaces outside the reductionist confines of their own administrative, governmental and professional boundaries. If we are to progress towards a phase two of decentralisation, that will take account of territorial differences, multiple identities, civic responsibility, and rouse public participation in place creating – and on the terms of those living within territories, rather than on the terms of the central state – we believe we shall be moving in the right direction. It will not be an easy or straightforward transition, and there will be much contention on the way. But that should not detract us from the objective at hand: to create new governance processes, embedded within their territories, which inspire civic engagement and build confidence and ownership.

References

Agnew, J. (1997) 'The dramaturgy of horizons: geographical scale in the "reconstruction of Italy" by the new Italian political parties, 1992–1995', *Political Geography*, 16(2): 99–121.

Allmendinger, P. (2001) 'The future of planning under a Scottish parliament', *Town Planning Review*, 72(2): 121–148.

Allmendinger, P. (2002) 'The death of the sovereign state? Prospects for a distinctly Scottish planning', *European Planning Studies*,10(3): 359–81.

Allmendinger, P. and Haughton, G. (2006) 'The fluid scales and scope of spatial planning in the UK', *Environment and Planning* A (forthcoming).

Allmendinger, P. and Tewdwr-Jones, M. (1997) 'Post-Thatcherite urban planning and politics: a Major change?', *International Journal of Urban and Regional Research*, 21(1): 100–16.

Allmendinger, P., Tewdwr-Jones, M. and Morphet, J. (2003) 'New order: planning and local government reforms', *Town and Country Planning* 72(9): 274–7.

BBC (2005) BBC Multiculturalism Poll, Final Results 10 August 2005, MORI25982, MORI, London (available at news.bbc.co.uk).

Beer, S. (2001) 'New Labour: Old Liberalism?', in S. White (ed.) *New Labour: The Progressive Future?*, London: Palgrave.

Berry, J., Brown, L. and McGreal, S. (2001) 'The planning system in Northern Ireland post-devolution', *European Planning Studies*, 9(6): 781–91.

Blair, T. (1998) *The Third Way. New Politics for the New Century*, London: Fabian Society.

Boyer, R. and Hollingsworth, J.R. (1997) 'From national embeddedness to spatial and institutional nestedness', in J.R. Hollingsworth and R. Boyer (eds), *Contemporary Capitalism: The Embeddedness of Institutions*, Cambridge: Cambridge University Press, 433–84.

Brenner, N. (2004) *New State Spaces: Urban Governance and the Rescaling of Statehood*, Oxford: Oxford University Press.

Brivati, B. and Bale, T. (eds) (1997) *New Labour in Power: Precedents and Prospects*, London: Routledge.

Buckler, S. and Dolowitz, D. (2000) 'New Labour's ideology: a reply to Michael Freeden', *Political Quarterly*, 71(1): 102–9

Callinicos, A. (2001) *Against the Third Way*, Cambridge: Polity Press.

Castells, M. (1996) *The Rise of the Network Society*, Oxford: Blackwell.

Coates, D. and Lawler, P. (eds) (2000) *New Labour in Power*, Manchester: Manchester University Press.

Cooke, P. (1983) *Theories of Planning and Spatial Development*, London: Hutchinson.

Counsell, D., Haughton, G., Allmendinger, P. and Vigar, G. (2003) 'From land use plans to spatial development strategies: new directions in strategic planning in the UK', *Town and Country Planning*, January: 15–19.

Driver, S. and Martell, L. (1998) *New Labour: Politics After Thatcherism*, Cambridge: Polity.

Driver, S. and Martell, L. (1999) 'New Labour: Culture and economy', in L. Ray and A. Sayer (Eds) *Culture and Economy After the Cultural Turn*, London: Sage.

Driver, S. and Martell, L. (2000) 'Left, right and the third way', *Policy and Politics*, 28(2): 147–61.

Easthope, A. (1999) *Englishness and National Culture*, London: Routledge.

Finlayson, A. (1999) 'Third Way Theory', *Political Quarterly*, 70(3): 271–9.

Freeden, M. (1999) 'The Ideology of New Labour', *Political Quarterly*, 70(1): 52–61.

Fukuyama, F. (1989) 'The End of History?', *National Affairs*, Washington DC:

Gamble, A. and Wright, T. (1999) 'Introduction: The new social democracy', in A. Gamble and T. Wright (eds) *The New Social Democracy*, London: Blackwell, 1–9.

Giddens, A. (1998) *The Third Way*, Cambridge: Polity Press.

Giddens, A. (2000) *The Third Way and its Critics*, Cambridge: Polity Press.

Gray, J. (1996) *After Social Democracy*, London: Demos.

Hague, C. (2005) 'Planning and place identity', in C. Hague and P. Jenkins (eds), *Place Identity, Participation and Planning*, London: Routledge.

Hall, S. (1998) 'The Great Moving Nowhere Show', *Marxism Today* special edition interactive version, http://www.ge97.co.uk/mt/

Harris, M. (1999) 'New Labour: government and opposition', *Political Quarterly*, 70(1): 42–51.

Harris, N., Hooper, A. and Bishop, K. (2002) 'Constructing the practice of "spatial planning": a national spatial planning framework for Wales', *Environment and Planning C: Government and Policy*, 20(4): 555–72.

Haughton, G. and Counsell, D. (2004) *Regions, Spatial Strategies and Sustainable Development*, London: Routledge.

Hay, C. (1997) Blaijorism: towards a one vision polity, *Political Quarterly*, 68(4): 372–8.

Hay, C. (1999) *The Political Economy of New Labour: Labouring under False Pretences?*, Manchester: Manchester University Press.

Healey, P. (1997) *Collaborative Planning*, Basingstoke: Macmillan.

Heffernan, R. (2001) *New Labour and Thatcherism: Political change in Britain*, London: Palgrave.

Hewitt, M. (1999) 'New Labour and social security', in M. Powell (ed.) *New Labour, New Welfare State? The Third Way in British Social Policy*, Bristol: Policy Press.

Hutton, W. (1996) *The State We're In*, London: Vintage Press.

Jacobs, M. (1999) 'Environmental Modernisation: The New Labour Agenda', *Fabian Society Pamphlet 591*, London: Fabian Society.

Jayasuriya, K. (2000) 'Capability, freedom and the new social democracy', *Political Quarterly*, 71(3): 282–99.

Jones, T. (1996) *Remaking the Labour Party: From Gaitskell to Blair*, London: Routledge.

Kenny, M. and Smith, M. (1997) '(Mis)understanding Blair', *Political Quarterly*, 68(3): 220–30.

Kerr, P. and Marsh, D. (1999) 'Explaining Thatcherism: Towards a multidimensional approach, in D. Marsh, J. Buller, C. Hay, J. Johnston, P. Kerr, S. McAnulla and M. Watson (eds), *Postwar British Politics in Perspective*, Cambridge: Polity Press.

Le Grand, J. (1998) 'The Third Way Begins with CORA', *New Statesman*, 6 March: 26–27.

Lloyd, M.G. and Illsley, B. (1999) 'Planning and developed government in the United Kingdom', *Town Planning Review* 70(4): 409–32.

Lloyd, M.G. and McCarthy, J. (2000) 'The Scottish Parliament, regulation and land use planning', *European Planning Studies*, 8(2): 251–6.

Lowe, P. and Ward, S. (1998) *British Environmental Policy and Europe: Politics and policy in transition*, London: Routledge.

McKinsey Global Institute (1998) *Driving Productivity and Growth in the UK Economy*, London: McKinsey Global Institute.

Marquand, D. (1999) *The Decline of the Public: The Hollowing Out of Citizenship*, Polity Press: Cambridge.

Massey, D. (2005) *For Space*, London: Sage.

Naughtie, J. (2001) *The Rivals*, London: Fourth Estate.

Neill, W.J.V. (2004), *Urban Planning and Cultural Identity*, London: Routledge.

ODPM (2004) *Creating Sustainable Communities: Planning Policy Statement 1*, London: ODPM.

O'Neill, M. (ed.) (2004) *Devolution and British Politics*, London: Pearson Longman.

Planning (2000) 'End of Term Report Chastises Planners', 15 December: p.3.

Plant, R. (2001) 'Blair and ideology', in A. Seldon (ed) *The Blair Effect: The Blair Government 1997–2001*, London: Little, Brown and Company.

Powell, K. (2001) 'Devolution, planning guidance and the role of the planning system in Wales', *International Planning Studies*, 6(2): 215–22.

Powell, M. (ed.) (1999) *New Labour, New Welfare State? The Third Way in British Social Policy*, Bristol: Policy Press.

Rawnsley, A. (2001) *Servants of the People* (2nd Edition), London: Penguin.

Reade, E. (1987) *British Town and Country Planning*, Milton Keynes: Open University.

Reeves, D. (2005), *Planning for Diversity: Policy and Planning in a World of Difference*, London: Routledge.

Rouse, J. and Smith, G. (1999) 'Accountability', in M. Powell (ed) *New Labour, New Welfare State? The Third Way in British Social Policy*, Bristol: Policy Press.

RTPI (2003) *A New Vision for Planning*, London: RTPI.

Rubinstein, D. (2000) 'A new look at New Labour', *Politics*, 20(3): 161–7.

Sandercock, L. (2004) *Mongrel Cities: Cosmopolis II*, London: Continuum.

Savage, S. and Atkinson, R. (eds) (2001) *Public Policy under Blair*, Basingstoke: Palgrave.

Scholte, J.A. (2000) *Globalization: A Critical Introduction*, London: Macmillan.

Scott, A.J. (1996) 'Regional motors of the global economy', *Futures*, 28(5): 391–411.

Scottish Executive (2001) *Review of Strategic Planning*, Edinburgh: Scottish Executive.

Seldon, A. (ed.) (2001) *The Blair Effect*, London: Little Brown Books.

Shaw, E. (1994) *The Labour Party Since 1979: Crisis and transformation*, London: Routledge,

Smith, N. (1995) 'Remaking scale: competition and cooperation in prenational and post-national Europe', in H. Eskelinen and F. Snickars (eds), *Competitive European Peripheries*, Berlin: Springer, 59–74.

Social Exclusion Unit (1998) *Bringing Britain Together: A National Strategy for Neighbourhood Renewal*, London: HMSO.

Stoker, G. (ed.) (2000) *The New Politics of British Local Governance*, London: Macmillan.

Tewdwr-Jones, M. (1999) 'Discretion, flexibility, and certainty in British planning: emerging ideological conflicts and inherent political tensions', *Journal of Planning Education and Research*, 18(2): 244–56.

Tewdwr-Jones, M. (2001) 'Planning and the National Assembly for Wales: Generating distinctiveness and inclusiveness in a new political context', *European Planning Studies*, 9(4): 553–62.

Tewdwr-Jones, M. (2002) *The Planning Polity: Planning, Government and the Policy Process*, London: Routledge.

Tewdwr-Jones, M. (2004) 'Spatial planning: principles, practice and culture', *Journal of Planning and Environment Law*, 57(5):

Thatcher, M. (1993) *The Downing Street Years*, London: Harper Collins.

Thornley, A. (1993) *Urban Planning Under Thatcherism: The Challenge of the Market* (2nd Edition), London: Routledge.

Vincent, A. (1998) 'New ideologes for old?' *Political Quarterly*, 69(1): 48–58.

Welsh Assembly (2001) *Strategic Statement On The Preparation Of 'Plan For Wales 2001'*, Cardiff: Government of the National Assembly for Wales.

White, S. (1999) '"Rights and Responsibilities": A social democratic perspective', *Political Quarterly*, 70(3): 166–79.

White, S. (2001) 'The ambiguities of the Third Way', in S. White (ed.) *New Labour: The Progressive Future?*, London: Palgrave.

Young, S. (2000) 'New Labour and the environment', in D. Coates and P. Lawler (eds) *New Labour in Power*, Manchester: Manchester University Press.

2 Nationality, devolution and policy development in the United Kingdom

Michael Keating

Introduction – devolution in the United Kingdom

Devolution in the United Kingdom, as in other European states, responds to three sets of needs. First is the effort to accommodate national and cultural diversity within a reformed constitutional order, a task that has confronted British statecraft for over two centuries. Second, there is the pressure for democratization and accountability in the modern complex state in which policy-making systems seem to have escaped the purview of elected institutions of government. Third is a set of functional needs related to the changing spatial scale of economic and social systems and the challenge of steering and managing the process of economic growth and adaptation. While these have all driven policy in the direction of strengthening the meso or intermediate level of government, they have not all pointed to the same solutions and have had different impacts in the different parts of the United Kingdom. The result is a patchy, incremental and asymmetrical territorial settlement, which can be judged in two ways. For some, it is an inherently unstable arrangement, which neglects the needs of the state as a whole and makes impossible coherent spatial planning or territorial equity. Citizenship is fragmented and an overarching sense of British national identity thrown away in a reversion to ancient loyalties. For others, it is an appropriate adjustment to the realities of a multinational union embedded in a broader European system of regulation and policy making. Like the British constitution itself, it is piecemeal and gradually evolving without a preordained plan; and it reflects accurately the emerging European order of overlapping authority and mixed sovereignty. In this view, the new constitution is both old and new. Indeed, the British may have achieved the unique feat of going from a premodern to a postmodern constitution without passing through modernity in the form of the consolidated, homogeneous nation state.

The United Kingdom as a multinational state

The United Kingdom is one of the few European states explicitly to accept that it is multinational. This is reflected in the name of the state, the flag, the use of the term 'national' for the institutions of its component parts including the monarchy, which adapts itself to the various component territories even to the point of changing its religion when passing between England and Scotland. It has been characterized as a 'union state' or simply a political union (Rokkan and Urwin, 1983; Mitchell, 2003; Keating, 2001). This is to distinguish it from a unitary state, with a single set of governing institutions and a project for political homogenization; and from a federal state, in which there is a formal division of power between centre and constituent units, with all the latter having the same relationship to the former. The union state, rather, is characterized by its compound form, in which different units have been added on, keeping some of their old institutions and practices, but without a formal federal division of powers. Until 1999, this recognition of the distinctiveness of the component parts of the UK was accompanied by a highly centralized system of government, in which the Westminster Parliament enjoyed a legislative monopoly and claimed absolute and undivided sovereignty. The circle was squared through an elaborate system of territorial management in which peripheral interests were brought to the centre for resolution and economic and other concessions made to the constituent nations in return for maintaining loyalty to the state and the state parties. A Secretary of State for Scotland, dating from 1885, had the dual role of representing Scottish interests in the Cabinet and representing London government in Scotland. While he had little independent policy discretion, he was able to put a Scottish face on British policy measures (Midwinter *et al.*, 1991). In 1964, a similar arrangement was introduced for Wales although with less power. While Secretaries of State for Scotland had by convention to be Scottish MPs, Conservative governments from the 1980s were able to appoint English-based MPs as Secretary of State for Wales. Integration was further served by the British parties, which until the 1970s monopolized the representation of Scotland and Wales. Ireland had been governed until 1922 in a more 'colonial' mode in which it sent representatives to Westminster but was run by a Chief Secretary who never represented an Irish seat. From the 1870s, the British parties largely disappeared from Ireland, leaving the field to varieties of nationalism and unionism. After 1922, the larger part of Ireland became independent, while six counties in the north were insulated from British politics by devolution to the Stormont Parliament. Following the Second World War, national integration was furthered by the welfare state, based on principles of individual equity and universal access and by regional economic policies aimed at distributing prosperity as evenly as might be possible.

The prevailing mode of territorial management was repeatedly challenged on several fronts. There were those, particularly on the left, who regarded all concessions to nationality as inherently reactionary and who favoured unity, homogeneity and centralization. In the UK, doctrinal Jacobinism insisting on the unity of the state was weaker than in countries such as France, but was expressed in other ways, such as the insistence by the Labour Party on the unity of the working class, or the theme of Britishness as essentially bound up with social justice. This contributed to the uniformity of the main welfare and economic policy measures. Second, there were nationalists, believing in the independence of the constituent nations. In Ireland, this position became hegemonic after the First World War, but in Scotland was not a serious contender before the 1970s and in Wales has been a marginal force. Third, was the option of Home Rule, first elaborated for Ireland, adopted by William Ewart Gladstone in the 1880s and evolving into the idea of Home Rule all Round, the conversion of the United Kingdom into a federation of self-governing nationalities. The Irish settlement of 1922 took the momentum out of the Gladstonian option and during the twentieth century there was a strong trend to centralization. From the 1970s, however,

Home Rule came back on the agenda in the form of devolution and the settle-ment of 1999 is recognizably Gladstonian. Home Rule or devolution is an adaptation of the union state, recognizing the distinct characteristics or needs of each territory and accepting that each might have a different relationship to the centre and its own form of government; yet, at the same time, maintaining the principle of parliamentary sovereignty and insisting on the right of Westminster unilaterally to alter the constitutional order. This is not, however, how it is seen in the smaller nations, where devolution tends to be seen as a concession to their right of self-government. In Scotland, in particular, there is a countervailing doc-trine to that of Westminster sovereignty, in the form of popular sovereignty, an argument that was deployed to some effect in the devolution campaigns of the 1980s and 1990s (Campaign for a Scottish Assembly, 1988).

The United Kingdom has thus balanced nationality and sovereignty claims in it own unique way. It has never confused the nation with the state, as is often the case elsewhere, and accepts that there can be a number of different formulas link-ing the two. This has placed it in an unusual position in the face of state transformation in the early twenty-first century. Europe is an evolving political order in which sovereignty is shared among various levels of government. It is complex and asymmetrical. By providing an overarching political order, it has transformed the politics of nationalism by defusing the question of independence and blurring the distinction between independence and home rule (Keating, 2004). Scottish and Welsh nationalists, like their counterparts in most of Europe, now accept European integration as a framework for their national aspirations thus giving up, in practice if not in theory, dreams of absolute independence. Home rulers, for their part, have incorporated the European dimension to accept that the nations and regions of the UK need to have access not only to state but also to supra-national levels of government and regulation. In this they have joined the Europe of the Regions movement, which for some twenty years has been exploring the possibilities for sub-state entities to act within European insti-tutions. The effect has been to blur the practical differences between independence and home rule as independence supporters accept shared sover-eignty within Europe, while home rulers accept that devolution is not merely an internal reorganization of the state, but can also bring possibilities for limited self-determination within the wider order. Public opinion surveys show that voters in the smaller nations of the United Kingdom, like their counterparts elsewhere, do not make a sharp distinction between independence and devolution, rather see-ing self-government as a continuum of options and powers (Keating, 2001).

All this might seem to make it easier to accommodate the nationalities of the United Kingdom within the state and the broader European order. On the other hand, the party systems in Scotland, Wales and Northern Ireland are now all aligned on the national question, pitting nationalists against unionists and so polarizing around the more pure options of integration and independence. Nor have the British parties absorbed the implications of European integration or shar-ing sovereignty within the European Union, despite the precedents of the UK union. It is true that the nationalist parties, with the exception of Sinn Féin, are all pro-European and have made the relevant connections, but they have only been partially successful in carrying their own supporters. Even in Scotland, where there is an elite consensus in favour of Europe and public opinion has consistently been less hostile to Europe (albeit only by small margins), it is not the nationalist voters who are the most pro-European (Keating, 2005). The British Conservative and Labour parties, for their part, insist on setting up a fundamental conflict between the sovereignty of the United Kingdom and any form of European federalism. This creates a series of tensions in the new politics of the United Kingdom.

Democratization and decentralization

There are traditionally two ways of conceptualizing the relationship between democracy and decentralization. The Jacobin tradition regards national unity as a prerequisite to democracy and citizenship and sees governments representing only part of the people as an obstacle to civic equality and possibly a bastion of reaction. A related liberal tradition, visible in the work of John Stuart Mill (1972), identifies shared nationality as a necessary condition for democracy, providing the necessary demos, shared norms and mutual trust. Modern writers including David Miller (1995) and Ralph Dahrendorf (1995) have similarly seen the nation state as the foundation of democracy and social solidarity. Another tradition, however, sees decentralization as an essential foundation of liberal democracy, with power exercised as close as possible to the citizen and divided among different levels of government. Others have pointed out that in complex multinational states, the nation state version of democracy is simply not viable and that democracy entails adapting government to the communities of identity that actually exist, which may be at various levels, including the transnational, the state, the component nations of a plurinational state, or the local community (Keating, 2001; Tully, 2001).

Political changes over recent decades have brought back these debates. There has been a strengthening of identities in the nations of the United Kingdom and, as important, they have become politicized (which they always were in Northern Ireland). The relevant community for the articulation of democratic citizenship for many Scots is now Scotland; a similar trend in Wales has been less complete. In Northern Ireland, there is a conflict precisely over which should be the relevant community, the United Kingdom, Ireland as a whole, or the province of Northern Ireland.

Functional restructuring (see below) has at the same time taken much of policy making and regulation away from elected and accountable institutions and into the hands of appointed bodies or into complex networks spanning tiers of government and the public and private sectors. A large part of the new state apparatus emerged at the meso level, between central and local government. This includes the old Scottish and Welsh Offices as well as the Northern Ireland Office, and the array of quangos and government offices in the regions of England. In spite of periodic efforts at suppression, the latter have regularly resurfaced since the 1960s, giving the impression of a new tier of unelected government. This has fed the demand for democratic institutions at this level.

Functional change

The functional importance of the regional level in Europe grew steadily in the latter half of the twentieth century. In the post-war years, it was a critical level for spatial planning and infrastructure provision. National anti-disparity policies, often starting out with a local focus, took on a regional dimension, where they were integrated more or less successfully with spatial planning and urban development. From the 1980s, economic development to some extent took over from planning as the dominant theme of regional level policies. With the decline of national diversionary policies, regional policy has increasingly been decentralized to the regions themselves. The new regionalist thinking emphasizes the importance of factors specific to regions and localities as the key to success, often focusing on the social construction of the region and institutional matters (Storper, 1997; Cooke and Morgan, 1998). This has stimulated institutional reform, although it has taken varied forms. One is a narrowly defined functional regionalism, dedicated to creating the conditions for economic growth and entrusted to specialized institutions, often dominated by business interests, 'social partners' or the state. This form of regionalism has its origins in the 1960s, where it was applied in France, Italy and the United Kingdom, and has recently been revived in England. Experience shows, however, that efforts to depoliticize economic development in this way and to insulate it from other social and political forces come under challenge and that the field rapidly becomes politicized. This leads to demands for elected regional government, as occurred in France and Italy and as we have seen in England. This is not to say that regional governments then take over the economic development function, as economic interests may remain outside or be organized at different spatial levels, as is the case in Italy (Trigilia, 1991).

There are also some trends to regionalization of welfare state functions. National health systems in Europe, unmanageable at state level, have in various ways been regionalized. Vocational education and training has also been regionalized and increasingly been aligned with welfare in the form of active labour market policy. Higher education has taken on a regional dimension, partly for managerial reasons and partly because of the connection with economic development policies. The main income support functions remain a national responsibility in most European welfare states, although there is some evidence that social solidarity, at least in regions and stateless nations with a strong identity, has taken on a territorial dimension.

Region and nation building

It would be wrong, as some excitable observers (e.g. Ohmae, 1995) have done, to extrapolate directly from functional changes to politics and institutions. Functional pressures to regionalization are mediated by political factors and have very different impacts on territories, depending on the local conditions. We see the emergence of regions across much of Europe, but they take very different forms. In some cases, the region is an economic space, with a well-articulated network of economic interactions and may constitute a local system of production. This is the strongest form of the 'new regionalism' that has been so prominent in thinking about economic development. A region may constitute a political space, in which citizens identify with it, articulate their demands within that spatial framework, and aspire to self-government. There may or may not be a regional civil society in the form of regionalized interest groups, associations and private institutions. Another variable is institutional. A region may form a level of government, be managed by agencies or not have any distinctive institutions at all. These dimensions are interrelated but in principle independent of each other. So a region might be an economic space without any political significance or a political space without any economic significance. It might be a political space but without its own institutions or have its own institutions but lack citizen identification with them. The territory might have a certain institutional identity, but the population remain divided on their national identity. Whereas in others, the various dimensions coincide with a strong form of regionalism or stateless nationality, as is the case in Scotland. In other cases, there is a weaker regionalism, as in most of England. Wales would be somewhere in between, while Northern Ireland is a clear case of a divided society.

Yet the extent of coincidence of the various dimensions is not given by nature. It is the result of political entrepreneurship, of social and political mobilization and of institution building. Historic nationalities like Scotland and Wales clearly have something to build on here, but only if the past can be interpreted and used as a guide to the future. Developments in the UK and in Europe have permitted this, by opening up questions of sovereignty and encouraging the rediscovery of identity, but this has been pursued by political movements, including the nationalist and home rule parties. Functional restructuring has also encouraged the rebuilding and redefinition of nationality, by pointing to the opportunities for economic development and competitiveness at the meso level. New regionalist theories show how, in a world where autarky is no longer possible, territorial groups and institutions can gain a measure of control over their own economic future and their insertion into the European and global trading order; and they do not need to become states to do so. These same theories point to the importance of social capital, co-operation and trust as factors in economic success so that the very elements that in the past were often seen as obstacles or evidence of retarded modernization, can now be invoked as assets.

The cultural, political, economic and social constitution of the region in turn will affect the nature of the development project pursued within it. Some regions are disorganized and subject only to state policies or dependent on the vagaries of the international market. Others are organized in the form of economic development coalitions including selected interests and excluding others. Where there are elected institutions and a strong sense of territorial political identity, then the political agenda is broadened. Interests may be brokered at the territorial level and social compromises made there. Institutional structures, patterns of interest articulation and political culture will condition policy choices as well as the process of policy making. It may be that in present-day Europe the great ideological divisions of the past no longer structure politics but, within the parameters of the market economy and welfare state, there are critical choices to be made and these are often being made at the meso level.

Nation and region in the United Kingdom

The United Kingdom has become something of a laboratory for constitutional change because, within its borders, it has territorial entities of so many kinds. The impact of functional change and the various principles of constitutional reform are therefore very different from one part to another.

England has a long unitary history, which it has retained even as the United Kingdom was built around it. Many English people treat the UK and England as interchangeable in the same muddled way that foreigners often do. Yet, when they do contemplate the smaller nations of the union, they tend to take a rather indulgent view of their rights to govern themselves. There is almost no popular opposition in England to Scottish and Welsh devolution or even independence, or to the prospect of Irish unity (Keating, 2001). Efforts to arouse them on the West Lothian anomaly, that Scots can vote on their domestic affairs while they (the English) cannot vote on the domestic affairs of the Scots, have been singularly unsuccessful. More broadly, it seems that the ideology of unionism, once such an important force in UK politics, has more or less died, an interesting but unexplained phenomenon that seems to have few counterparts in any modern polity. And yet this has been accompanied by only a faint revival of a distinct English identity, most being content to muddle through as in the past.

The territorial question in England, as a result, has concerned the regional level, with the driving forces being decentralization and democratization, and functional needs related to economic development and planning. At the regional level, there is a rather weak identity, especially outside the northern part of the country (Bond and McCrone, 2004) and interest groups are weakly regionalized (Keating and Loughlin, 2002). Regional planning structures established by the Labour government in the 1960s involved a mixture of central state direction and regional-level corporatism in an effort to co-opt local social and economic actors into the national development effort. Since they did not rest on firm foundations of territorial identity and did not provide power bases for significant political elites, they were rather easily swept aside by the incoming Conservative government in 1979. They had, in any case, lost much of their rationale with the collapse of the National Plan in 1966–7. The regional development machinery established by Labour after 1997 bears a strong family resemblance to the 1960s model, although there is now provision for private sector dominance of the Regional Development Agencies, in accordance with the new political orthodoxy shared by Labour with the Conservatives. As in other cases, this narrowly based functional regionalism has been under some challenge by those wanting to move to elected regional government. Although the government's failed schemes for English regional government would have essentially limited it to the same co-ordinating, planning and development role that the RDAs and Regional Assemblies have done since 1999, their political agenda would inevitably have been widened, and they would have become the focus of interest group mobilization. In this way, they could have stimulated the development of distinct regional civil societies and the growth of regional political spaces. These, however, would still be regions of England and not component parts of the United Kingdom in the way that Scotland, Wales and Northern Ireland are.

Scotland is a historic nation, with a strong sense of identity. Unlike many stateless nations in Europe, it has rather stable and uncontested borders. Identity is not based on language or ethnic criteria, but on the maintenance of important pre-Union institutions and a civil society. These have made possible a rather inclusive or 'civic' nationalism that has largely overcome historic divisions between Highlanders and Lowlanders, Scots and Irish or Catholics and Protestants. Since the 1970s, Scottish national identity has become more important at the expense of British identity (Rosie and Bond, 2003) and has become more politicized, linked increasingly to demands for home rule. It is not clear just why this is so. One theory is that Britishness was never a dominant identity in Scotland, merely an instrumental one, or that it was really tied to the Empire rather than the UK state. With the weakening of instrumental support for the union and the end of Empire, an underlying Scottishness thus reasserted itself. Such an explanation is difficult to test, and underplays the strong sense of British identity that did seem to be fostered in the two world wars and by the welfare state in the twentieth century. A more institutional explanation would point to the way in which the British state itself has sustained a distinct Scottish identity through progressive extensions of administrative devolution, themselves intended to assuage separatist threats or at least the rise of territorial parties. Around the Scottish Office and its associated agencies, there grew an array of interest groups, encouraging the articulation of sectoral demands in a territorial framework. This created a distinct Scottish political arena, which could be accommodated within the British political system as long as Scottish electors voted more or less the same way as those in England, and Scottish questions reduced to matters of economic advantage, which could be addressed through spending policies and regional policy initiatives. When Scotland was ruled by political parties for which it had not voted, however, the legitimacy of the system was put in question. This happened in the late nineteenth century, in the aftermath of the First World War, between 1970 and 1974 and – the most prolonged example – between 1979 and 1997.

During this last phase, Scottish political identity was reconstructed around opposition to Thatcherism and defence of the welfare settlement. It is not that Scots are naturally or massively more favourable to collectivism and welfare than the English. They are marginally so but the difference is not enough to explain the divergences in voting behaviour over recent decades (Keating, 2005). In any case, Thatcherite welfare policies never gained majority support in England. What happened in Scotland is that nationality and territory were used as a basis to mobilize resistance to neo-liberalism at a time when neither British national solidarity nor class-consciousness appeared up to the task. In the process, historical memories, myths and cultural norms were invoked and re-interpreted, presenting Scotland as a naturally progressive country with a strong sense of community. There was a Thatcherite counter-mythology, deployed by the Iron Lady herself in the famous 'sermon on the mound' in which she strove to present Adam Smith and the Scottish Enlightenment generally as a form of proto-Thatcherism, but to no avail. The cross-party campaigns for home rule in the 1980s and 1990s also fostered an ethos of consensus and a belief that, freed from the grip of London, Scotland could embark on a 'new politics'. Such ideas were unlikely to survive unscathed the rough and tumble of post-devolution politics but have left some mark on Scottish political society. With this preparation, the proposals for a Scottish Parliament brought forward by the Labour government in 1997 faced little serious opposition and won the support of three out of four voters in the referendum of 1998.

National identity has been weaker in Wales. The country was united with England earlier than Scotland and largely absorbed administratively. National identity was carried by the language, by religion (Protestant nonconformity) from the eighteenth century, and by a radical Liberal voting tradition from the nineteenth (Morgan, 1980). Language, however, was increasingly divisive as Welsh retreated to the northern heartlands, while nonconformity and radicalism declined in importance in the modern state and the main British parties took over. In the referendum of 1979, Welsh devolution was massively rejected, gaining significant support only in the Welsh-speaking heartlands, now dominated politically by Plaid Cymru. Industrial south Wales, English speaking but still Welsh, was uninterested; while a third Wales had emerged of English incomers with little connection to Welsh traditions. Welsh civil society was weak in comparison with its Scottish counterpart, and there were few distinctively Welsh interest groups. There was no cross-party movement for home rule in Wales during the 1980s and 1990s, and yet there were some developments comparable to those in Scotland. After a short phase of apparent convergence, Welsh voting behaviour from 1987 diverged again from England. Interest in devolution revived in some political and intellectual circles, and Wales began to re-emerge as a more integrated national region, especially in the context of the Europe of the Regions. Labour was persuaded to include Welsh as well as Scottish devolution in its electoral programme and, while it only gained the support of half the voters on a 50 per cent turnout, this represented a massive advance on 1979.

Nationality in Northern Ireland is the main political and social divider and, since the collapse of the Unionist hegemony in the early 1970s, governments have sought a compromise solution that would allow for recognition of both traditions and a sharing of power between the two communities. The main items have been a power-sharing or consociational devolved government for Northern Ireland, and an articulation of this both with the United Kingdom and with the Republic of Ireland. The Sunningdale agreement of 1973 brought in moderate Unionists and the moderate nationalists of the Social Democratic and Labour Party (SDLP) against the opposition of republicans (Sinn Féin) and extreme unionists. It provided for a power-sharing assembly and executive and an advisory Council of Ireland but was brought down by a strike of loyalists in the Ulster Workers' Council. Twenty-five years later, the Good Friday Agreement produced a more elaborate scheme but whose principles were identical. The difference was that Sinn Féin and another section of the unionists (including David Trimble) had grown up sufficiently to accept the compromise, although this still left about half of Protestant opinion outside the fold. This has had a chequered existence but so far nobody has come up with any other way to accommodate the national question in Northern Ireland. Were it allowed to function, then the expectation could be that it would become a significant focus for interest group activity and mobilization in the province, creating a Northern Ireland political system in which parties could work without giving up their longer-term national aspirations. The functional requirements of government could thus be separated to some degree from the politics of national identity.

Devolution, nationality and policy performance

The strength and resilience of the devolved institutions depend to a large degree on the factors reviewed above, nationality, civil society and political trust. They also depend on the performance of the institutions themselves and the contribution they make to economic and social progress. Surveys at the time of devolution show that expectations in Scotland were high that devolution would improve economic performance and public services and give Scotland a stronger voice in the United Kingdom (Paterson *et al.*, 2001). In the other cases, where there had been less mobilization, expectations were not set as high.

The Scottish Parliament rests on a wide and deep sense of national identity, reflected in the large popular majority for its establishment. Scottish identity, together with the existence of Scots law and the existing machinery of administrative devolution, determined that it should have broad legislative and executive powers, extending to all items not expressly reserved for Westminster. This, in turn, ensured that it would become a central part of Scottish political life, with no interest groups able to ignore it entirely, although some are much more engaged than others. Its legitimacy does therefore depend on its policy performance. Polls over the years have shown that, while few voters think that devolution has done any harm, most feel that it has made little difference. But support for devolution remains as high as ever, even as expectations have been disappointed. In Wales, there is a similar lack of enthusiasm for what the Assembly has done, but a marked increase in support for it as an institution (Curtice, 2004) suggesting, again, that it rests upon a new sense of common identity and not just upon its output record. There is a Welsh political life that did not previously exist and the Assembly has become a central part of this. English regional governments are unlikely to be able to depend on this kind of identity or centrality, and will probably be judged strictly on their results. This would weaken them as institutions, should they fail to deliver on policy expectations, allow central government to dominate them and even permit their abolition by some future Parliament, as happened with the metropolitan counties in the 1980s. The Northern Ireland institutions have not had a long enough period of continuous existence to judge them, but the ease with which the central government was able to suspend them suggests that they are seen largely in instrumental terms.

It is difficult to judge the effect of devolution on public policies in the nations of the United Kingdom, since England, Scotland and Wales have been under Labour-dominated governments that have tended naturally to pursue similar policy lines. Spending priorities have been similar, with a strong emphasis on education and health. There is, however, a difference in policy style and there are important differences in the way public services are managed and delivered. Policy making in Scotland and Wales tends to be more consensual, subject to consultation and negotiated with groups and in networks. This is partly down to the pre-devolution commitment to consensus and new politics, and partly to the institutional arrangements, including the role of committees of the Scottish Parliament and National Assembly for Wales. It is also a product of the weaker policy capacity of the executive branches in the devolved territories, which were used before 1999 merely for adapting policy made in Whitehall. This forces them to work through professional networks and with groups rather than laying down policy from the centre. The legitimacy given to the devolved institutions by national identity and their endorsement by referendum in turn encourages groups to co-operate.

In the public services, there is a greater commitment to traditional social democratic principles of universalism and egalitarianism, in contrast to the emphasis on competition, choice and differentiation in England. So Scotland and Wales have rejected foundation hospitals, star-ratings for hospitals, selective schools and school league tables. Scotland has abolished up-front university fees and top-up fees and introduced free personal care for the elderly. Wales has tried, within its powers, to follow suit and has also resisted the Private Finance Initiative/Public Private

Partnerships. This is not a direct response to public opinion that, except for opposition to selective education in Scotland, tends to be rather close to that in England; but it does draw upon a particular sense of social solidarity that is tied to national identity and an emerging distinct social citizenship. It may also owe something to the strength of public sector professionals and their greater commitment to universalism and equality for which there is some sporadic evidence (Keating, 2005). There is also a difference in objective conditions. Labour has explicitly sought to keep the middle classes in the south of England within the welfare state by promising them their own niche within it. In Scotland, where there is less than half the amount of private medicine and a quarter of the amount of private education, this has proved unnecessary. Welsh First Minister Rhodri Morgan has made much of this too, marking out a distinct Welsh model of social democracy tied to community, and talking of putting 'clear red water' between Cardiff and London. The Scottish Executive, while following a similar line, has been altogether more reticent and it seems that, left to his own devices, First Minister Jack McConnell could be much more New Labour. Under pressure from London, steps have been taken in 2005 to bring Scotland more in line with the English system of health care management. However, unlike Tony Blair, McConnell faces more opposition to his left than to his right and is in coalition with the Liberal Democrats, who have tended to pull him away from New Labour ideas. Again, we should not exaggerate the natural difference from England. There was massive opposition among English MPs to top-up fees and foundation hospitals, which were only forced through with the support of loyalist Scottish Labour MPs. In Scotland, however, it has been possible to mobilize opposition to differentiation in the welfare state more effectively.

New regionalist theories of economic development suggest that institutions and norms (or 'social capital') are important elements in determining success or failure. We might expect, therefore, that the new and re-created forms of national identity and social interaction created by devolution will affect the constitution and working of economic development coalitions. Here again the scope for policy divergence is limited, not just within the United Kingdom but across Europe where there has been something of a convergence around a model of endogenous development, focused on human capital, research and development, and entrepreneurship. This has been adopted in Scotland and Wales, but perhaps with a stronger emphasis on social inclusion than in England. It has not been easy to mobilize an inclusive development coalition based on shared national interests. Although there has historically been a Scottish and Welsh lobby able to come together to defend certain material advantages, the business community has long been suspicious of political devolution and still tends to look to London rather than Edinburgh or Cardiff. Devolution has not re-created the local bourgeoisie that existed in Scotland, Wales and the English regions in the nineteenth century, or that was created in Quebec during the Quiet Revolution. Culture and shared identity are sometimes invoked in Wales as development assets, but the first major Scottish Executive statement on economic policy expressed some bemusement about the theme of culture, recognizing that it might be important but confessing that it did not know why (Scottish Executive, 2000). Over time, more of a consensus has developed on economic policy, recognizing the limitations of the devolved institutions and bringing in the major stakeholders, but Scotland and Wales are a long way from the sort of concerted action and mobilization that has proved possible in other small European nations.

Conclusions

The United Kingdom is an asymmetrical union kept together by a certain sense of common nationality and by the principle of parliamentary sovereignty. Both have come under some challenge in recent decades but demands for constitutional reform from the constituent nations have been very different. The response has been an asymmetrical constitution, with distinct forms of government for the component parts and provision for differentiation within England. The functional requirements for modern territorial government and spatial development may be similar across the UK and Europe, but the impact is distinct. Nationality and regionalism are affecting not only the institutions of government in the component parts of the United Kingdom, but the policy process and the substance of policy itself.

Acknowledgement

The research on which this paper is based was supported by the ESRC programme on Devolution and Constitutional Change, and by the Leverhulme Trust programme on Nation and Region in the United Kingdom.

References

Bond, R. and McCrone, D. (2004) 'The growth of English regionalism? Institutions and identity', *Regional and Federal Studies*, 14(1): 1–25.

Campaign for a Scottish Assembly (1988) 'A claim of right for Scotland', in O.D. Edwards (ed.) *A Claim of Right for Scotland*, Edinburgh: Polygon.

Cooke, P. and Morgan, K. (1998) *The Associational Economy: Firms, Regions, and Innovation*, Oxford: Oxford University Press.

Curtice, J. (2004) 'Restoring confidence and legitimacy? Devolution and public opinion', in A. Trench (ed.), *Has Devolution Made a Difference? The State of the Nations 2004*, Exeter: Imprint Academic.

Dahrendorf, R. (1995) 'Preserving prosperity', *New Statesman and Society*, 15/29 December, pp. 36–41.

Keating, M. (2001) *Plurinational Democracy: Stateless Nations in a Post-Sovereignty Era*, Oxford: Oxford University Press.

Keating, M. (2004) 'European integration and the nationalities question', *Politics and Society*, 31(1): 1–22.

Keating, M. (2005) *The Government of Scotland: Public Policy Making after Devolution*, Edinburgh: Edinburgh University Press.

Keating, M. and Loughlin, J. (2002) *Territorial Policy Communities and Devolution in the United Kingdom*, Working Paper SPS 2002/1, Florence: European University Institute.

Midwinter, A., Keating, M. and Mitchell, J. (1991) *Politics and Public Policy in Scotland*, London: Macmillan.

Mill, J.S. (1972) *Utilitarianism, On Liberty and Considerations on Representative Government*, London: Dent.

Miller, D. (1995) *On Nationality*, Oxford: Oxford University Press.

Mitchell, J. (2003) *Governing Scotland: The Invention of Administrative Devolution*, London: Palgrave.

Morgan, K. (1980) *Rebirth of a Nation: Wales, 1880–1980*, Oxford: Oxford University Press.

Ohmae, K. (1995) *The End of the Nation State. The Rise of Regional Economies*, New York: Free Press.

Paterson, L., Brown, A., Curtice, J., Hinds, K., McCrone, D., Park, A., Sproston, K. and Surridge, P. (2001) *New Scotland, New Politics?*, Edinburgh: Polygon.

Rokkan, S. and Urwin, D. (1983) *Economy, Territory, Identity: Politics of West European Peripheries*, London: Sage.

Rosie, M. and Bond, R. (2003) 'The personal and political significance of feeling Scottish', in C. Bromley, J. Curtice, K. Hinds and A. Park (eds) *Devolution – Scottish Answers to Scottish Questions?* Edinburgh: Edinburgh University Press.

Scottish Executive (2000) *A Framework for Economic Development*, Edinburgh: Scottish Executive.

Storper, M. (1997) *The Regional World: Territorial Development in a Global Economy*, New York and London: Guildford.

Trigilia, C. (1991) 'The paradox of the region: economic regulation and the representation of interests', *Economy and Society*, 20(3): 306–27.

Tully, J. (2001) 'Introduction', in A.-G. Gagnon and J. Tully (eds) *Multinational Democracies*, Cambridge: Cambridge University Press.

3 The theoretical challenge of devolution and constitutional change in the United Kingdom

Mark Goodwin, Martin Jones and Rhys A. Jones

Introduction

This chapter is concerned with the geographies of devolution and state restructuring, particularly in the UK. Devolution is a key policy agenda currently being pursued by the Blair administration and 'may yet come to rank as its most significant achievement' (Gamble, 2002: 22). As part of a comprehensive UK-wide agenda, and perhaps the biggest change to the UK state since the Acts of Union, the Labour Party has established the Scottish Parliament, elected Assemblies for Wales, Northern Ireland, and London, and Regional Development Agencies within England's regions (see detailed discussions in Bogdanor, 1999; Hazell, 2000, 2003; Tomaney, 2000; Trench, 2001, 2004). Through these developments, a unitary system of government is gradually being replaced by a quasi-federal system and this is certainly bringing the UK closer to other European countries in terms of its governing structure.

Although there is a range of historically embedded cultural and political reasons for the territorial and scalar restructuring of the UK's national state/space (see especially Chaney *et al.*, 2001; Jones and Osmond, 2002; Morgan and Mungham, 2000; Taylor and Thomson, 1999), common to these institutional developments is a concern to secure economic prosperity through supply-side policies in order to create a competitive advantage under globalisation (compare DED, 1999; DTLR, 2002; Labour Party, 2001; National Assembly for Wales, 2001; Scottish Executive, 2001). Economic development is a devolved policy across the United Kingdom (see Keating, 2002, for full details) and, in terms of these academic issues, devolution has the potential to alter the UK's institutional architecture and set in train changes to frameworks supporting economic governance in each of the territories. However, the reverse is also true, and the performance of the new institutions of economic development also holds considerable implications for the process of devolution.

Now that we are more than six years or so into the devolution and constitutional change settlement in Britain, it is evident that new political geographies are emerging – particularly in relation to the organisations used to govern each of the four devolved territories, their structures of power, and future capacities to deliver policy. For instance, with regard to economic governance, it is clear that economic development now takes place at different scales, involves different actors, covers different territories, and includes different sets of intergovernmental linkages (see Goodwin *et al.*, 2002; Jones, 2004; Jones *et al.*, 2004). This is more the case in some devolved territories than others. A good example of this is (post-devolution) skills training in Britain – one of the key policy frameworks for boosting economic performance and ensuring post-national economic prosperity. The Training and Enterprise Councils have been abolished in recent years and merged with other branches of the local state apparatus to form new local Learning and Skills Councils in England and ELWa (Education and Learning Wales) and its partners in Wales (see Ainley, 2001, 2003; Jones *et al.*, 2004). ELWa has been subsequently scrapped, as part of the 'bonfire of the quangos', and its powers transferred to the expanded Welsh Assembly Government. By contrast, the network of Local Enterprise Companies in Scotland has remained more or less intact (see Scottish Executive, 2001) and a new Department of Employment and Learning has been created in Northern Ireland, which superseded the Training and Employment Agency side of the Department for Economic Development. Such changes have not occurred solely with respect to economic governance but are rather a feature of all aspects of territorial governance post-devolution (see Keating 2002). Clearly, there is a high level of asymmetry here with regard to the various territories' response to devolution: in terms of the adaptation of new organisations of governance, along with their concomitant policies and strategies.

Outside these introductory comments though, the chapter is not concerned with detailed policy concerns. We are instead concerned with developing conceptual frameworks to inform our understandings of the role of devolution in the contemporary reorganisation of the UK state by addressing the ways in which we can begin to theorise the changes taking place within the devolved territories. Commenting on emerging work in the UK on devolution and constitutional change, the political scientist Fred Nash suggests that it suffers from a 'lamentable lack of theoretical and conceptual grounding' (Nash, 2002: 30). Indeed, in many ways this work is very rich in empirical analysis (see *Local Economy*, 2002; *Regional Studies*, 2002; Tomaney and Mawson, 2002), but less developed conceptually and theoretically. This is something of a lost opportunity, and we would argue that devolution presents geographers and planners with an ideal circumstance in which to develop geographically informed theoretical debates on contemporary state restructuring.

A number of popular approaches to understanding the institutions of economic development are discussed in the chapter, but considered to be inadequate for *explaining* the variable geopolitical situations emerging within the UK – such as that briefly highlighted through the example of skills training. The chapter then takes forward academic debate by extending Jessop's (especially 1990; 2001a; 2002) strategic-relational approach to the state. We argue that it is no longer enough to simply refer to a multivariate 'hollowing out' of the nation state in an era of economic and political restructuring. Instead, devolution presents economic and political geographers and spatial analysts with a unique opportunity to theorise the state as an institutional-materialist territorial and scalar matrix, which is currently being 'filled in' unevenly across the four territories, thus leading to an increasingly complex spatial division of the state.

Theorising the political geographies of state restructuring

One entry point into this new theoretical venture is to take stock of previous debates on the political geographies of state restructuring in the *after*-Fordist era.

For although we might lack theoretical work on UK devolution, we do have considerable conceptual insights into state restructuring at a more general level. Indeed, in recent years there has been a growing concern in the social and political sciences to theorise the many ways in which economic and political geographies are being actively remade in the context of globalisation and the erosion of the Fordist–Keynesian national welfarist state settlement. Human geographers have been contributing much to a number of literatures that are specifically concerned with mapping the emergence of new sub-national territories, scales, and places of regulation and governance (see *Environment and Planning A*, 2001). In contrast to 'boosterist' accounts of globalisation, which emphasise the death of the nation state and the end of geography, human geographers have emphasised the need to explore the complex connections between the institutionalisation of political activity, the changing role of the capitalist state, and the importance of territorially embedding continued capital accumulation within the context of economic competitiveness and social cohesion (see Jones and Jones, 2004). Such analysis suggests the need to consider the ways in which social, economic and cultural processes are institutionally mediated and actively produced through struggles occurring within and between different spatial scales. Some of these concerns are being shared by political scientists, whose work on multilevel governance (MLG) seeks to capture the complex dynamics taking place with respect to territorial restructuring and political change in Western Europe (compare, for example, John, 2000; Marks, 1996; Pierre and Stoker, 2000; Scharpf, 1997). MLG theorists, for instance, see state power and authority as 'dispersed' rather than 'concentrated' and political action occurs 'at and between various levels of governance' (Jones and Clark, 2001: 206).

We would suggest that key literatures on the new regionalism, on multilevel governance and on the political economies of scale, hold some useful pointers for those concerned with the contemporary economic and political geography of the UK, but we would also claim these bodies of thought are incapable of fully understanding the processes of state restructuring that underpins much of these new geographies. This work has been well summarised, and extensively critiqued, elsewhere (see the accounts offered by Jessop, 2002; Jones, 2001; Lovering, 1999; MacLeod, 2001a, 2001b; MacLeod and Goodwin, 1999a, 1999b; MacLeod and Jones, 1999; Martin and Sunley, 1998) so we will restrict ourselves to drawing out some of the concerns which arise when these frameworks are applied to issues of devolution and state rescaling.

First, whilst the literature on the 'new regionalism' emphasises the central roles played by institutions, networks, trust and social capital in the maintenance of successful economies, it is less strong on analysing the continued role of the national state in nurturing and sustaining these institutional norms and networks. This may not be surprising given the heavy emphasis within this research on supply-side innovation and the replacement of formalised government with less formal networks of partnership and governance, but we still need to understand how the state helps to produce, reproduce and articulate these new sites and scales of economic governance. Second, multilevel governance is stronger on the politics of the new sub-national institutions, but overplays the vertical nesting of discrete policy competencies, at the expense of analysing the dense network of 'tangled hierarchies' (Jessop, 2001b) which mesh together to produce and implement policy horizontally across any one scale. Multilevel governance also tends to reify the different scales within these hierarchies – when in practice scales of governance are relative and are actively produced (not least by the national state). Third, the emerging literature within geography on the political economies of scale does stress the active constitution of spatial scale, but in seeing scale as the product of relational processes there is the danger of not appreciating how territorial fixity is created partly through processes of government and governance. Thus, in many ways the process-based strength of this approach can become a weakness – a point confirmed by devolution research, which highlights how the production, or fixing, of scale is occurring through a number of context-specific institutions which are themselves bound up with the reconfiguration of national state competencies (Jones, 2004; Jones and Jones, 2004). Collectively, then, we would argue that these debates on the importance of sub-national institutions of economic governance need to heed Swyngedouw's (still valid) argument that: 'Although the thesis of state re-scaling has been advanced by a number of authors ... the *actual mechanisms* through which processes take place remain vague and under theorised' (Swyngedouw, 1996: 1500, emphasis added). We would suggest that a focus on such mechanisms – of seeing the state as a 'political process in motion' (Peck, 2001: 449) – leads us to pay more attention than hitherto to the continuing role played by political strategy in the production of new sites and scales of state governance. This is clearly the case in a post-devolution UK, where the central state, together with each devolved administration, clearly has a key influence in structuring the scales and actions of a range of emergent new sub-national institutions.

One theoretical approach that does stress state strategy is the strategic-relational approach (SRA) to state theory (especially Jessop, 1985, 1990, 1995, 1997, 1999, 2001a, 2002). Developed as a means of analysing the shifting contours of economic and political restructuring in Western Europe over the past twenty years, the SRA is particularly useful for beginning to conceptually position the current round of variable state restructuring in the UK. It suggests that the nation state's dominance is being undermined by the three interrelated processes of 'de-statisation', the 'internationalisation of policy regimes', and 'denationalisation' in its long-run evolution from a Keynesian Welfare National State (KWNS) towards a Schumpeterian Workfare Post-National Regime (SWPR). In Jessop's account, the *de-statisation* of the political system is most clearly reflected in a functional shift from government to governance. This is associated with a relative decline of the central state's direct management and sponsorship of economic and social projects, and an analogous engagement of quasi-state and non-state actors in public–private partnerships. *Internationalisation* refers not only to the more significant role of international policy communities and networks, but also to the heightened strategic significance of the international and global contexts within which state actors now operate, and to the processes of international policy transfer as in, for example, the UK's adoption of 'workfare' from the USA. The *denationalisation* of statehood is reflected in a structural 'hollowing out' of the national state apparatus (cf. Holliday, 2000; Rhodes, 1994). As the nation state's capacity to promote economic development is being weakened through the processes of globalisation, international trade, financial deregulation, and space-shrinking technologies, so old and new state capacities are being reconfigured territorially and functionally along a series of spatial levels – sub-national, national, supra-national, and trans-local.

Geographers have tended to focus on denationalisation at the expense of the other two interrelated processes, attracted perhaps by the spatial metaphor of 'hollowing out'. However, Jessop is very careful not to imply the inevitable decline of the nation state. In a deliberate echo of Ohmae's notion of the 'hollow corporation', Jessop points to its continuing importance:

> The 'hollow state' metaphor indicates two trends: first, that the national state retains its 'headquarters' (or crucial political) functions – including the trappings of central executive authority and national sovereignty as well as the discourses that sustain them and overall responsibility for maintaining social cohesion; and, second, that its capacities to translate this authority and sovereignty into effective control are becoming limited by a complex displacement of powers upwards, downwards, and outwards. This does not mean that the national state loses all importance: far from it. Indeed, it remains a crucial site and discursive framework for political struggles; and it even keeps much of its sovereignty.
>
> (Jessop, 2002: 212)

Significantly, Jessop developed this thesis on state reorganisation through empirical work on economic governance during the 1990s, drawing the conclusion that these transformations have resulted in various sets of 'tangled hierarchies' (Jessop, 1998). The recent processes of constitutional change in the UK, discussed below, have altered this picture in two key ways. First, the 'tangled hierarchies' of governance, operating at various interlinked spatial scales, have, if anything, been made more complex. We can witness this growing level of complexity with regard to economic governance in the UK. Even before devolution, a number of reports pointed out that the multiplicity of agencies involved in economic development, and their often overlapping responsibilities, proved to be a constant source of inefficiency and confusion (see Audit Commission, 1989). Devolution has further complicated this picture. The UK now has four (or five if we include London) distinct institutional structures for the promotion of economic development, which hold important implications for the processes of de-statisation and internationalisation. Second, as we empirically chart below, devolution *has* begun a critical process of formally building new architectures of governance *within* the hollowed out structure of the UK state. Jessop hinted at this conceptually, when he wrote that the 'hollowed out' nation state:

> *has a continuing role in managing the political linkages across different territorial scales*, and its legitimacy depends precisely on doing so in the perceived interests of the social base ... Moreover, just as multinational firms' command, control, communication and intelligence functions are continually transformed by the development of new information and communication technologies and new forms of networking, bargaining and negotiation, so, too, as *new possibilities emerge*, are there changes in how hollowed out' states exercise and project their power.
>
> (Jessop, 2002: 212, emphasis added)

Devolution, of course, has at once altered how the state manages its 'political linkages across different scales' and opened up 'new possibilities' in the exercise and projection of power. However, 'hollowing out' is perhaps not the most appropriate metaphor to explore the shifting political geographies of the British state. For as particular elements of the UK state (at the national scale) are 'hollowed out', the process of devolution has meant that at other (sub-national) scales they are 'being filled-in'.

New states, new theories: from hollowing out to 'filling in'

We feel that the hollowing out metaphor is *not* an appropriate one for two reasons. First, hollowing out is often advanced as the explanation; that is, we have witnessed devolution because the state is being hollowed out, when in fact it is precisely the hollowing out which needs explaining. Second, hollowing out is not unidimensional: as one element of the state is being hollowed out, others are indeed being *'filled in'*. Under devolution we would suggest that hollowing out and 'filling in' are twin aspects of the same process: you can't have one (and understand the state's shifting spatial architecture) without the other. And yet the very popularity of the hollowing out concept leads to an analytical focus that directs attention away from this dual aspect. Using existing SRA accounts we therefore have a somewhat one-sided treatment of a dialectical process, which is resulting in simultaneous de-statisation *and* re-statisation. All that is solid, including the state apparatus and its political institutions, does not melt into air because the state is 'incompressible': it can re-emerge at a different scale and with a different territorial form at different points in time (Poulantzas, 1978).

To develop strategic relational accounts on the state, then, it is important to think spatially. Hollowing out has become a popular metaphor because it is deployed (usually implicitly) at a national scale, but as we discuss below, the situation within the UK's devolved territories is far more complex and illustrative of asymmetrical 'filling in'. And so there is an interesting paradox within the SRA. Although hollowing out is a spatial metaphor, it is not sufficiently spatially sensitive, at least not in the way it has hitherto been deployed in Jessop's original thinking. This argument is acknowledged in Jessop's more recent work, which stresses the role of space and territory in the reproduction of state power (see Brenner *et al.*, 2003; Jessop, 2002). Hollowing out, however, still remains central to this ongoing analysis and tends to be seen as a top-down – rather than a dialectical, iterative and negotiated – process. But as we discuss below, devolution and constitutional change is a multifaceted and multi-scalar process of state restructuring and remaking involving 'complex changes in the relations *between* different levels/scales and branches/departments of the state' (Peck, 2001: 447, emphasis added). Hence, as Peck adds, in an echo of Poulantzas' argument on incompressibility, we 'need to see "hollowing out" as a qualitative process of state restructuring, not a quantitative process of state erosion or diminution' (ibid.). Thus, with the advent of devolution and constitutional change, it is no longer enough to simply refer to the hollowing out of the state – future research also needs to concentrate on assessing the processes, practices, and resulting geographies of 'filling in'.

This still leaves us with the problem of deploying (or reworking) adequate conceptual frameworks to grasp these complex qualitative processes of state (and territorial) restructuring. It might be argued that the concept of 'hollowing out' implies 'filling in', and that the latter term is therefore unnecessary. After all, 'hollowing out' does refer to the ways in which state capacities are refigured functionally *and* territorially. But it is the implication that is the problem. 'Hollowing out' refers to a potential rescaling away from the national state, both upwards and downwards, at a series of levels from the local to the supranational. In this sense it lacks specification. The use of 'filling in' as a concept draws attention to such specification, but focuses on the manner in which power is being transferred, and on the scales it is being transferred to. In other words, the very process of 'filling in' is geographically constituted and spatially constructed – in contrast to 'hollowing out' which implies an abstract sense of restructuring at any number of potential scales. 'Filling in' also allows a specification of the relationship between different tiers of the state, and stresses the active and contested process of new state formation. As Swyngedouw (1997: 141) suggests, our theoretical priority should be concerned with 'the process through which particular scales become (re)constituted' – which necessitates a dialectical emphasis on filling in as well as hollowing out.

Modifying strategic-relational state theory to take account of scale

One of the great strengths of strategic-relational state theory is that it conveys an understanding of the state 'not as some lumbering bureaucratic monolith, but as a (political) process in motion' (Peck, 2001: 449). Moreover, in analysing this process, as its name suggests, the SRA is at once strategic and relational. It stresses that state power, and indeed activity, can only be assessed relationally, as a set of social relations. As Jessop has famously argued:

> The state as such has no power – it is merely an institutional ensemble; it has only a set of institutional capacities and liabilities which mediate that power; the power of the state is the power of the forces acting in and through the state. These forces include state managers as well as class forces, gender groups as well as regional interests, and so forth.
>
> (Jessop, 1990: 269–70)

Following this statement, it is possible to focus our analytical lens on the state as what we might term a 'peopled organisation', not a set of anonymous institutions referred to in some of the literatures identified above. We would maintain that there is more to this: this theoretical framework deployed also highlights the notion of strategy – the state is viewed as the site, generator and product of particular strategies. In other words, the state, and the policies and activities pursued through it, 'constitutes a terrain upon which different political forces attempt to impart a specific direction to the individual or collective activities of its different branches' (Jessop, 1990: 268). And in viewing the state as the site of strategy, it can be analysed as a territorially and functionally dispersed system, whose structural form and mode of operation are more permeable and suitable to some types of political agents than to others. For Offe, this often entails a tendency for the state to favour certain interest groups (usually capital) – a process referred to as 'structural selectivity' (Offe, 1974). However, rather than being the result of some essentialist logic, whereby the state functions as if it were an 'instrument of the interest of capital' (Offe, 1984: 51), the structural selectivity of any state needs to be analysed as contingent upon the actions and strategic endeavours of a whole plethora of agents peculiar to specific places and times. The complex articulation of these processes of *strategic selectivity* means that:

> Particular forms of economic and political system privilege some strategies over others, access by some forces over others, some interests over others, some spatial scales of action over others, some time horizons over others, some coalition possibilities over others. Structural constraints always operate selectively: they are not absolute and unconditional but always temporally, spatially, agency, and strategy specific. This has implications both for general struggles over the economic and extra-economic regularization of capitalist economies and specific struggles involved in securing the hegemony of a specific accumulation strategy.
>
> (Jessop, 1997: 63)

This relational character of strategic selectivity also implies that the differential ability of social forces to pursue their interests through different strategies 'is not inscribed in the state system as such but in the relation between state structures and the strategies which different forces adopt' (Jessop, 1990: 260). This thinking can shed much light on the seemingly incessant rounds of post-devolution restructuring that we briefly referred to above because this theoretical framework is inherently spatio-temporal – considerations of space and time are etched into its formulation, although Jessop has perhaps been reluctant to pursue this in concrete research. We would suggest that this etching takes on three forms (cf. Jessop, 2001a, 2002; MacLeod and Goodwin, 1999b).

First, like all structures, and indeed institutions, states have a definite spatio-temporal extension: they emerge in specific places at specific times, they operate over particular scales and territories, and they have certain temporal horizons. States are in fact complex 'geographical accomplishments', which are forged through the interaction of economic and extra-economic systems and their social relations (Philo and Parr, 2000). Using a modified strategic-relational state theory can make us appreciate how devolution has served to make more complex the scalar and territorial geographies of the UK, leading to the creation of territorial mismatches and policy overlaps. Put simply, the modified SRA can help us focus on the spatial tensions inherent in devolution, most specifically in terms of collaboration and competition between the various new organisations of governance.

Second, some practices and strategies are privileged, and others made more difficult to realise, according to how they match the spatial and temporal patterns inscribed in the structures in question. Critically important here are the relationships between state structures and the strategies which given forces adopt. Once again, the modified SRA can enable us to tease out the relationships between new post-devolution strategies and particular territories and certain scales. It can also help us to highlight the efforts being made to dovetail strategies between territories and scales, but also the instances in which strategies – promoted within particular territories and at certain scales – compete with each other.

Third, because of its peculiar territorial nature as an institution, the very forces that can legitimately gain access to particular parts and branches of the state are spatially bounded and to a certain extent scale dependent. In this context, the modified SRA can draw our attention to the differential access of certain groups to the plural institutions of the state, post-devolution. Once again, access is territorial and scalar dependent.

The SRA, then, has the ability to reveal how the power of the social forces acting in and through the state, and their interplay with the state's institutional form, are dependent on sets of relations, which are geographically constituted and contested. And it is this brand of relational strategic and process-orientated analysis that social scientists can draw on in the immediate future as we seek to uncover the shifting sites and scales of state activity within an actively devolved system. Moreover, in addition to helping us to explore the newly devolved territories of the UK state, we argue that the modified SRA is useful as a way of examining the political geographies of *all* states that contain regional and/or local tiers of government. Most especially, it can aid our understanding by encouraging us to focus on a number of important themes. First, whilst one might need to focus on institutions in order to conduct concrete research, there is a need to appreciate that one is not studying institutions per se, for their own sake, but rather one is studying them to uncover political, social and cultural dynamics they entail (and the way these are scaled). Second, there is a critical need to think about the processes that operate in between institutions not just inside them. Institutions do not operate in a vacuum and, therefore, we must explore the dynamic relationships that exist between them. Third, there is a need to appreciate how scale is constructed horizontally (through territory and the connections between scales) as well as vertically through notions of multilevel governance. These horizontal relationships were especially apparent in the empirical material discussed earlier and should represent a key object of enquiry within geographical studies of the contemporary state. Finally, there is a need to develop an analysis, which can fuse the spatial and the political and a suitable modified strategic-relational state theory has the potential to help us do this. As the later chapters of this book indicate, devolution provides us with an ideal empirical opportunity for addressing these conceptual challenges.

Conclusions

This chapter has sought to provide a contribution to contemporary debates on state restructuring in the context of devolution and constitutional change in the United Kingdom. At a basic level devolution has certainly been followed by a remaking, and indeed a rescaling, of the institutions responsible for formulating a variety of policies and strategies. These changes have not just been the result of any simple transference of power and functions from the UK scale to each individual territory. What we are witnessing is a very complex rescaling of governance, both vertically between scales, and horizontally between institutions operating over the same territory, leading to what we might label as an increased 'spatial division of the state'. We would contend that understanding this complexity requires an appreciation of filling in as well as hollowing out. This is partly because the asymmetrical devolution witnessed within the UK has meant that these new spatial divisions are themselves *uneven* across the four devolved territories. Attempts at rationalisation have occurred, both at a formal governmental level, and at a more local delivery level. But these have proceeded very differently in each territory. Whilst the concept of hollowing out can be drawn on to situate the beginnings of these processes, notions of fill-ing in can provide an understanding of their differential and uneven outcomes. Moreover, the twin processes of hollowing out and filling in are recursive – each influences and is influenced by the other. Thus, the next round of hollow-ing out from central government is likely to be conditioned by the political, social, cultural and economic impacts of the current processes of filling in within the locality.

Given that devolved organisations' structures, policies and strategies will have a direct impact on their ability to act and to achieve their stated goals, then a clear understanding of the conceptual and empirical character of the various processes of filling in becomes imperative. If this is the case, then the exact con-figurations of filling in within the various UK territories post-devolution may well possess serious implications for the whole devolution project.

Acknowledgements

We are grateful to the Economic and Social Research Council for funding a research project – entitled 'Constitutional Change and Economic Governance: Territories and Institutions' (Grant L219252013) – as part of its Devolution and Constitutional Change Research Programme, which has informed the arguments presented in this chapter.

References

Ainley, P. (2001) 'From a national system locally administered to a national system nation-ally administered: the new Leviathan in education and training', *Journal of Social Policy*, 30: 457–76.

Audit Commission (1989) *Urban Regeneration and Local Economic Development: The Local Government Dimension*, London: HMSO.

Bogdanor, V. (1999) *Devolution in the United Kingdom*, Oxford: Oxford University Press.

Brenner, N., Jessop, B., Jones, M. and MacLeod, G. (eds) (2003) 'Introduction: state space in question' in Brenner *et al.* (eds) *State/Space: A Reader*, Oxford: Blackwell.

Chaney, P., Hall, T. and Pithouse, A. (eds) (2001) *New Governance, New Democracy? Post-Devolution Wales*, Cardiff: Cardiff University Press.

DED (1999) *Strategy 2010: Report by the Economic Development Strategy Review Steering Group*, Belfast: Department of Economic Development.

DTLR (2002) *Your Region, Your Choice: Revitalising the English Regions*, Cm 5511, London: HMSO.

Environment and Planning A (2001) 'Theme issue: Reflections on the "institutional turn" in local economic development', *Environment and Planning* A, 33: 1139–1241.

Gamble, A. (2002) 'Divided, different, but not a disaster', *The Times Higher*, 4 January, p. 22.

Goodwin, M., Jones, M., Jones, R., Pett, K. and Simpson, G. (2002) 'Devolution and eco-nomic governance in the UK: Uneven geographies, uneven capacities?', *Local Economy*, 17: 200–15.

Hazell, R. (ed.) (2000) *The State and the Nations*, Exeter: Imprint Academic.

Hazell, R. (ed.) (2003) *The State of the Nations 2003: The Third Year of Devolution in the United Kingdom*, Exeter: Imprint Academic.

Holliday, I. (2000) 'Is the British state hollowing out?', *The Political Quarterly*, 71: 167–76.

Jessop, B. (1985) *Nicos Poulantzas: Marxist Theory and Political Strategy*, London: Macmillan.

Jessop, B. (1990) *State Theory: Putting Capitalist States in their Place*, Cambridge: Polity Press.

Jessop, B. (1995) 'The future of the nation state: Erosion of reorganisation?' *Lancaster Regionalism Working Papers (Governance Series) Number 50*, Lancaster: Department of Sociology, University of Lancaster.

Jessop, B. (1997) 'A neo-Gramscian approach to the regulation of urban regimes', in M. Lauria (ed.), *Reconstructing Urban Regime Theory: Regulating Urban Politics in a Global Economy*, London: Sage.

Jessop, B. (1998) 'The rise of governance and the risks of failure: The case of economic development', *International Social Science Journal*, 155: 29–45.

Jessop, B. (1999) 'Globalisation and the nation state', in S. Aaronowitz and P. Bratsis (eds), *Rethinking the State: Miliband, Poulantzas and State Theory*, Minneapolis: University of Minnesota Press.

Jessop, B. (2001a) 'Institutional re(turns) and the strategic-relational approach', *Environment and Planning A*, 33: 1213–35.

Jessop, B. (2001b) 'Multi-level governance and multi-level meta-governance', *Mimeograph*, Lancaster: Department of Sociology, Lancaster University.

Jessop, B. (2002) *The Future of the Capitalist State*, Cambridge: Polity Press.

John, P. (2000) 'The Europeanisation of sub-national governance', *Urban Studies*, 37: 877–94.

Jones, A. and Clark, J. (2001) *The Modalities of European Union Governance: New Institutionalist Explanations of Agri-Environmental Policy*, Oxford: Oxford University Press.

Jones, B. and Osmond, J. (eds) (2002) *Building Civic Culture: Institutional Change, Policy Development and Political Dynamics in the National Assembly for Wales*, Cardiff: Welsh Governance Centre, Institute of Welsh Affairs.

Jones, M. (2001) 'The rise of the regional state in economic governance: "Partnerships for prosperity" or new scales of state power?', *Environment and Planning A*, 33: 1185–1211.

Jones, M. (2004) 'The regional state and economic governance: regionalized regeneration, or territorialized political mobilisation?' in A. Wood and D. Valler (eds) *Governing Local and Regional Economies: Institutions, Politics and Economic Development*, Aldershot: Ashgate.

Jones, M. and Jones, R. (2004) 'Nation states, ideological power and globalisation: Can geographers catch the boat?' *Geoforum*, 25: 409–424.

Jones, R., Goodwin, M., Jones, M. and Simpson, G. (2004) 'Devolution, state personnel, and the production of new territories of governance in the United Kingdom', *Environment and Planning A*, 36: 89–109.

Keating, M. (2002) 'Devolution and public policy in the United Kingdom: Divergence or convergence?' in J. Adams and P. Robinson (eds) *Devolution in Practice: Public Policy Difference in the UK*, London: IPPR/ESRC.

Labour Party (2001) *Ambitions for Britain: Labour's Manifesto 2001*, London: Labour Party.

Local Economy (2002) 'Special issue: Devolution, regionalism and local economic development', *Local Economy*, 17: 186–252.

Lovering, J. (1999) 'Theory led by policy: The inadequacies of the "new regionalism" (illustrated from the case of Wales)', *International Journal of Urban and Regional Research*, 23: 379–95.

MacLeod, G. (2001a) 'Beyond soft institutionalism: Accumulation, regulation, and their geographical fixes', *Environment and Planning A*, 33: 1145–67.

MacLeod, G. (2001b) 'New regionalism reconsidered: Globalization and the remaking of political economic space', *International Journal of Urban and Regional Research*, 25: 804–29.

MacLeod, G. and Goodwin, M. (1999a) 'Space, scale and state strategy: Rethinking urban and regional governance', *Progress in Human Geography*, 23: 503–27.

MacLeod, G. and Goodwin, M. (1999b) 'Reconstructing an urban and regional political economy: On the state, politics, scale and explanation', *Political Geography*, 18: 697–730.

MacLeod, G. and Jones, M. (1999) 'Reregulating a regional rustbelt: Institutional fixes, entrepreneurial discourse, and the "politics of representation"', *Environment and Planning D: Society and Space*, 17: 575–605.

Marks, G. (1996) 'An actor-centred approach to multilevel governance', *Regional and Federal Studies*, 6: 20–40.

Martin, R. and Sunley, P. (1998) 'Slow convergence? The new endogenous growth theory and regional development', *Economic Geography*, 74: 201–27.

Morgan, K. and Mungham, G. (2000) *Redesigning Democracy*, Bridgend: Seren.

Nash, F. (2002) 'Devolution dominoes' *The Times Higher*, 1 February 2, p. 30.

National Assembly for Wales (2001) *The National Economic Development Strategy*, Cardiff: National Assembly for Wales.

Offe, C. (1974) 'Structural problems of the capitalist state', *German Political Studies*, 1: 31–57.

Offe, C. (1984) *The Contradictions of the Welfare State*, London: Hutchinson.

Peck, J. (2001) 'Neoliberalizing states: Thin policies/hard outcomes', *Progress in Human Geography*, 25: 445–55.

Philo, C. and Parr, H. (2000) 'Institutional geographies: Introductory remarks', *Geoforum*, 31: 513–21.

Pierre, J. and Stoker, G. (2000) 'Towards multi-level governance' in P. Dunleavy, A. Gamble, I. Holliday and G. Peele (eds) *Developments in British Politics 6*, London: Macmillan.

Poulantzas, N. (1978) *State, Power, Socialism*, London: New Left Books.

Regional Studies (2002) 'Special issue: Devolution and the English question', *Regional Studies* 36: 715–810.

Rhodes, R.A.W. (1994) 'The hollowing out of the state: The changing nature of public services in Britain', *The Political Quarterly*, 65: 138–51.

Scharpf, F. (1997) 'The problem-solving capacity of multi-level governance', *Journal of European Public Policy*, 4: 520–38.

Scottish Executive (2001) *A Smart, Successful Scotland: Ambitions for the Enterprise Networks*, Enterprise and Lifelong Learning Department, Edinburgh: Scottish Executive.

Swyngedouw, E. (1996) 'Reconstructing citizenship, the re-scaling of the state and the new authoritarianism: Closing the Belgium mines', *Urban Studies*, 33: 1499–521.

Taylor, B. and Thomson, K. (eds) (1999) *Scotland and Wales: Nations Again?*, Cardiff: University of Wales Press.

Tomaney, J. (2000) 'End of the empire state? New Labour and devolution in the United Kingdom', *International Journal of Urban and Regional Research*, 24: 675–88.

Tomaney, J. and Mawson, J. (2002) *England: The State of the Regions*, London: Policy Press.

Trench, A. (ed.) (2001) *The State of the Nations 2001*, Exeter: Imprint Academic.

Trench, A. (ed.) (2004) *Has Devolution Made a Difference? The State of the Nations 2004*, Exeter: Imprint Academic.

4 Towards a transnational space of governance?

The European Union as a challenging arena for UK planning[1]

Ole B. Jensen and Tim Richardson

Introduction

The chapter develops a critical analysis of how European Union spatial policies are shaping new policy narratives at multiple spatial scales and within new complex governance settings or – alternatively – the 'spatialisation of the European project'. We find a Europeanisation of spatial narratives, where notions of a seamless flow space of Europe are reproduced, as part of the reproduction of a 'Europe of Monotopia'. Within EU territories, this process of Europeanisation is manifested in the shaping of policies for mobility, infrastructure and urban and regional development, as well as in the general expression of becoming more European (alongside other 'scales' of territorial identity). In the chapter, these 'hard' and 'soft' policy issues are linked, to identify a new policy field towards which territories within the UK are reorienting their policy narratives of territory, identity and space through a complex engagement with Europe.

Britain in a fragile Europe: an ambivalent relationship

In the last year, the confidence that might have been expected by the EU at the historic moment of enlargement to EU-25 has been shattered by the impact of world events: the political differences between the leaders of its member states over war in Iraq cast a shadow over the EU summit which had hoped to reach consensus on a new European Constitution; and the threat and shocking reality of global terrorism has been mobilised within the debating chambers as a new reason to find a European consensus. Yet red lines and disputes over voting rights reflect an at best ambivalent attitude to the EU that seems increasingly pervasive; for example, Britain's (or perhaps Tony Blair's) position in the Constitution debate has been characterised as anti-European, and about achieving 'victory over' rather than 'victory for' Europe (Cook, 2004: 39). And the citizens of Europe remain disinterested or else sceptical, as the recent 'No' vote results in the referenda for a new European Constitution in France and the Netherlands demonstrated most acutely. The participation rate in the 2004 elections to the European Parliament hit its lowest level to date. For many people, the EU remains a remote yet somehow threatening entity, as illustrated also by the unprecedented success of the UK Independence Party, returning among others new MEPs, one of whom stated that his mission will be to 'wreck' the European Parliament. Today, the European project can be seen as a pale shadow of the strident institution of a decade ago. The EU appears fragile, striated by a multitude of lines of weakness along which further fractures might occur.

Understanding the Europeanisation of planning

Planning researchers have noted that it is easy to miss even the most obvious influences of European policy and legislation on planning in Britain:

> for example, with 34 British implementing regulations and 27 guidance documents issued since 1988, the practice of environmental assessment has become so embedded in the British planning system that it is easy to forget its parenthood in the European Commission's EIA Directive 85/337.
>
> (Tewdwr-Jones and Williams, 2001: 11)

Tewdwr-Jones and Williams note that, as a consequence, British planners are frequently unaware of the European origin of the frameworks and ideas they work with(in). Researching the Europeanisation of planning in Britain is therefore likely to be difficult, and this difficulty is made worse by the fact that the Europeanisation of spatial planning takes place in a range of ways which include but extend far beyond its legal framework. This chapter concentrates on some of the less apparent ways in which Europeanisation is taking place, through the subtle shaping of ideas in spatial strategy making processes.

The 'spatialisation of ideas'

Here we first briefly set out the core concepts of the 'spatialisation of ideas', and the notion of the spatialisation of the European project. But what do we really mean by this notion of 'spatialisation of ideas'? First of all, we recognise the fact that spaces and places are linked to policy ideas and concepts, even without the articulating subject being fully aware of this. Thus, we would argue for the basic understanding of the 'European Project' as being one of inherent spatial quality. This comes to the fore in multiple fields of activity from Common Agricultural Policies, through trans-European transport networks, to the mechanics of the Internal Market. However the policy ideas and conceptualisations at work in this institutionalisation of European integration also carry notions of space and place, and most importantly of the organisation of European territory. Part of such 'spatialisation of policy ideas' has to do with constructing a particular narrative of Europe through the creation of place images, which on their side comes about in a social process of simplification and stereotyping (Shields, 1991: 47). Such a process of collectively creating places through the construction of place images amounts to a set of ideas in currency, gaining value through conventions circulating in a discursive economy. When a whole set of place images are brought together they form a 'place myth' (1991: 61). On a very general level, we would argue that the rational ordering of the European territory is the spatialisation of the political notion of 'Europe as the cradle of Western Civilization'.

The political and economic discourses of European integration create very specific demands on mobility, necessitating what Maarten Hajer, following Castells, has described as a discourse of a Europe of flows (Hajer, 2000). This discourse is characterised by the development of technologies such as Just in Time logistics, which have enabled a more footloose approach to economic development, reliant on the fast, low cost movement of goods by road. The increasing demand for personal mobility is also characteristic of this discourse.

The spatialisation of the European project

If the European Project is inherently spatial in its character, then the grounding ideas behind the project may also be said to be spatial in their underlying rationale. Evidently, whether one stresses the importance of the peace keeping agenda, the new economic zones of global competitiveness, or even the notion of European values and identity, we would hold that a vital dimension is the way that the spatial entity called Europe is imagined. In times of enlargement, this issue is even more evident. However, the ordering of European space is also present within different policy fields. Here we shall argue that the multiple policies aiming at producing a European Union with as little friction to the flow of people, goods, ideas, and capital is the ordering principle that most predominantly expresses the spatialisation of the European project. Thinking about such a smooth space of European flows can be understood as an agenda for producing a Europe of monotopia. By this is meant the idea of a one-dimensional (mono) discourse of space and territory (topia/topos) (Jensen and Richardson, 2004) where the basic spatial organising principle is a discourse of Europe as monotopia set as an organising set of ideas that looks upon the EU territory within a single overarching rationality of making a 'one space', made possible by seamless networks enabling frictionless mobility. It is, in other words:

> an organised, ordered and totalised space of zero-friction and seamless logistic flows ... we will argue that, though the word 'monotopia' will not be found in any European plan, policy document or political speech, this idea of monotopic Europe lies at the heart of the new ways of looking at European territory. We will argue that a rationality of monotopia exists, and that it is inextricably linked with a governmentality of Europe, expressed in a will to order space, to create a seamless and integrated space within the context of the European project, which is being pursued through the emerging field of European spatial policy. The future of places and people across Europe seems closely linked to the possibility of monotopia ... A vision of monotopic Europe centres on the idea of a 'zero-friction society' (Hajer, 1999; Flyvbjerg *et al.*, 2003) based on an increasing harmonisation of mobilities of people, goods and information, leading to a new dimension of ambivalence.
>
> (Jensen and Richardson, 2004: 3)

However, we need to take this notion of spatialisation of policy ideas into a more concrete and specific mode in order to argue the case of a Europe of monotopia. This will be done by probing the spatialisation of two key policy ideas: cohesion and polycentricity.

The spatialisation of ideas – cohesion

The notion of cohesion policy has, as we briefly stated before, an inherent spatial logic to it since it concerns the coherence not only in economic, social and political terms, but also in a spatial and territorial sense. In the official EU policy discourse, the issue of cohesion is one of a spatial/territorial redistributive logic; however, contrary to this, Rumford interprets the concept as one of subsumed economic competitiveness:

The EU has, over the past decade, increasingly come to view cohesion not as an objective in its own right, but as a contributor to other aims, notably competitiveness ... Cohesion is not to be understood as the levelling out of disparities through programmes of wealth distribution. The EU embraces a market economy not a redistributive one.

(Rumford, 2002: 179)

However, this simply means that seen from the perspective of 'spatialisation of ideas', we should think of the ordering of a single space of European flow as the spatial expression of an economic rationale of global competitiveness. Similar conclusions can be reached when the European Spatial Development Perspective (ESDP) is considered in its relation to cohesion:

The overwhelming emphasis on economic development within the ESDP suggests that the EU's spatial strategy will be played out in competition between cities and regions, between urban and rural, between core and periphery, and along growth corridors, with the marginalisation of social and environmental concerns. It is in response to these risks of fracture and injustice that the EU's cohesion agenda has emerged. And it is here that we find the issue of territorial identity reflected in the new vocabulary. Springing from cohesion as a response to a territory in danger of breaking apart, the new language of territorial identity itself subtly articulates an idea of European unity.

(Jensen and Richardson, 2004: 98–9)

The policy idea of cohesion thus links to territorial identity as well since the 'imagined community of monotopic Europe' has to have a platform for articulating the vision and idea of a level and coherent playing field.

The spatialisation of ideas – polycentricity

The other key policy idea that we need to think about in order to grasp the idea of a spatialised European project is the notion of polycentricity. This has become a key concept in the general EU spatial policy area (Böhme, 2002). However, this is also an example of a policy idea with a very wide semantic reach – making it ideal for less precise policy visions. Taken at face value, the concept refers to many centres as opposed to the classic one-centre model ('monocentric'). The crucial issue is, however, the question of scale. Thinking about a network of cities as forming a poly-nucleated system gives the impression of a more equal spatial distribution of urban nodes or agglomerations. But, there is a big difference as to whether one thinks of a region within a nation state in Europe, or the issue of the whole system of cities and urban spaces in the European Union. Thus, the concept is rather vague and polyvalent in itself.

The core ESDP policy goal centres on a policy triangle of economic and social cohesion, sustainable development and balanced competitiveness, iterated in the final document (CSD 1999, 11) as:

• development of a balanced and polycentric city system and a new urban–rural partnership;

- securing parity of access to infrastructure and knowledge; and
- sustainable development, prudent management and protection of nature and cultural heritage.

Although the importance of social cohesion and sustainable development is highlighted, the rationale of economic competitiveness is nevertheless dominant (Davoudi, 1999). This can be seen, for example, in the way that the notion of a balanced regional development is linked to the issue of global economic competitiveness (CSD, 1999: 24). The powerful 'core region' of Europe (the so-called 'Pentagon') is framed as a model for other EU regions. This is pursued by the concept of 'dynamic global economy integration zones' (1999: 24), which should be created in other regions to imitate and duplicate the prosperous core. This is in spite of severe problems associated with traffic congestion, for example, which the ESDP recognises do not contribute to the sustainability objective. For weaker regions, outside the proposed 'dynamic global economy integration zone', the approach is to widen the economic base and carry out economic restructuring.

Risky and smooth narratives of Europe

Mobility technologies like trans-European transport networks (TENs) are torn between the risky storylines of network failure and the smooth storylines of unrestricted flows, parity of access, and the possibility of harmonious, balanced development. The risky storyline relates to the imminent vulnerability of infrastructures, as we for instance find them in terrorist attacks or the 'culture of congestion'. Terrorists seem potentially attracted to these technological networks and nodes partly due to their vulnerability, but also due to the dependency that the 'flow society' has on these technical systems. On the other hand, TENs are the materialisation of the smooth storyline of Europe of monotopia, of the frictionless society and of accessibility and mobility enhancing technologies as promises of a better world.

In the light of the era of 'liquid Modernity' (Bauman, 2000), the very attempt to transcend human mortality is expressed in the fight against network failure. The battle can be seen as an expression of the 'death denial' surfacing in smooth storylines of monotopic Europe, epitomising not only the dream of the frictionless society but also the denial of friction, breakdown and catastrophe – and ultimately the denial of death: 'The dominant trend in our societies, as an expression of our technological ambition, and in line with our celebration of the ephemeral, is to erase death from life, or to make it meaningless ...' (Castells, 1996: 454).

It is the real and imagined negotiation of the policy space between the smooth and the risky storylines that makes up for the potential spaces of alternative mobility practices. The risky storylines of devastating network failure challenge the reason and rationality that we use under the auspices of the smooth storylines of flow. From these more abstract issues, let us now return to the policy fields dealing with regional growth and flows of people and goods, where the spatialisation of the European project is becoming a dance with frictionlessness and the denial of death.

Territorial cohesion, enlargement and the Constitution: Europe at the crossroads?

Elsewhere we have analysed how spatial ideas are reproduced through the contested EU policy spaces of transport and regional development (Jensen and Richardson, 2004). In our analysis, we have tried to articulate the idea of a policy agenda to build a monotopic space that explicitly seeks to tackle the dangers of increasing urban sclerosis and political uncertainty, offering the binding power of new mobility infrastructures that promised to restore economic flows and enable further convergence and harmonious development. The analysis suggests the importance of close attention to the knowledge claims that are deployed in multi-level struggles to assert smooth futures in the face of dysfunction. Here, though, we explore one aspect of this research agenda by considering the implications of the expansion of the borders of the European Union, by considering how attempts to smooth out European spatial dynamics can be related to issues of cohesion and belonging (and therefore to the wider constitutional debate). With the view on both future European enlargement and the debate over a new constitution, issues of whether Europe has come to a threshold or a crossroads cannot be ignored.

So how is the EU spatial project changing shape in response to questions of enlargement? In particular we focus on the imposition of a spatial agenda on the newcomers through the emerging concept of 'territorial cohesion', which has embedded the 'spatial turn' in the political integration process through its inclusion in the 2004 draft Constitution through the embedding of a spatialised idea of cohesion. The result is the policy idea of 'territorial cohesion', considered to be the single defining concept which embodies the spatialisation of the European project, and which underpins many of the other more applied spatial concepts of networks and connectivity, urban relations, environmental governance, and natural resource management. This key policy concept builds on economic and social cohesion (Articles 3 and 158 of the EC Treaty). The policy objective is to contribute to the balanced and harmonious development of the European Union as a whole (CEC, 2003: and see Box 4.1). It identifies geographical discontinuities:

- Excessive geographic concentration of innovative economic activities
- Great territorial diversity ... a challenge for development and planning activities
- Significant socio-economic disparities between the regions.

It builds on existing cohesion policy by:

- Encouraging cooperation and networking between areas
- Greater focus on development opportunities
- Greater attention to strengths of areas (better targeting of instruments and investments)
- Full incorporation of environmental and sustainable development principles including prevention of natural risks that are potentially damaging to regional development prospects
- Promoting greater coherence and coordination between regional policy and sectoral policies with a substantial territorial impact.

Territorial cohesion was included in the draft EU Constitution (Article 3), and its importance was acknowledged in Article 16 (Principles) in which it was recognised that citizens should have access to essential services, basic infrastructure and knowledge, by highlighting the significance of 'services of general economic interest for promoting economic and social cohesion' (CEC, 2004).

Box 4.1 Extract from the amended EU Constitution (European Council, 2004)

Economic, social and territorial cohesion (article III-116)

In order to promote its overall harmonious development, the Union shall develop and pursue its action leading to the strengthening of its economic, social and territorial cohesion.

In particular, the Union shall aim at reducing disparities between the levels of development of the various regions and the backwardness of the least favoured regions.

Among the regions concerned, particular attention shall be paid to rural areas, areas affected by industrial transition, and areas which suffer from severe and permanent natural or demographic handicaps such as the northernmost regions with very low population density, and island, cross-border and mountain areas.

The creation of EU-25 poses new threats to social and economic cohesion as new gaps in welfare appear. There is evidence of a number of cross-border 'welfare gaps' with socio-economic disparities doubling and the average GDP of the Union decreasing by 12.5 per cent (CEC 2004: XXV). The enlargement event thus creates a substantial challenge of completing the homogenisation of the transnational flow space in the face of the clear risk of regional GDP 'drop outs'. The discourse of monotopic Europe must now find a way to level out not only 'missing links' of infrastructure, but also 'missing growth' of the newcomers. New smooth narratives are required:

> Areas with low population density have reduced attractiveness because of low level of infrastructure and services. The modernization and further development of this infrastructure and services raises difficulties for public and private decision-makers. This explains why population is still declining in a number of these regions.

> (ESPON, 2004)

This becomes more evident when the urban system is considered in the post-enlargement situation. Looking at the urban system based upon analysis of 1595 Functional Urban Areas (FUA) combined with their linkage to the transnational infrastructure network, it can be seen that: 'not one accession country has a transport node of European significance' (2004: 18). In other words, the previously acknowledged centre-periphery relations seem only to widen within the EU-25 (2004: 12). However, if the issue of accessibility is broken down to individual modes of travel, a clear and very important picture emerges:

Analysed on the basis of transportation of persons, Europe-wide accessibility shows a clear centre-periphery pattern as far as the accessibility of road and rail is concerned. Regions with highest Europe-wide accessibility are located within the Pentagon. Accessibility by air shows a quite different pattern, with a number of regions on the periphery having high accessibility levels, provided they have a well-developed airport.

(ESPON 2004: 56)

The accession countries have better accessibility and linkage to the rest of Europe in air traffic terms. This is a conclusion that might have serious contra-productive repercussions for the attempts to orchestrate a more sustainable organisation of European flows, if this is followed up by an increase in air infrastructure. The problem identification of 'missing links' is thus still very much on the agenda, even though the phrasing today is one of 'serious deficits in connectivity' (2004: 70).

Evidently, territorial cohesion is the subject of continuing debate. The precise links between territorial cohesion, and economic and social cohesion – two fundamental aims of the EU (Article 16 of the Treaty) – remains to be further clarified. However, in the light of the Constitutional debate, we find clear evidence of the 'spatialisation of the European project'. Thus in Article III-116 of the Constitutional Draft it was clearly indicated that the European Union should promote economic, social and territorial cohesion (see also European Convention, 2003). Here, Faludi has argued that the adoption of territorial cohesion in the EU Constitution may finally bring about the empowerment of the EU to intervene in spatial planning (Faludi, 2004).

Policy implications for member states

Clearly the policy development described in this chapter is an example of a complex policy field in the making. The full effects of these policies are therefore not yet known. Neither in terms of what this means for the Community in general nor for individual member states. Thus, the policy implications for the UK are needless to say not easy to 'read off' these discursive representations. Nevertheless, we shall try to stipulate what the potential policy challenges for the UK might be said to be. Here, our discussion will be divided into two themes; the first being mobility, and the second being the complex relationship between democracy, identity and citizenship. These are, following our analysis of the European spatial policy discourse and its 'monotopic' spatial representation, some of the more important themes that need to be addressed even though they might not normally be related within mainstream planning and policy analysis. However, in accordance with the description of the overarching theme of this book, there is a need to think about the relationship between territory, identity and space in novel terms. At the end of the day, answering these types of questions inevitably leads to notions of more 'ordinary' issues such as the importance of these policy fields to the sovereignty of the member state, in this example the UK.

Mobility – the core dimension of monotopic Europe

Within the ESDP mobility is framed as accessibility, and accessibility is framed in economic rather than social or environmental terms. This rhetorical construction appears to ignore the rather different ways that accessibility is being used in transport policy debates. In the UK, for example, accessibility has rapidly become a core focus of policy, but has concerned quality of life and social inclusion as much as economy growth. Here, access to employment, services, leisure, etc. are considered to be important policy concerns. The second core element of the discourse is efficiency. The problem of mobility, framed here, is the growth in road and air transport with resulting environmental and efficiency problems. The need to promote alternative modes is emphasised, but with several strong caveats:

> this objective must be achieved without negative effects on the competitiveness of both the EU as a whole and its regions … [and] both road traffic for passengers and freight will remain of great importance, especially for linking peripheral or sparsely populated regions.
>
> (CSD, 1999: 28)

Similarly, while the potential for high speed rail is recognised as a competitor to air travel in the denser regions, 'in sparsely populated peripheral regions, particularly in insular locations, regional air transport including short-haul services has to be given priority' (1999: 28). Here, once again, arises a question of the extent of harmony between EU and national policy discourse. Drawing again from the UK example, where policy discourse has shifted towards demand management and integration, efficiency of networks is certainly an increasingly important objective. However, in the UK, policy shifted away from road building in the 1990s, whilst road building remained the major component of overall spending on TENs. Elsewhere, in the accession countries, increasing road traffic levels – car ownership in particular – are positively welcomed as signs of freedom in the post-Soviet era. Indeed, the rhetoric of the ESDP suggests that growth in overall traffic movements will be the key to improving accessibility.

Therefore, the UK will face difficult challenges and choices when it comes to 'picture its place' not only within the complex transnational policy field of the European Union, but also when it comes to decide how the UK 'links' in a physical and material sense to the 'rest' of the European Union. The question of how to fit within the 'monotopic Europe' trickles down to divergent parts of the UK, and to regions and cities, all of which need to figure out how to fill gaps and missing links in the new transnational 'flow space'.

Democracy, identity and citizenship

It is clear that practice in policy making across the arenas of European spatial policy making reflect the wider crisis of democracy in the EU. The general problem of the lack of democratic legitimacy of EU decision making is reproduced and even amplified in the informal, infra-national nature of much of the work of spatial policy making, and the non-constitutional nature of many of the new cross-border and trans-European institutions that have been established. This, then, is one of the most important questions for the future development of this policy field: while questions of competency remain, how can EU spatial policy develop as a field of action which can attain high standards of legitimacy, transparency and accountability?

Policy making for European spatial planning, from the TENs to the ESDP, provides anything but exemplary cases of participatory and transparent European democracy in action. In making this claim, it is important to bear in mind the nature of 'democracy' being adhered to. Certainly, the policy discourse clearly demonstrates a claim to legitimacy based on the democratic nature of its construction. However, we have found many cases of elitist and semi-closed policy and decision making. There are difficult unresolved issues here: will the ESDP – as both planning framework and strategic spatial vision – institutionalise a spatial discourse capable of delivering spatial justice within the EU (and at what cost to those outside it?) or will the discourse of monotopic Europe preclude such possibilities?

The issues of democracy and citizenship are obviously not to be seen as a finished project in terms of European integration. Furthermore, many analysts point to the crucial fact that apart from the lack of a common mother tongue, the real challenge to European democracy is the lack of a public arena or sphere for democratic discussion (Morley and Robins, 1995; Habermas, 1996; Weiler, 1999; Shore, 2000). Weiler, and Morley and Robins, variously explain this as a vacuum of citizenship, the absence of a 'people' from which democratic legitimacy can flow, due to deficiencies in the political culture of Europe:

> Citizenship is not only about the politics of public authority. It is also about the social reality of peoplehood and the identity of the polity. Citizens constitute the *demos* of the polity … Simply put, if there is no *demos*, there can be no democracy.
>
> (Weiler, 1999: 337, emphasis in original)

> The European Community has so far failed to develop an adequate political culture or a basis for European citizenship. Questions of identity and of citizenship have become dissociated.
>
> (Morley and Robins, 1995: 19)

Certainly, the EU seeks to construct a common identity, of which the territorial dimension is highly significant. Part of this process is clearly to create a form of citizenship, and here lies the problem. Citizenship creates subjects of government since it is both a 'way of ordering people as well as being an "identity-marker"' (Shore, 2000: 71):

> European citizenship is not so much a discourse of rights as a discourse of power. It is a political technology for making the EU more visible to its newly constituted subjects … Its function is to generate a new category of European subjectivity and thereby enable the new European state to communicate directly with its citizens.
>
> (Shore, 2000: 83)

Obviously one needs to be very careful not to reproduce a nationalist sense of belonging and assert the nation state model as the only model for democracy. Having said that, it seems evident that any polity marketing itself as a democratic entity needs to think carefully about the question of legitimacy. Thus, at the end of the day, there must be an answer to who is going to decide policies on behalf of whom? And how can the loyalty of those governed be manufactured and maintained? In Weiler's analysis, democracy and identity are linked via the question of 'multiple loyalties', which we will return to below. These are clear challenges to politicians, policy makers and institutional designers. However, in the absence of European democracy, what options are left to those who are in the shadow of European spatial policy? And crucially, can spatial justice be secured in a monotopic Europe?

The Europeanisation of British planning space

What, then, does the Europeanisation of spatial policy making and planning mean in the UK? Tensions potentially exist between the emergence of overarching regional spatial policy objectives at different scales, and the individual programmes of national planning systems within them. The existence of many national systems of land ownership, planning control and building regulations, working with separate and potentially exclusive objectives, could be argued as a counter measure to the single market, which depends on member states and other interests pursuing policies and actions which are in harmony with the overall EU integration project. This certainly creates problems in the new transnational and other multilevel spatial planning situations, and challenges the possibility of constructing a planning framework within which these conflicting objectives can be pursued equitably (Williams, 1996). So, the momentum towards EU spatial planning can be seen to be linked with the need to deal with not only spatial problems, but also with conflicting spatial objectives at different levels and in different regions.

As a further attempt to facilitate the institutionalisation and legitimation of the Europeanisation of spatial policy, the ESDP proposes that member states take into account its policy aims and options in their respective national spatial policies (CSD, 1999: 44):

> The member states should also take into consideration the European dimension of spatial development in adjusting national spatial development policies, plans and reports. Here, the requirement for a 'Europeanisation of state, regional, and urban planning' is increasingly evident. In their spatially relevant planning, local and regional government and administrative agencies should, therefore, overcome any insular way of looking at their territory and take into consideration European aspects and inter-dependencies right from the outset.
>
> (CSD, 1999: 45)

This is an 'exercise' that countries such as The Netherlands and Denmark have already progressed quite a long way with (Faludi, 1998; Jensen, 1998). However, such a message has not been universally accepted as a neutral and objective statement, and opens up the question of how divergent interests, agendas and power relations will be played out across scales of policy making. Needless to say, the Europeanisation of spatial planning will play itself out differently in the different member states. As part of a 'new regionalism', the sub-national scale is where the European agenda is being most actively pursued in England. This is explicitly acknowledged as a new direction within national planning policy guidance (Jensen, 1998):

> Widening the spatial planning scope of RPG is in keeping with the trends elsewhere in Europe. Moreover, both the European Spatial Development Perspective (ESDP) and the Community initiative on transnational cooperation on spatial planning – INTERREG IIC and INTERREG IIIB – programmes will provide a European context for the preparation of RPG. So too, will other European funding regimes, in particular EU Structural Funds.
>
> (DETR, 2000, para. 3.1)

This is nicely illustrated in research on local planning in England which reveals a context of Britain simultaneously becoming more integrated in the EU while devolving powers within the state, and where central government is failing to keep up with local government's increasing relationships with other European regions as part of the emerging transnational planning practices (Tewdwr-Jones and Williams, 2001). From case studies of urban and rural planning in Kent, Northamptonshire, Strathclyde, Mid Glamorgan, Leicester and Gwynedd, they conclude that:

> In most of the case study areas there were explicit references in the development plans to: the requirements of particular items of EU legislation; the eligibility of particular areas for EU financial assistance; and/or the impact on the local economy of developments in particular EU policies ... As well as influencing the general context of development plan preparation, EU membership was found to have influenced the formation of individual policies.
>
> (Tewdwr-Jones and Williams, 2001: 150)

Interestingly, in this study the regional and national levels were found not to be as affected by the EU as the local level, with evidence of a general transformation within the planning community towards networking in new transnational settings. The authors end the analysis by regretting the 'dismissive' attitudes within British planning towards European policies – an 'isolated view' which the authors suggest limits the actors' 'capacity for thinking in terms of EU space and spatial relations' (2001: 162). What is being pointed to here is the Europeanisation of local planning both through a top-down emphasis on the need to think European when preparing local development plans, but also a more informal exchanging and spreading of ideas through transnational structures and programmes.

Attending a meeting of the North of England Assembly of Local Authorities (NEA) on the INTERREG IIC North Sea Programme[2] illustrated that the motives and rationales that drive the local authorities do have a certain European flavour. Though it has to be admitted that the basic reasoning seemed to be more to do with 'bidding for funds' than about European integration. The whole exercise was very much about how to 'tune in' on the programme by fitting project descriptions to the goals and intentions of INTERREG IIC. At the NEA meeting, this was expressed in terms of what one participant called 'using the right EU buzz words'. At this meeting at least, there was apparent consensus about making an application that stressed 'empowerment, sustainability and social and economic cohesion'. The feeling was, that this was the necessary vocabulary to use in order to be seriously considered for funding. This is part of a wider trend, where many local and regional authorities are engaged in making their own 'foreign policy' (Williams, 1996: 250). It is also in accordance with the general 'bypassing the capital' trend that seems to be sweeping European local authorities (Rometsch and Wessels, 1996: 343–6). And finally, it is a sign that local authorities work together to construct agreed 'storylines', marketable to external audiences such as the EU (Healey, 1995: 267).

New spatialities in the devolved UK territories

What are the implications for emerging devolved patterns of spatial governance in the UK? With devolution, the UK has begun to *look* spatially fragmented. The preparation of new spatial strategies across the devolved regions seems to articulate a new sense of different spatial conditions, challenges, opportunities and trajectories, within a spatial context that is as European as it is British. Scotland, Northern Ireland and Wales have moved quickly to prepare separate spatial strategies, while in England, the regions are preparing spatial strategies. Though each of these strategies recognises the ESDP, they respond differently, and to a different degree, to the EU spatial agenda. Further research is required to understand how this new mosaic of spatial strategies across the UK territories embodies a Europeanised spatial perspective, but several tentative observations can be made. Each of the strategies sets out a particular context, but this extract from the draft Wales Spatial Plan exemplifies in straightforward terms the Europeanised context for this new wave of activity:

> The concept of spatial planning gained momentum from the publication of the European Spatial Development Perspective (ESDP) in 1999. This sets a framework for spatial planning at national and regional scales within the European Union. Many of our neighbours are taking spatial planning processes forward for their areas. Northern Ireland, Ireland and Scotland have, for example, recently developed spatial strategies or are currently working on them, whilst Regional Planning Guidance for English regions is being replaced by Regional Spatial Strategies. Furthermore, networks of European regions are developing spatial visions for their territories. Wales is included in both the North West Europe and Atlantic Arc areas.
> (Welsh Assembly Government, 2003: 3)

Is it possible, then, to trace regional and national identity differences in these new strategies with respect to the Europeanisation of their spatial thinking? The Welsh plan, for example, strongly emphasises a sense of place distinctiveness (see also the chapter by Neil Harris and Alan Hooper in this book):

> Distinctiveness, sense of identity and pride in place are important elements of successful communities and countries. Respecting community identity is intrinsic to our approach… A distinctive approach, building upon industrial heritage, culture or local environment, can offer economic opportunities and future identity for communities.
> (Welsh Assembly Government, 2003: 5)

Yet, under scrutiny, the plan's recognition of a wider European spatial perspective seems limited to a recognition of the importance of Welsh transport corridors to Ireland's international accessibility. Northern Ireland's strategy, in contrast, shows a stronger response to the EU agenda, with a strategy identifying polycentric networks of rural settlements and new growth poles, and a strong expression of connectivity in relation to EU space (see the chapter elsewhere by Geraint Ellis and Bill Neill). The emerging strategies play a role of mediating new regional identities between local distinctiveness and EU connectivity, between traditional spatial hierarchies and a new logic of functional relations.

As things stand, there remains no common spatial vision for the UK, or for England. The strategies under preparation by the English regions set out sharply defined boundaries in relation to their UK territorial neighbours, but articulate strong links – actual and desired – to mainland Europe and global markets. The exception is the border between Northern Ireland and Eire, which is clearly recognised on both sides as a space of integration, corresponding to current political tensions and opportunities. Overall, however, with no spatial vision or strategy at the English and UK levels, concerns are raised both about a lack of spatial integration across this new scale of activity, and about the diversity of ways in which the spatial thinking in these strategies may be Europeanised.

This changing context of UK spatial governance is further complicated by the increasing array of new spaces at different scales, as actors respond to the unstable conditions of Europeanised space. Such spaces include the organisational spaces of: the EU Structural Funds; the cross border spaces which link South East England with North West France or the North European Trading Axis, linking the trans-Pennine corridor with Ireland and the Benelux countries; the more abstract spaces of cooperation between non-contiguous international partners under the INTER-REG programme; and the transnational spaces such as North West Europe and the North Sea Region. Each of these configurations draws actors together in new ways, each within an explicitly Europeanised spatial context, as mobilities, polycentric city networks and environmental risks are re-articulated at new scales which do not always sit easily with the UK's new territorial configuration. The emerging regional spatial strategies must establish how they will respond to this unstable spatial context. The North West of England regional spatial strategy sets out a clear context of being on the edge of the North West European area. Strategies in Northern Ireland and northern England declare the significance of the North East Trading Axis.

It is clear that scale still matters, since spatial ideas are being organised and articulated at particular scales of analysis and action. But rather than simply seeing a devolved spatial structure which foregrounds the regional scale, UK spatial governance needs to be understood within a contingent and dynamic context where scale has become unpredictable, multi-scalar, fluid and unstable (Brenner, 2004).

Towards a progressive sense of place/spaces of resistance

The question is, then, what could be done in the face of the monotopic Europe? And further, is there a future for the local, the peripheral and for (new) places of friction spread across the European territory? Is there scope for 'spaces of resistance' to the overarching notion of monotopia? Shore identifies the 'agents of European consciousness' and sees these as forces and objects through which knowledge of the European Union is embodied (Shore, 2000: 26). Such 'agents' can thus both be human agents, artefacts, bodies, institutions, policies, and representations that serve as an instrument for promotion and acceptance of the 'European idea' (2000:26). As the making of identities uses materials from history, geography, biology, institutions, collective memory and personal fantasies, power apparatuses and religious revelations (Castells, 1997: 7), the European idea reveals a contested field for political action. Or in the words of Chantal Mouffe: 'What we commonly call "cultural identity" is both the scene and the object of political struggles' (Mouffe, 1994: 110). However, if the European spatial policy discourse is part of the effort to picture European unity and identity, one might ask, what sort of identity?

Some analysts are quite dismissive of the potential for developing a European identity due to the situation of more or less permanent flux in the vital elements of its building blocks: 'The prerequisite for a specific socio-spatial consciousness and identity are weak in Europe, since both the territorial, symbolic and institutional shapes of this entity are unclear' (Paasi, 2001: 21). However, according to Castells, it is not as much a question of whether a European identity can be constructed as on what premises:

> If meaning is linked to identity, and if identity remains exclusively national, regional or local, European integration may not last beyond the limits of a common market, parallel to free-trade zones constituted in other areas of the world. European unification in a long-term perspective requires European Identity ... So, by and large, there is no European identity. But it could be built, not in contradiction, but complementary to national, regional, and local identities.
>
> (Castells, 1998: 332–3)

The interesting question here, particularly in the aftermath of the new European Constitution referenda results of 2005, is how these more general notions of identity and territory link to the European spatial policy discourse. Needless to say, more in-depth case based studies are needed in order to excavate the complex relationships between the UK at the regional and national level and the European Union's spatial policies.

In this chapter, we have offered a critical analysis of how European Union spatial ideas are shaping new policy narratives at multiple spatial scales and within new complex governance settings. We argue for the existence of a Europeanisation of spatial narratives, where notions of a seamless flow space of Europe are reproduced, as part of the reproduction of a Europe of Monotopia. Within EU territories, this process of Europeanisation is manifested in the shaping of policies for mobility, infrastructure and urban and regional development, as well as in the general expression of becoming more European (alongside other 'scales' of territorial identity). These 'hard' and 'soft' policy issues are linked, making the territories within the UK in need of reorienting their policy narratives of territory, identity and space, through a complex engagement with Europe. We are witnessing cases of Europeanisation occurring in different ways, as a transnational institutional creation, as the penetration of national planning, as the reproduction of a top-down political project, but also as the creation of new spaces and potentials for the content of Europeanisation – for the policy ideas – to be reconstructed and contested, at different scales and in diverse institutional settings.

Apart from the understanding of the European Union as another level of policy institutions that UK spatial planners and policy makers now deal with, we would argue that these spatial policies and representations are more than that. They are new ways of organising space from the scale of the 'big Europe' to the smallest village in the English countryside. They are new ways of picturing space and its relationship to social agents. Thus, behind the 'hard' policies of regional development and infrastructure looms 'softer' issues of belonging and notions of how to juggle the new nested identities of the Millennium. It is here, perhaps, that policy makers, planners and citizens of the UK are facing even bigger challenges.

Notes

1   This chapter is a revised version of the paper 'The Spatialisation of the European Project' presented at the XVIII AESOP Congress, 1–3 July 2004, Grenoble, France.
2   The meeting took place on 5 March 1997 in Sunderland.

References

Bauman, Z. (2000) Liquid Modernity, Cambridge: Polity Press.
Böhme, K. (2002) Nordic Echoes of European Spatial Planning, Stockholm: Nordregio.
Brenner, N. (2004) New State Spaces: Urban Governance and the Rescaling of Statehood, Oxford: Oxford University Press.
Castells, M. (1996) The Information Age: Economy, Society and Culture, Vol. I: The Rise of the Network Society, Oxford: Blackwell Publishing.
Castells, M. (1997) The Information Age: Economy, Society and Culture, Vol. II: The Power of Identity, Oxford: Blackwell Publishing.
Castells, M. (1998) The Information Age: Economy, Society and Culture, Vol. III: End of Millennium, Oxford: Blackwell Publishing.
CEC (2003) 'Structural policies and European territories: competitiveness, sustainable development and cohesion in Europe – from Lisbon to Gothenberg', Brussels: CEC.
CEC (2004) 'A new partnership for cohesion: convergence competitiveness cooperation: Third report on economic and social cohesion', Brussels: CEC.
Cook, R. (2004) 'This Brussels summit will do nothing to offer the people a European vision', The Independent, London: 35.
CSD (1999) 'European spatial development perspective – towards balanced and sustainable development of the territory of the EU', Presented at the Informal Meeting of Ministers Responsible for Spatial Planning of the Member States of the European Union, Potsdam May 10/11 1999, Committee for Spatial Development.
Davoudi, S. (1999) 'Making sense of the ESDP', Town & Country Planning, 68(12): 367–9.
DETR (2000) Planning Policy Guidance Note 11: Regional Planning, London: HMSO.
ESPON (2004) ESPON in progress: preliminary results by autumn 2003, Denmark: ESPON.
European Convention (2003) Draft Treaty Establishing a Constitution for Europe, E Convention – CONV, 2003, http://european-convention.eu.int/bienvenue.asp?lang=EN&Content.
European Council (2004) Presidency note on IGC 2003 following Meeting of Heads of State or Government, Brussels, 17/18 June 2004, CIG 85/04, PRESID 27, Brussels: CEC.
Faludi, A. (1998) 'Polynucleated metropolitan regions in Northwest Europe', European Planning Studies, 6(4): 365–77.
Faludi, A. (2004) 'Territorial cohesion: old (French) wine in new bottles?' Urban Studies, 41(7): 1349–65.

Flyvbjerg, B., Bruzelius, N. and Rothengatter, W. (2003) *Megaprojects and Risk: An Anatomy of Ambition*, Cambridge: Cambridge University Press.

Habermas, J. (1996) 'Citizenship and national identity' in J. Habermas (ed.) *Between Facts and Norms: Contributions to a Discourse Theory of Law and Democracy*, Cambridge: Polity Press: 491–515.

Hajer, M.A. (1999) 'Zero-friction society: the cultural politics of urban design', *Urban Design Quarterly*, 71, 29–34.

Hajer, M.A. (2000) 'Transnational networks as transnational policy discourse: some observations on the politics of spatial development in Europe', in A. Faludi and W. Salet (eds) *The Revival of Strategic Planning*, Dordrecht: Kluwer.

Healey, P. (1995) 'Discourses of integration: making frameworks for democratic urban planning', in P. Healey (ed.) *Managing Cities: The New Urban Context*, Chichester: John Wiley & Sons: 251–72.

Jensen, O.B. (1998) *Polycentrisk balance eller hierarkisk konkurrence? – Byerne og det Globale i Danmarks og EU's planlægnings diskurser*, København, Konference om Storbyens Forvandlinger og Kortlægninger, Kunstakademiets Arkitektskole, Center for Tværfaglige Urbane Studier, 25–26 May.

Jensen, O.B. and Richardson, T. (2004) *Making European Space: Mobility, Power and Territorial Identity*, London: Routledge.

Morley, D. and Robins, K. (1995) *Spaces of Identity: Global Media, Electronic Landscapes and Cultural Boundaries*, London: Routledge.

Mouffe, C. (1994) 'For a politics of nomadic identity' in G. Robertson, M. Marsh, L. Ticker *et al.* (eds) *Traveller's Tales: Narratives of Home and Displacement*, London: Routledge: 105–13.

Paasi, A. (2001) 'Europe as a social process and discourse: considerations of place, boundaries and identity', *European Urban and Regional Studies*, 8(1): 7–28.

Regional Development Strategy for Northern Ireland 2025, at http://www.drdni.gov.uk/shapingourfuture.

Rometsch, D. and Wessels, K. (eds) (1996) *The European Union and its Member States: Towards Institutional Fusion?* Manchester: Manchester University Press.

Rumford, C. (2002) *The European Union: A Political Sociology*, Oxford: Blackwell Publishing.

Shields, R. (1991) *Places on the Margin: Alternative Geographies of Modernity*, London: Routledge.

Shore, C. (2000) *Building Europe: The Cultural Politics of European Integration*, London: Routledge.

Tewdwr-Jones, M. and Williams, R.H. (2001) *The European Dimension of British Planning*, London: Routledge.

Weiler, J.H.H. (1999) *The Constitution of Europe: 'Do the New Clothes Have an Emperor?' and Other Essays on European Integration*, Cambridge: Cambridge University Press.

Welsh Assembly Government (2003) 'People, Places, Futures: The Wales Spatial Plan' .Consultation draft. Cardiff: National Assembly for Wales.

Williams, R. H. (1996) European Union, Spatial Policy and Planning, London: Paul Chapman Publishing.

5 Territory, integration and spatial planning

Patsy Healey

Introduction

> So we need to move again towards the idea of planning as the integration of strategies and policies as they impact on use of land, but with a different approach ...
>
> (LGA, 2000: 17)

> Successful spatial planning is integrated
>
> (RTPI, 2001: 3)

In contemporary discussions in the UK on the purpose and organisation of the planning system and the practices surrounding it, a recurring theme is the role of the system in 'integrating' disparate agendas, activities and actors. An 'integrated' approach is put forward: as a way of linking diverse policy objectives (the search for a beneficial economic, social and environmental 'balance'); as a way of connecting issues as they play out spatially (for example, housing and economic development, or land use and transport); as a way of linking different types of government intervention (especially regulatory power and investment power); as a way of overcoming the fragmentation of area- and development-based policy initiatives and the competition between individual projects; or as a way of linking policy with 'implementation'. An 'integrated' approach may also be used to refer to increasing the connections between levels of government, or to linking multiple stakeholders in pursuit of an agreed framework or strategy.

In these discussions, the focus is sometimes on 'joining up' or 'holistic' government in some general sense (6, Leat *et al.*, 2002; Wilkinson and Appelbee, 1999). This may become more specific through an emphasis on 'joining up' at a sub-national level of government – regions, local authorities, or town, parish and neighbourhood administrative units. These concerns to connect different governance initiatives are focused in the arena of the planning system, through the system's concern with land, its use and development, with spatial organisation and the qualities of places. With this remit, the system is forced to connect economic, social and environmental arguments, and address intense conflicts over the qualities of places and the impacts of development projects.

The advocacy of the role of the planning system in 'joining up' diverse governance initiatives rests on the idea that 'territory', 'place' and 'spatial organisation' have some particular value as a focus of attention in the 'search for integration'. This echoes a longstanding claim in planning thought that a policy focus on the qualities of place and territory promotes the effectiveness of different policies as they play out in specific places. In contrast, in economic and social policy, the emphasis has been on the delivery of social welfare and economic support services, with complex policy traditions and communities of practice building up around different sectors (education, health and social support; policies for economic development; policies for agriculture, forestry, industry and trade; policies for specific infrastructures). It has been left to the planning system to work out how these policies interlink at the local level, through development investment and land use regulation. In practice, rather than a critical arena for the articulation of place-focused territorial development policies, the planning system's integrative task was a local job of finding space for, and fitting in, diverse public and private development initiatives.

In the years of neo-liberal policy, even this interest in policy co-ordination faded. The remit of planning was increasingly focused on land use regulation, with conflicts being resolved through contestation and challenge over planning frameworks that expressed norms and criteria for development proposals, and through conflict over specific development proposals. The scale of contestation increased rapidly, as the environmental movement acquired significant leverage in public policy and in social movement activism. The planning system became a critical site where the tensions between economic interests and environmental values were played out (Grove-White, 1991). During the 1990s, as the urban policy agenda evolved into an area-focused urban regeneration agenda, pressures developed to make clearer connections between regeneration initiatives and policies embedded in land use regulation planning practices. Increasingly, from many perspectives, demands emerged for a more integrated approach to a wide range of disparate strategic initiatives arising from different policy sectors. The call for more 'joined up' government was taken up vigorously in the 1997 Labour government's initiatives to 'modernise' local government (Department of the Environment, 1998; see also Janice Morphet's discussion elsewhere in this book). In a somewhat ambiguous relation with the modernising initiative, the government was also committed to a regionalisation initiative, devolving power to regional bodies. One result has been the emergence of an array of partnerships and agencies focused on delivering services at regional and local levels, linking together policy communities which formally proceeded separately (Wilson and Game, 2002; Imrie and Raco, 2003). In England, where formal political regional jurisdictions do not exist, and are now unlikely to for many years, the urban and regional governance landscape is splattered with an array of regional and urban agencies and partnerships, charged with developing strategies on various topics and encouraged to 'integrate' around new foci of attention and to develop policy frames in new multi-agency groups.

Within this array of initiatives, the planning system has been faced with an ambiguous challenge. On the one hand, it is presented as a key site for 'integrating' disparate policies and strategies around a place focus:

> We have sought a better way to integrate planning and a whole range of local government activities. We have also explained ways to secure community involvement in the processes that shape the future pattern of development investment and service delivery. This requires a wide-angle, joined-up approach at county, district, borough and neighbourhood level.

However, local planning cannot be seen in isolation. The benefits we hope to achieve can only be realised if the regional and national context facilitate more effective, holistic and accountable planning at the local level.

(LGA, 2000: 4)

Most people are looking for the planning system to do the joining-up between a whole range of public and private activities and services that have implications for the way land is used.

(LGA, 2000: 7)

On the other hand, the practices of the system had become deeply embedded in an increasingly legalised conflict resolution process. The first part of the 2000s has seen the emergence of a reform movement as regards the planning system and its practices, the aim of which has been to widen the remit and perspective of the system, to take on an integrative and strategic role in territorial development, at regional, sub-regional, urban and neighbourhood/village scale. This 'spatial planning movement', pushed by forces within local government (LGA, 2000) and the planning profession (Royal Town Planning Institute, 2001), has had a significant role in shaping a new English government initiative in the planning area, heralded as a 'fundamental reform' of the system (DTLR, 2001).

However, this movement is highly ambiguous and is being promoted in a context where, in the recent past, planning practice has been discouraged from, and lost skills in, the strategic spatial thinking which could deliver a strategic imagination with focusing power which could promote territorial integration (Vigar *et al.*, 2000). It bumps up against the institutionally embedded power of sectoral policy communities and demands significant changes in the political currency used by local politicians when addressing issues to do with development projects, local environments and economic, social and environmental conditions. There are also major uncertainties over what actually is to be integrated with what, where integration is to take place and through what processes, what old 'integrations' will be challenged, and how far spatial organisation and place qualities are central to the 'integrations' being sought.

This chapter examines the concepts of integration mobilised through this 'spatial planning' reform of the planning system, as expressed in recent national policy statements and the potential for a territorially based integrative focus to policy development in the English context. The next section reviews the meanings of 'integration' emerging in government statements about the newly reformed English planning system. The chapter then reviews the 'opportunity structure' for a focus on space, place and territory as an integrative force at local and regional level. The chapter concludes with comments on the potentials and limits of such a focus in the developing English context.

Multiple meanings of 'integration'

The ODPM (Office of the Deputy Prime Minister), the government department responsible for local government, housing policy, urban policy and planning, has produced a stream of legislation, initiatives, policy statements and guidance documents in all these areas in the period 2000–4. In 2003, an important initiative was taken to link these fields of policy, each with their own accretion of practices and policy communities, in the context of a general approach, the *Sustainable Communities* action plan (ODPM, 2003). This strategy, focused particularly on housing issues, has since been used to provide an overarching purpose to all the areas of the ODPM's work, including the planning field. Through the concept of sustainable development, the strategy emphasises that environmental, economic and social issues must be addressed together:

> It will be essential for all development, especially new housing develop-ments, to respect the principles of sustainable development and address potential impacts on the environment alongside social and economic goals.
> (ODPM 2003, para. 3.11: 33)

Sustainable development has become the overarching purpose of the English planning system, as articulated in the Planning and Compulsory Purchase Act 2004. Drawing on an earlier government statement,[1] these principles were expressed in guidance on national planning policy (PPS1) as four aims:

- social progress which recognises the needs of everyone
- effective protection of the environment
- the prudent use of resources
- maintenance of high and stable levels of economic growth and employment.
> (ODPM, 2005, para. 4: 2)

In the *Sustainable Communities* strategy, the principles, and more particularly a statement about the qualities of sustainable communities (p. 5), are intended as a strategic, integrating platform upon which linkages between the ODPM's own pol-icy areas can be developed and from which campaigns to encourage more co-ordination with other government departments on regional and local develop-ment issues can be mounted. This is particularly important, as the strategy defines areas of low demand (linked to funding from housing and urban regeneration bud-gets) and growth areas, the latter requiring major investments from other national government departments, especially, transport, health and education. However, the struggle to focus attention and co-ordinate regulatory and investment activity between government departments and within the ODPM remains a difficult one, as the new regional agencies stress (National Audit Office, 2003). Within the ODPM itself, while all fields make reference to the others, it is within the planning policy area that the rhetoric of 'integration' is most strongly developed. The rhetoric of integration, rather hesitant in the initial proposals, has become increasingly stri-dent in the policy statements about the new planning system. The arenas and practices of the system are now expected to develop to provide the key integrative force to draw together the different policy frames and programmes of action as they affect territories and places:

> Planning for sustainable development should ensure that these four aims are tackled in an integrated way, in line with the principles for sustainable development ...
> (ODPM, 2004a (draft PPS 1) para. 1.14: 8)

These aims should be pursued in an integrated way through a sustainable, innovative and productive economy that delivers high levels of employment, and a just society that promotes social inclusion, sustainable communities and personal well-being, in ways that protect and enhance the physical environment and optimise resource and energy use.

(ODPM, 2005 (final PPS 1), para. 4: 2)

The core strategy [of the Local Development Documents] should draw on any strategies of the local authority and other organisations that have implications for the development and use of land ... Where appropriate, the core strategy should provide an integrated approach to the implementation of these aspects of other strategies.

(ODPM, 2004c (PPS 12), para. 2.10: 7)

But, as these quotations reflect, within the planning rhetoric, many meanings and foci of integration coexist. Integration is a relational word. It implies some kind of synergetic linkage and interconnection. It can only be understood in terms of what is to be linked and merged. One example is the interlinking of functional policy 'sectors' around a specific issue or area focus, such as the development of a 'transport corridor' or a multi-agency crime prevention initiative or an 'integrated area development' strategy. But a momentum for 'integration' does not merely generate a co-ordinating force. It also creates new divisions, as linkage in one direction closes off opportunities for linkage in another direction. The result could be that while old boundaries are broken down to create new, rich and intertwined connectivities, new boundaries and divisions are created. The term 'integration' is thus one of a family of words – joined-up, holistic, co-ordinated, interrelated, sometimes including older terms like comprehensive and balanced, which focus attention on connectivities and relations. It is therefore important to look carefully at the specific connectivities being emphasised in particular contexts.

Table 5.1 summarises the meanings of integration to be found in recent English government statements about the planning system, drawing out the connectivities implied. The focus here in particular is on a key forerunner report from the Local Government Association (LGA, 2000), the Green Paper proposing a 'fundamental reform' to the system (DTLR, 2001), the *Sustainable Communities* strategy produced by the ODPM in 2003, three Planning Policy Statements produced by the ODPM in 2004, the consultation draft of PPS 1, *Creating Sustainable Communities* (ODPM, 2004a), PPS 11 on *Regional Spatial Strategies* (ODPM, 2004b), PPS 12 on *Local Development Frameworks* (ODPM, 2004c) and the final version of PPS 1 produced in early 2005, now entitled *Delivering Sustainable Development* (ODPM, 2005).

A core meaning of integration centres on *co-ordination* between policy fields, sometimes expressed as 'bringing together', usually linked to some concept of co-alignment, and making policies in different fields mutually consistent. In this co-ordinative meaning, the planning system is recalled to a role intended for it in the 1940s, which was to give spatial expression to a range of policy activities. Within this co-ordinative meaning, the most frequent reference is to align planning policies with other policies (DTLR, 2001). The role of planning strategies and policies is to ensure land is available for the development needs and opportunities identified by other government programmes by 'clearing' them through the regulatory processes of the planning system. As the momentum for strategic

Table 5.1 Meanings of 'integration' in recent policy statements about the English planning system

		Type of integration	Meaning of integration
A		*Co-ordination*	
	A1	Aligning	Fitting in to other policies and strategies
	A2	Co-aligning	Mutual adjustment among diverse strategies
	A3	Multilevel co-aligning	Mutual adjustment both vertically and horizontally
B		*Framing*	
	B1	Widening a policy frame	Extending an existing frame to encompass a new dimension
	B2	Creating a new frame or vision	Developing a new policy focus and discourse
	B3	Creating a place-focused policy frame	Focus specifically on place qualities and spatial organisation
C		*Linking policy and action*	
	C1	Policy and delivery	Connecting policy assertions to specific delivery mechanisms
	C2	Regulation and investment	Linking principles governing land use regulation to those governing development investment
D		*Linking multiple actors*	
	D1	Involving	Drawing the community and key stakeholders into plan-making processes
	D2	Sharing knowledge and ideas	Drawing on and developing knowledge with stakeholders
	D3	Sharing ownership	Developing a shared commitment to the content and legitimacy of a plan/strategy

spatial planning increased during the early 2000s, this adaptive meaning has moved to greater emphasis on co-alignment in interactive processes, particularly between community strategies under regeneration initiatives and local development frameworks under the planning legislation (see the quotations above from LGA 2000 and draft PPS 1). Local Development Frameworks are expected to perform an even more complex co-ordinative feat, mutually adjusting both vertically and horizontally. In this emphasis, the arenas of the planning system are promoted as key sites for the performance of multilevel and multi-actor governance. Yet this is to be done in a situation of competing arenas and multiple policy networks and coalitions, all encouraged to develop horizontal or multi-scalar relations in contexts where national power and traditions of nested hierarchy remain strong.

In this context, a second meaning of integration refers to the formation of forceful policy *frames*, usually referred to as 'visions', which are expected to lie at the heart of regional and local planning strategies. For example:

> The RSS should articulate a spatial vision of what the region will be like at the end of the period of the strategy and show how this will contribute to achieving sustainable development objectives …
>
> (ODPM, 2004b: 2, para. 1.7)[2]

> The core strategy should set out the key elements of the planning framework for the area. It should be comprised of a spatial vision and strategic objectives for the area; a spatial strategy; core policies; and a monitoring and implementation framework with clear objectives for achieving delivery.
>
> (ODPM, 2004b: 7, para. 2.9)

The vision provides a synthesising force, capable of drawing together key priorities and expressing them in ways that act as a shared policy frame. The *Sustainable Communities* strategy (ODPM, 2003) is designed to have this kind of force at national level. The work of policy framing, here and in PPS 1, focuses on developing the four aims of sustainable development in an integrated way. Sometimes the old vocabulary of 'comprehensive' planning creeps in, when faced with these competing foci around which an integrated 'vision' could develop (LGA, 2000: 20). But in PPS 1, the overarching policy statement about the planning system, another old idea of 'balancing' competing claims is used in a discussion of how to justify giving more weight to one consideration than another (ODPM, 2005, para. 29). There are other references which emphasise the importance of linking planning and transport policies (PPS 12), planning and economic development (PPS 11), and planning and housing (the *Sustainable Communities* strategy, and the Barker Review (Barker, 2004)). Integration is also sometimes used in a quite different way, to refer to the mixed tenure housing developments and mixed use development projects which are supposed to characterise 'sustainable communities' (see the *Sustainable Communities* strategy and PPS 1, both showing the influence of the earlier report of the Urban Task Force (1999)).

This confusion of foci for policy frames and visions reflects the complexity of the task facing planning system practices in engaging in 'integrative' policy framing work. The planning system has developed its own baggage of policy discourses around specific policy topics and spatial organising ideas, notably around housing, environmental protection and contained (compact) settlements surrounded by green belts (Vigar *et al.*, 2000). One way forward is to widen an established frame to include concepts developing in another policy community. This is already happening in the linkages between the multi-modal strategies developing in the transport field and planning strategies emphasising development at public transport nodes; the longstanding link between housing policy and planning strategy has been reinforced by the *Sustainable Communities* strategy and the Barker Review. But creating such a linkage is an institutionally complex process. Local Strategic Partnerships are encouraged to produce *new* strategic frames when they articulate the Community Strategies intended to focus urban regeneration investment for particular parts of cities and sub-regions. These then should feed into, and potentially change, strategies for land allocation and development investment which focus on the *whole* of a particular administrative area (Bailey 2003; Entec, 2003). Although this is the ambition of protagonists of the 'new' reformed planning system (LGA, 2000; RTPI, 2001), developing such new policy frames requires very substantial institutional effort, underpinned by the mobilising power to manage the coalition building and intellectual imagining which developing a new policy frame involves. The same challenge has to be met at regional level too, where any efforts in developing new policy frames have to pull in the already well-developed framing work undertaken through the Regional Development Agencies, along with several other regional strategy initiatives.

Effective integration between planning policies and other policies thus demands substantial institutional work in co-aligning the policies of diverse policy communities, each with their own traditions, pressures and innovation dynamics as they focus on regional and local scales. Co-framing efforts are promoted as a way to achieve such co-alignment. Behind this is an important objective of the *Sustainable Communities* agenda. Through the integrative work achieved through the planning system, economic, environmental and social considerations are, the rhetoric proposes, kept in balance, avoiding the crowding out of environmental and social considerations by economic ones, or of housing development considerations at the expense of environmental ones. The planning system, therefore, is expected to be the carrier of multiple policy objectives into the sectoral heartlands of other policy communities. This is where the spatial planning agenda gets its leverage. It is not just about co-ordinating and aligning the spatial aspects of the policies of other sectors. It is about the 'nature of places and how they function' (ODPM, 2004a, para. 1.29). This implies that the integrative policy frames and visions focus on qualities of places and principles of spatial organisation. The next section examines the opportunity structure for such a policy framing effort at regional and local scales in England. Before doing this, a brief comment is provided on two more meanings of integration to be found in recent policy statements about the planning system.

The first concerns the relation between *policy* and *action*. This is emphasised particularly in the *Sustainable Communities* strategy, and in the advice on *Regional Spatial Strategies*; the concern in both is with investment project 'delivery vehicles'. Here the aim is to link national strategies to mechanisms to carry particular projects forward, notably the various regeneration initiatives, now combined with the investment projects needed in the 'growth areas'. But another dimension of the link between policy and action is the connection between development investment strategies and land use and development regulation. The planning system is stridently criticised for its narrow focus on 'traditional land use [meaning: regulatory] planning' (ODPM, 2004a, para. 1.29; LGA, 2000). Yet this regulatory activity is critically important as it provides a check and balance on the state's development role, to ensure that property rights and any other principles locked into regulatory norms and policies are properly respected. Investment projects can be pursued through all kinds of formal and informal arrangements, but regulatory practices achieve their legitimation through semi-judicial and judicial judgements about the limits of state power. There are thus inherent tensions between the state as entrepreneurial enabler and facilitator, and the state's regulatory role in protecting the public interest.

The policy ambition is that overarching policy frames can be produced through a *multi-actor and multilevel collaborative effort*, which can be perceived as, and defended as, legitimate in both development investment and regulatory arenas and practices. This final meaning of integration in the current policy rhetoric about the planning system emphasises the complexity of the processes through which substantive integrations of policy content are to be arrived at. Drawing key stakeholders and the 'community' into plan-making processes will, it is hoped, help to produce new shared understandings and policy frames, achieve greater co-ordination, reduce conflict once plans and projects reach detailed specification, and generally buttress planning strategies with robust legitimation (ODPM, 2004a).[3] But it is a big step from involving stakeholders and the community in the arenas where key policy frames are produced, which is difficult enough, to sharing knowledge among parties, and developing the ideal of co-ownership promoted in some statements.

It is no wonder that the PPS 1 talks of the need for a 'culture change' in planning. This implies major changes in the practices of all those who operate in the planning area, that is, the planning policy community – the lobby groups and environmental activists, as well as local authority politicians and planning officials, developers and their planning and legal advisers. But the range of the 'culture change' implied is in reality much wider than this. It anticipates the development of a governance culture in England that challenges the established landscape of British governance and public policy, with its history of strong centralism and sectoralism. It implies the emergence of powerful policy discourses and arenas focused around regional and local coalitions, to challenge and reconstruct the overall policy landscape in more decentralised and territorially focused ways. Where does the mobilisation force for such a struggle come from, how strong is it likely to be, and what is its potential to lead to policy discourses, arenas and governance practices which focus policy attention on spatial organisation and place quality?

Space, place and territory as an integrative focus

The search for a stronger governance focus on spatial organisation and place quality is not confined to the UK. In Europe generally, the idea and practice of 'strategic spatial planning' has taken hold at both EU level[4] and in urban and regional planning practice in many EU countries (Healey *et al.*, 1997; Salet and Faludi, 2000; Albrechts *et al.*, 2001; Tewdwr-Jones and Williams, 2001; Albrechts *et al.*, 2003). Le Gales has argued that European traditions of city and regional identity create a significant cultural underpinning to these endeavours (Le Gales, 2002), producing a social grounding for the development of political coalitions to articulate place-focused policy frameworks and position city policy agendas in national and EU policy space. However, studies of the experience of metropolitan regions in building common agendas emphasise the tensions of coalition building around place-focused agendas where the relational dynamics which generate place qualities do not match up with political territorial jurisdictions, as is commonly the case in contemporary urban and regional geographies (Salet *et al.* 2003; Gualini, 2004a). To carry integrative force across a multi-actor governance landscape, political coalitions need to develop which can mobilise and sustain policy attention through time.

The literature and policy debate on urban and regional policy in recent years has emphasised that an opportunity structure exists which favours such mobilisation (Amin and Thrift, 1994; Cooke *et al.*, 2000; Salet and Faludi, 2000; Le Gales, 2002). Some ground their arguments in the claim that, in an era where economic forces emphasise global markets and economic networks, the qualities of cities and sub-regions have become key factors in attracting and sustaining economic investment, and hence in the creation of national GDP. Others emphasise the interdependence of environmental systems on various scales, and the significance of local action in contributing to global environmental sustainability. A socio-cultural argument also supports the claim that qualities of places become more noticed and better defended as people become increasingly conscious of how places of significance to them (as living places, visiting places, culturally significant places) are positioned in a global context. The unfolding crisis of rising expectations of state spending and limited state resources adds to pressures on national states to offload and download service delivery and investment responsibility. Finally, increasing concern about the distance between national politics and policy making and citizens, and the alienation from the political and policy-making classes that results, generates a search for greater 'subsidiarity' in political organisation, with a stronger role for cities and sub-regions. All of these forces are creating pressures for some reformulation of the governance arrangements of European welfare states (Gualini, 2004a).

These pressures generate a structuring force to encourage greater decentralisation and devolution to sub-national jurisdictions. But this does not necessarily imply that political mobilisation and policy agendas will focus around place qualities and 'integrated' territorial development agendas. Nor does this mean that the arenas of the planning system will necessarily become significant in articulating decentralised agendas. In England, escaping from hyper-centralism is a hard and ongoing struggle. The national level remains important, but exactly what central government should do to encourage and monitor regional and local performance is highly contested. From a national government perspective, there have been attempts to create a geography of north and south, with the development of the idea of a 'Northern Way' as a banner under which political interests outside the powerful South East can cluster. At the regional level, as other chapters in this book show (see especially that by Iain Deas on the North West), there are political struggles over where power should lie between regional bodies and municipalities, particularly the strong cities, which have mobilised under a 'core cities' umbrella (Jonas *et al.*, 2004). Policy development can all too easily remain structured at local and regional levels around the coexistence of sectoral policy agendas, loosely coupled with each other, and often prone to domination by a particular sector, such as, for example, economic considerations. Some business groups may promote an integrated place-focused agenda, usually focused on sub-regions defined in terms of critical economic linkages (land and labour markets, transport and other infrastructures), as do many citizens concerned about their daily living environments. But it is difficult to translate such perceptions into political mobilisation, even in territories with a strong cultural tradition of distinctiveness. Meanwhile, the political capital locked into the landscape of pressure groups, from the business sector and from civil society, tends to emphasise 'single issue' politics, continually contesting policy agendas pursued from other perspectives, rather than searching for a multi-actor, integrated perspective.

This suggests that, while an opportunity structure exists for an integrated, place-focused policy agenda in England's cities and sub-regions, capturing this opportunity will require long-term and hard political work to create a new landscape of governance practices, underpinned by a governance culture which exchanges a predominantly national focus on the construction of policy attention for a much more horizontal focus. This means that many politicians, policy experts (including planners), pressure groups and lobbyists will have to disembed themselves from their current discourses, arenas and practices, and change their landscape of institutional cores and boundaries, in order to make these new, more horizontal integrations. This disembedding process is not just a question of exchanging one policy focus for another. It involves struggles over competing logics and legitimacy principles (Tewdwr-Jones, 2002: 279), and the formation of new communities of policy practice around different arenas and discourses, with different logics. The logic of development project investment and the logic of land use regulation invoke different practices (entrepreneurial energy and management skill versus fairness and reasonableness) and different legitimacy principles (the logic of results versus the logic of authoritative norms). In the business sector, different kinds of company relate to different labour market time–space relations, and place identities may invoke many different meanings and locales of place (Healey, 2002). The logics of *ad hoc* governance partnerships are not coterminous with those of city politics. The decentralisation movement in England is thus likely to proceed in a highly uneven and 'jerky' way, as moves towards greater decentralisation in one part of the governance landscape are counteracted by countermovements elsewhere. This means that any attempt to change practices and cultures is a risky business, as there is no certainty that change initiatives reinforce each other.

In this context of multiple, often conflicting, governance change dynamics, the efforts to shift the agendas and practices of the planning system are but one among many initiatives. The role of spatial concepts as integrative devices is much less well-developed than, for example, in the Netherlands (Faludi and van der Valk, 1994; Hajer and Zonneveld, 2000; van Duinen, 2004). In Wales and Northern Ireland, and in the work on the London *Spatial Development Strategy*, a new vocabulary of spatial concepts is beginning to be invoked (see the chapters on Wales, Northern Ireland and London in this book). In some cases, these are providing a focus for developing investment strategies and agendas of key development projects. Elsewhere, there is a competition over where, if anywhere, in the local and regional landscape, the most powerful integrative arenas are likely to be. In some places, the Local Strategic Partnership (LSP) have managed to produce an integrative Community Strategy (CS), within which the planning system activity can be located. Sometimes, the community strategy provides the core to a local authority's overall strategic management. In other areas, the LSP/CS process is confined largely to the spending of urban regeneration budgets (Counsell *et al.*, 2003; Entec, 2003). The final version of PPS 1 exhorts local actors to 'integrate the wide range of activities relating to development and regeneration' (ODPM, 2005, para. 32 (iii)).

In these struggles over discourses, arenas and practices, the arenas of the planning system and its practices are caught up in many complex tensions. On one dimension, there is tension between which territory is in focus in place-focused development strategies: is it that of a formal politico-administrative jurisdiction, a territory of material functionality such as a land or labour market, a place of identity such as a historic city or ancient shire county? On another dimension are the struggles over the logics of legitimacy for a strategy: is it that of a political mobilisation movement which has recognised critical connectivities within a sub-region? Is it the logic of functional relations, such as a travel-to-work area? Is it the logic of elected politicians? Is it the logic of the construction of arguments that have standing in judicial arenas?

In situations which lack strong political mobilisation around place-focused agendas, it is unlikely in the short term that the integrative ambitions of the new planning reforms will get much beyond aligning planning policies with the action programmes proposed in other investment and regulation strategies. Where those responsible for plan-making tasks are able to mobilise attention more forcefully, a stronger role in co-alignment may be possible. This has been the outcome of earlier attempts at strategic spatial planning in the British context (Healey *et al.*, 1988; Brindley *et al.*, 1996). But in this co-ordinative work, conceptions of place and territory, although asserted, have typically had a weak role with little political meaning beyond the planning policy community.

To move beyond this, and to generate mobilising force behind concepts of place and territory which could develop, over time, into widely shared understandings with the power to motivate action across an urban or regional governance arena, efforts are needed to build the intellectual resources and debates about place qualities through which places and their qualities can be 'seen' (Healey, 2002). This then may help to build horizontal social capital around qualities of places. Generating such intellectual and social capital may help to both build political coalitions and construct governance arenas with the force to convert such capital into political capital, with enduring power to reform as well as co-align governance policy attention around critical place qualities to defend, develop and maintain. To achieve this, planning strategies need to connect to rich debates about place qualities, in terms which link to identities as well as material assets and conditions. An enduring mobilisation force needs sufficient political and intellectual robustness to command attention among multiple actors necessary to make a planning strategy contribute to more efficient co-ordination and more legitimate conflict resolution than current practices. Creating a vision and building an integrated core strategy is thus not a task merely for specialist planning teams. An integrative focus for planning strategies, centred around place qualities, does not exist 'out there' to be articulated and used in specific plan-making episodes; it has to be grown, and to take root, in a governance culture.

The prospects for an 'integrative' role for the planning system

The challenge for the recent planning reforms is that they assume a new governance landscape that has yet to arrive. It is possible to imagine a new landscape in which they would flourish. Regions would have significant investment and regulatory power across key services and activities. Within regions, local mobilisation would produce broadly based coalitions across business, citizens, pressure groups and all kinds of civil society groups which would promote city and sub-regional place qualities and identities, and express widely shared ownership of locally evolved concepts of spatial organisation. These principles in turn would shape investment projects, action programmes and regulatory practices. The national role would remain important, but confined to articulating spatial strategies for key infrastructures and nationally important projects, for safeguarding critical social and environmental norms and for redistributing resources to promote social justice between citizens and even out economic inequalities between regions.[5] In such a context, which is opening up to a limited extent in Scotland, Northern Ireland and Wales already, there is a chance that territorial policy communities might evolve to intersect with and re-frame the agendas and practices of the sectoral ones. Even in this scenario, major tensions are still likely to arise.

The planning system is acutely positioned between the logic of legal authority, which needs some reference to 'sovereign authority' for defining limits to private interests in land use and development, and the logics of functional materiality and identity that motivate many sub-regional political and partnership mobilisations. In the wider landscape, intense struggles are inevitable between different arenas of informal mobilisation and formal administrative jurisdictions and between local and regional levels of jurisdiction where more than one tier exists (Gualini, 2004a, b).

The planning system reform project is thus set to co-evolve in a complex, dynamic and conflictual governance landscape. In this context, the project could be undermined by a continuation of the hierarchical practice of 'looking up' to the national level to sort out the many tensions. This is very evident in the repeated calls from practising planners for 'more guidance' from the national level.[6] There is also a danger that the national level, fearful that the reform project will not be vigorously pursued locally, will end up being over-prescriptive; this is also a likely response to the demand for more guidance. But over-prescription also arises because the national level still holds the regulatory power and is continually concerned that planning principles as expressed in plan documentation have the robustness to meet the logic of legal challenge. As a result, over-prescription could reinforce the upward glance of planning officials and impede the 'culture change' that the spatial planning reform movement so earnestly seeks.

The planning reform project, and the devolution project generally, faces a further danger. The content of local and regional policy agendas, where they refer to place qualities and spatial organisation, are currently located primarily within the thoughtworlds of the existing planning policy community, and/or those local and regional policy activists who get involved in territorial stakeholder communities of practice. Instead of developing a broadly based discussion and debate about place identities, qualities and functionalities, which could ground a place-focused perspective in popular consciousness and political mobilisation, such a perspective could become just another competing policy logic in the complex urban and regional governance landscape. The result might be either that the focus is largely ignored or, if it attracted attention among many other policy groups, its content might end up narrow and exclusionary because of the lack of a broadly based debate and relation to popular consciousness. It could then be open to all kinds of challenges from those not part of the established policy nexus. Building a broadly based, place-focused agenda, capable of commanding legitimacy across the many dimensions of local governance and politics, needs institutional space and time to evolve.

In conclusion, the rhetoric of integration currently mobilised within the statements about the reformed planning system could be dismissed as just another plea for better co-ordination in the face of a continually evolving, fragmented landscape of governance initiatives. But it is also a call to the planning policy community to realign itself with the making of place development strategies, rather than merely anchoring itself to the logic and practice of land use regulation. It is a call to be part of the remoulding of the governance landscape, rather than merely acting as the guardian of a particular, though very important, function. However, in this repositioning momentum, it would be helpful for protagonists to be explicit about what specific integrations are being sought, where in the institutional landscape these are to be developed, what moment of political opportunity they expect to exploit, and how initiatives are to be developed and sustained. It is also important to indicate what linkages are likely to be weakened as a result of integrative efforts.

The contribution of the spatial planning movement and the planning policy community is not, as imagined in the past, to produce a 'comprehensive', all-encompassing, strategy for the evolution of a place or territory. Instead, much greater attention is needed to feeding debates and building connectivities. Already, some of the planning efforts in producing Regional Spatial Strategies and Local Development Frameworks show what this involves. This kind of strategic practice acknowledges that what is identified as important in policy terms about place and territory is an emergent property of multiple imaginative efforts, mobilised through political processes of coalition building and discourse formation. It is political mobilisation, not planning technique, which will have the power to carry the place-focused decentralisation movement into the remoulding of the landscape of urban and regional governance:

> A *New Vision of Planning* is required which seeks to build the capacity within society and its institutions to take effective and relevant decisions. This challenges us to think beyond the scope of statutory systems and to take a broader view of what society needs through planning. It also challenges us to see planning as an activity which professional planners facilitate, but do not own or monopolise.
>
> (RTPI, 2001: 2, emphasis in original)

Notes

1   *A better quality of life, a strategy for sustainable development in the UK*, Cm 4345, May 1999.
2   Note, also, in this statement, the idea of a strategy as depicting a state to be arrived at rather than a revisable trajectory.
3   'Sustainable development needs the community to be involved with developing the vision for their areas. Communities should be able to contribute ideas about how that vision can be achieved and have the opportunity to participate in the process for drawing up specific plans and policies and to be involved in development proposals' (ODPM, 2004a, para. 1.34: 15). 'Community involvement should ... enable the local community to say what sort of place they want to live in at a stage when this can make a difference' (ODPM, 2004d, para. 1.3: 4).
4   in the evolution of the *European Spatial Development Perspective* (CEC, 1999) and its subsequent development in the SPESP and ESPON programmes.
5   There is some evidence from the Netherlands and Italy for the link between significant decentralisation in unitary states and the emergence of strong place-focused policy agendas.
6   This emerges in comment in the weekly *Planning* journal.

References

6, P., Leat, D., Selzer, K. and Stoker, G. (2002) *Towards Holistic Governance: The New Reform Agenda*, Houndmills, Basingstoke: Palgrave.

Albrechts, L., Alden, J. and Kreukels, T. (eds) (2001) *The Changing Institutional Landscape of Planning*, Aldershot: Ashgate.

Albrechts, L., Healey, P. and Kunzmann, K. (2003) 'Strategic spatial planning and regional governance in Europe', *Journal of the American Planning Association* 69(2): 113–29.

Amin, A. and Thrift, N. (eds) (1994) *Globalisation, Institutions and Regional Development in Europe*, Oxford: Oxford University Press.

Bailey, N. (2003) 'Local strategic partnerships in England: the continuing search for collaborative advantage, leadership and strategy in urban governance', *Planning Theory and Practice* 4(4): 443–57.

Barker, K. (2004) *Review of Housing Supply: Securing our Future Housing Needs*, London: The Stationery Office.

Brindley, T., Rydin, Y. and Stoker, G. (1996) *Remaking Planning: The Politics of Urban Change in the Thatcher Years*, London: Routledge.

Cooke, P., Boekholt, P. and Todtling, F. (2000) *The Governance of Innovation in Europe*, London: Pinter.

Counsell, D., Haughton, G., Allmendinger, P. and Vigar, G. (2003) 'New directions in UK strategic planning: from development plans to spatial development strategies: new directions in strategic planning in the UK', *Town and Country Planning* 72(1): 15–19.

DETR (1998) *Modern Local Government: In Touch with People*, London: DETR.

DTLR (2001) *Planning: Delivering a Fundamental Change*, Wetherby: DTLR.

Entec UK Ltd (2003) *The Relationship between Community Strategies and Local Development Frameworks*, London: ODPM.

Faludi, A. and van der Valk, A. (1994) *Rule and Order in Dutch Planning Doctrine in the Twentieth Century*, Dordrecht: Kluwer Academic Publishers.

Grove-White, R. (1991) 'Land, the law and the environment', *Journal of Law and Society* 18(1): 32–47.

Gualini, E. (2004a) 'Integration, diversity and plurality: territorial governance and the reconstruction of legitimacy', *Geopolitics* 9(3): 542–63.

Gualini, E. (2004b) 'Politicising territorial governance: embedding the "political economy of scale" in European spatial policy', ECPR Workshop 14: *European Spatial Politics or a Spatial Policy for Europe?*, Uppsala, Sweden, April.

Hajer, M. and Zonneveld, W. (2000) 'Spatial planning in the Network Society – rethinking the principles of planning in the Netherlands', *European Planning Studies* 8(3): 337–55.

Healey, P. (2002) 'On creating the "city" as a collective resource', *Urban Studies* 39(10): 1777–92.

Healey, P., Khakee, A., Motte, A. and Needham, B. (eds) (1997) *Making Strategic Spatial Plans: Innovation in Europe*, London: UCL Press.

Healey, P., McNamara, P., Elson, M.J. and Doak, J. (1988) *Land Use Planning and the Mediation of Urban Change*, Cambridge: Cambridge University Press.

Imrie, R. and Raco, M. (eds) (2003) *Urban Renaissance? New Labour, Community and Urban Policy*, Bristol: The Policy Press.

Jonas, A.E.G., Gibbs D.C. and While, A. (FF) (2004) 'Uneven development, sustainability and city-regionalism contested: English city-regions in the European context', in

H. Halkier and I. Sagan (eds), *Regionalism Contested: Institution, Society and Territorial Governance*, Aldershot: Ashgate.

Le Gales, P. (2002) *European Cities: Social Conflicts and Governance*, Oxford: Oxford University Press.

Local Government Association (LGA) (2000) *Reforming Local Planning: Planning for Communities*, London: IDeA Publication Sales.

National Audit Office (2003) *Success in the Regions*, London: The Stationery Office.

ODPM (2003) *Sustainable Communities: Building for the Future*, London: ODPM.

ODPM (2004a) *Consultation Paper on Planning Policy Statement 1: Creating Sustainable Communities*, London: ODPM.

ODPM (2004b) *Planning Policy Statement 11: Regional Spatial Strategies*, London: ODPM.

ODPM (2004c) *Planning Policy Statement 12: Local Development Frameworks*, Norwich: The Stationery Office.

ODPM (2004d) *Community Involvement in Planning: The Government's Objectives*, London: ODPM.

ODPM (2005) *Planning Policy Statement 1: Delivering Sustainable Development*, Norwich: TSO.

Royal Town Planning Institute (RTPI) (2001) *A New Vision for Planning*, London: RTPI.

Salet, W. and Faludi, A. (eds) (2000) *The Revival of Strategic Spatial Planning*, Amsterdam: Koninklijke Nederlandse Akademie van Wetenschappen (Royal Netherlands Academy of Arts and Sciences)

Salet, W., Thornley, A. and Kreukels T., (eds) (2003) *Metropolitan Governance and Spatial Planning: Comparative Studies of European City-regions*, London: E&FN Spon.

Tewdwr-Jones, M. (2002) *The Planning Polity: Planning, Government and the Policy Process*, London: Routledge.

Tewdwr-Jones, M. and Williams, R. (2001) *The European Dimension of British Planning*, London: Spon Press.

Urban Task Force (1999) *Towards an Urban Renaissance*, London: E & FN Spon.

van Duinen, L. (2004) 'Planning imagery: the emergence and development of new planning concepts in Dutch national spatial policy', *Faculteit der Maatschappij en Gedragswetenschappen*, Amsterdam: Universiteit Amsterdam.

Vigar, G., Healey, P., Hull, A. and Davoudi, S. (2000) *Planning, Governance and Spatial Strategy in Britain*, London: Macmillan.

Wilkinson, D. and Appelbee, E. (1999) *Implementing Holistic Government*, Bristol: Policy Press.

Wilson, D. and Game, C. (2002) *Local Government in the United Kingdom*, London: Palgrave Macmillan.

Part 2

Studies of territorial and spatial planning

What a wondrous place this was – crazy as fuck, of course, but adorable to the tiniest degree ... What an enigma Britain will seem to historians when they look back on the second half of the twentieth century. Here is a country that fought and won a noble war, dismantled a mighty empire, created a far-seeing welfare state – in short, did nearly everything right – and then spent the rest of the century looking on itself as a chronic failure.

(Bill Bryson)

What is England, really? If you want to be brutal about it, England is shopping malls, staggeringly thick-witted and insensitive road schemes, lousy architecture, supermarkets, theme pubs and crowds of people wandering around, looking puzzled and disgruntled ... North to south, you find the same chainstores, the same eateries, the same cretinous planning fuck-ups.

(Charles Jennings)

6 The contested creation of new state spaces

Contrasting conceptions of regional strategy building in North West England

Iain Deas

Regional identity, policy and institution-building in North West England

The organisation of the state as a territorial entity has come back firmly onto the academic agenda in recent years. Initial ideas about the hollowing-out of the nation state began to give way from the late 1990s to more refined accounts which emphasise the contingency and complexity of the scalar division of the state and the importance of recognising the intricate multi-scalar geometry of the 'filled-in', re-territorialised state (see, amongst many examples, Brenner, 2004; Goodwin, *et al.*, this volume). These ideas have generated no little academic interest. The sometimes arcane intellectual arguments underpinning them – about the internationalisation of capital, the re-expressed pattern of state territoriality and, in some accounts, the changing centrality of the nation state – are invariably poorly understood amongst policymakers. The important point, however, is that a simplified reading of this challenging and varied set of ideas has helped to effect some significant shifts in the shape of institutional structures, and to embed some new orthodoxies amongst policymakers about urban and regional economic development.

This chapter explores the multifaceted ways in which the re-territorialisation process has unfolded in the context of the North West of England. It focuses specifically on strategic land-use planning, an area of public policy that has received relatively limited attention in the literature on questions of scale and territory. Most research on strategic land-use planning has tended to have a narrow emphasis, predominantly procedural in its focus, sometimes revolving around 'great men' historical accounts of policy evolution, and devoting limited attention to the links with other areas of policymaking or to wider considerations around regionalism and the changing territorial form of the state (Counsell and Haughton, 2003; Haughton and Counsell, 2004).

The North West of England provides an instructive context in which to explore these under-researched issues. With two of Britain's principal provincial cities, Liverpool and Manchester, the region provides a revealing illustration of the dynamics of scale and territory in the context of a large polycentric area of some 6.8 million residents. This is reinforced by the presence of a multiplicity of sub-regional territories and identities, and an array of agencies and policy initiatives that reflect the region's position at the epicentre of local economic development efforts in England. The chapter begins by charting the complex and conflicting ways in which the territorial reorganisation of the state has been expressed in the context of the North West. This provides the basis for an exploration of the specific experience of strategic land-use planning efforts in the region. The subsequent section considers regional land-use planning in the context of wider shifts in territorial governance in the North West. The chapter concludes by assessing the implications of territorial relations for the shape of strategic land-use planning in the English regions.

Re-territorialisation as a process in North West England

Part of the received wisdom stemming from 'new regionalist' thinking has centred on the widely held notion that regions such as the North West ought to acquire a clear and coherent institutional (and perhaps popular) identity if they are to induce improved economic performance. Such arguments have been given added salience by three factors of particular relevance to the North West. The first relates to the region's history of fragmentation and the limited effectiveness of many or most of the previous efforts to develop institutions with a North West base. Region-building efforts in the North West have been continually frustrated by a series of potent inter-jurisdictional tensions amongst its 46 local authorities (an upper tier of three counties with 25 lower tier districts, together with 15 metropolitan and three unitary districts) that, in turn, have limited the development of popular regional consciousness (Bristow, 1987; Tickell *et al.*, 1995). The most palpable tension has revolved around Liverpool and Manchester, the region's two principal cities, whose fraught relationship has oscillated over time between periodic enmity and reciprocal insouciance (Harding *et al.*, 2004). Alongside this most high-profile tension, the North West's cohesion has also been undermined by intermittent antagonism between the region's two principal cities and their Cheshire suburban hinterlands (Barlow, 1995; Deas, 2005; Hebbert and Deas, 2000; Lefevre, 1998), as well as with the free-standing Lancastrian industrial towns and with the rural surrounds. The uncertain status of Cumbria, alternately part of the North (East) and North West English regions, but never truly comfortable within either, has added a further impediment to region-building.

Second, in common with most of the other provincial regions of the UK, the North West has had over several decades a history of economic sluggishness relative to London and the South East, together with a preponderance of associated social problems. This has fuelled longstanding concern within the region about intractable, structurally embedded and politically sanctioned interregional inequalities which favour the economic heartlands of London and the South East over the provinces (see Amin *et al.*, 2003; see also Musson *et al.*'s chapter, this volume). Such concerns have been compounded by the contemporary view, drawing on emerging ideas from the 1980s about the need for inter-urban competition, that the major British cities lack the competitive dynamism of their principal provincial peers elsewhere in Western European (Parkinson *et al.*, 2004; Robson, 2001).

Running alongside and reinforcing these concerns about the North West's history of institutional fragmentation and its weak identity as a political actor, a third factor explaining the rise of regionalism has centred on reservations in the early 1990s about the region's (in)ability to exploit the perceived largesse of the EU Commission through its Structural Funds and an array of related Community Initiatives. As the recipient of more Structural Funds than any other Objective 2 area, and with Objective 1 status for Merseyside a possibility, developing effective mechanisms for attracting funding was unsurprisingly viewed as a key priority at that time (Williams and Baker, 2006). Such concerns might be seen as prosaic and straightforwardly managerial: the schisms that cross-cut the region had undermined its capacity to present a coherent and convincing case to the Commission. This was a view given added piquancy at the time by the perception that policy-making within the Commission was becoming ever more regionally based; and the North West, lacking the regional institutional superstructure evident elsewhere in the EU (including the 'Celtic' UK and, to a lesser extent, North East England), was at a significant disadvantage in attracting policy resources.

The combination of these interrelated concerns about institutional fragmentation and lack of political voice, economic underperformance and inability to procure EU resources created a powerful momentum in the early to mid-1990s for the development of regionally based governance in the North West. The result was a desire within the North West, shared by central government, for the region to 'get its act together': to develop new regionally based institutions and policy initiatives and, even if just for pragmatic reasons of improved resource procurement capability, to sacrifice supposedly parochial rivalries in favour of the regional good. This was initially reflected in the establishment of the North West Regional Association and then the North West Partnership, which laid the foundations for a regional institutional infrastructure subsequently formalised before the end of the decade with, respectively, the establishment of the North West Regional Assembly (NWRA) and the North West Development Agency (NWDA).

The Association was established in 1992 as a consortium of the region's constituent local authorities. Although there had previously been a regionally based association of four of the North West's then county councils (including the two metropolitan counties, but excluding Cumbria), this had withered in the early 1980s, partly because the bread-and-butter strategic planning issues around overspill which had fuelled previous regional planning efforts were superseded by a focus on intra-conurbation regeneration, and partly because central government interest in strategic planning had evaporated. The formation of the Association reflected the rise of some new issues that were seen, both within the region and to a lesser extent beyond it, as warranting a regionally based approach. The most significant spur for its formation was the concern about the region's standing in Brussels (Burch and Holliday, 1993). But there was also a view amongst its leading proponents – notably Louise Ellman, at the time leader of the previously unenthusiastic Lancashire County Council – that the region needed to develop a stronger identity, a series of regional structures and an increase in its funding if it was to begin to challenge the economic ascendancy of London and the South East, and compete on more even terms with other economically 'underperforming' territories like North East England, Scotland and Wales. Further emboldened by the Labour Party's growing embrace of regionalism in the early 1990s – and by the increasing prospect, from 1992, that a Labour government might be elected – the Association began to develop momentum. Its most significant act was to broker sufficient agreement within the region to underpin the publication, in 1993, of a regional economic strategy – the first for twenty years. Foreshadowing the collective agenda that was later to characterise the NWDA and NWRA, the Association's initial aims focused on image-building (largely geared towards inward investment and visitor attraction) and lobbying (directed at central government and the EU), with the overall goal of enhancing 'prosperity' and bolstering 'competitiveness' (North West Regional Association, 1993a).

While the establishment of the Association followed in a long line of earlier attempts to create differently configured regionally based alliances of local authorities, what characterised these efforts in the early 1990s was the increasingly instrumental role of business interests in influencing the shape and direction of change. The growth of private sector involvement in public policymaking had grown steadily since the early 1980s. At first, this was confined largely to particular areas of policymaking, like urban regeneration, for which central government sought to develop mechanisms directly to encourage business involvement. By the late 1980s, private sector involvement had begun to extend to regional affairs. 1989 saw the relaunch of the North West Business Leadership Team (NWBLT), a body originally established at government's behest in the wake of the 1981 riots in Manchester and elsewhere (Valler *et al.*, 2004). This provided a forum for the most committed regionalists from within the private sector to reflect on the scope for increased business involvement in public policymaking. Crucially, their deliberations mirrored (and themselves influenced) the emerging pro-growth agenda of the public sector. This was an agenda which privileged economic development concerns (and downplayed social and environmental ones) and which saw the region's future economic health as dependent on a set of interrelated, mutually reinforcing priorities: improvements to the region's road and airport (and to a lesser extent rail) infrastructure; the creation of more effective and better-resourced arrangements for attracting inward investment; the growth of flagship property development projects and events linked to urban regeneration efforts; and the enhancement of the service-based economies of the main city centres.

The establishment in 1994 of the North West Partnership (NWP), as a private–public counterpart to the local authority-dominated Association and the Business Leadership Team, reflected this emerging consensus about the ways in which economic growth could best be promoted. It drew inspiration not only from arguments about the need for a united regional voice, but from government exhortations for partnership working, and for coalitions of key public and private actors to develop strategies through which to stimulate economic performance. In retrospect, it might be viewed as an embryonic regional development agency. Although lacking any real resources and possessing no formal powers, its not unimportant symbolic significance was in signalling the view, shared by both the regionalist 'business leaders' and the local authorities, that a regionally based development agency for the North West was justified.

These developments, most of them emerging from within the region (Burch and Holliday, 1993; Deas and Ward, 2000), were to be reinforced by a series of regionally based institutional innovations authored by central government from the mid-1990s onwards. The first came in 1994 with the advent of Government Offices of the Regions (GORs), launched in recognition of the need for a clear and coordinated central government presence in the regions, again partly because of arguments about the need to present a coherent voice to Brussels (Mawson and Spencer, 1997). These new bodies also set out to challenge the persistent problem of departmentalism; the GORs, in this sense, were a newer version of the 'Little Whitehalls' to which Senior (1965: 82) refers in relation to economic development planning. However, the establishment of GORs was complicated in the North West context by the existence, uniquely amongst English conurbations, of a separate government office for Merseyside – a legacy of the decision by central government, in the wake of the urban riots in Liverpool in 1981, to award the area special status for regeneration purposes. The parallel existence of GORs for Merseyside and the remainder of the North West continued until 1998, their eventual amalgamation reflecting both pressures from within the region for a single identifiable voice, and also a desire from central government for symmetry across the English regions.

Region-building efforts continued to gather momentum under the Blair governments from 1997, reflecting growing (if not entirely enthusiastic) support for regionalism amongst some in the Labour Party over the previous decade. The Regional Development Agencies Act of 1998 enabled the creation of the NWDA in 1999. Located in Warrington – the region's would-be compromise capital – it became the best-funded of the RDAs, with a budget in 2003/04 of £334 million (in per capita terms approximately £49, second only to the North East). The NWDA was to be a business-led body, with eight of its 15 board members from a private sector background. A central part of its remit concerns the development of regional strategy. Its two regional economic strategies (NWDA, 2000, 2003a), as with those in other regions, were criticised by many as bland and lacking in detail (Bridges *et al.*, 2001; Nathan *et al.*, 1999). Their focus was on developing propulsive growth sectors (for example, biotechnology); promoting major inward investment on 'strategic sites'; assisting SMEs through managed workspace provision, business incubation and loan and grant funding; supporting flagship property projects and physical development; overseeing skills development, and 'learning' and 'knowledge' related to higher education research and technology transfer; and supporting and coordinating sub-regional regeneration partnerships (for example, through Urban Regeneration Companies established initially in Liverpool and Manchester, and subsequently in West Cumbria, Salford and Blackpool). This strategy was to be implemented partly by orchestrating other agencies within the region – a significant challenge in the context of an area with the policy richness and diversity of the North West. Strategy was also to be delivered by utilising funding streams inherited from other bodies (collapsed in 2002 into a more flexible 'single programme'). This was to help in delivering government commitments in relation to the Urban (DETR, 2000), Rural (DETR/MAFF, 2000) and Competitiveness (DTI, 1998) White Papers.

Alongside the NWDA, the 1998 Act also enabled the creation of the voluntary regional chamber, the NWRA. Its role is essentially to provide a forum for considering regionally significant issues; liaise with central government and the EU Commission to present the regional case; build identity within the region; and work collaboratively with other agencies. In contrast to the NWDA, its constitution allows for a membership drawn predominantly from the public sector, with 56 local authority members alongside 24 'economic and social partners' comprising representatives of the private and voluntary sectors, other public bodies and trades unions. With few powers, and a limited budget in 2003/04 of £2.3m, mainly raised through local authority subscription, the Assembly in its current form generally has limited capacity to extend beyond this relatively narrow remit of discussion, networking and lobbying. The main substantive function of its modest secretariat is to prepare Regional Planning Guidance and, from 2004, the North West's first Regional Spatial Strategy. The Regional Assemblies (Preparations) Act of 2003, which followed an earlier Regional White Paper (Cabinet Office/DTLR, 2002), provided for the possibility of a broader remit for directly elected Regional Assemblies, were there to be demonstrable demand within any given region. The budget of such an assembly in the North West would have amounted to an estimated £780 million of expenditure directly managed by Whitehall and associated government agencies, with 'influence' over an additional £1.6 billion controlled by other (quasi-)public bodies whose activities the Assembly would be expected to scrutinise (based on 2002/03 figures) (ODPM, 2004a). Additional resources could be raised in the longer term via Council Tax precepts. The North West was originally scheduled to be one of a first slew of three regions in which public support for elected assemblies would be gauged, but government – in the face of growing opposition from an influential group of the region's Labour MPs, and concerned about the ramifications of a 'no' vote for its national popularity in the lead-up to the General Election of 2005 – opted in mid-campaign to postpone indefinitely the referendum scheduled for late 2004. The likelihood of the North West gaining a directly elected assembly further diminished – and, in many eyes, ended for the foreseeable future – in November 2004, with the surprising but resounding rejection of proposals in a referendum in the North East, hitherto considered universally as the most supportive of all England's regions. For the North West, the prospect of a more expansive, formalised regional governance materialising in the immediate future seemed increasingly unlikely.

The renaissance of regional planning in the North West

Reflecting the fractured politics and city-based identities that characterise the region, work to develop land-use strategy which extends across the whole of the North West only began to take shape in the 1960s. This came some forty years after the North West saw the first efforts to construct strategic plans at the conurbation scale, with the initial meeting of the Manchester and District Joint Town Planning Advisory Committee in 1921 (Wannop, 1995). The first concerted effort to develop strategy for an area approximating to the North West region emerged in 1965 with the North West Study. Focusing on economic rather than land-use planning, this, and the later strategies of 1966 and 1968, embraced for the first time the city-regions of both Manchester and Liverpool, as well as the wider areas of Cheshire and Lancashire. From the outset, however, conflict between the constituent local authorities frustrated attempts to develop and take forward an agreed region-wide strategy for land-use planning. The difficulty, until at least the early 1960s, in brokering agreement between Manchester and Cheshire about issues like population overspill, new town designation and green belt delimitation meant that work on developing the Strategic Plan for the North West did not begin until 1971. Thereafter, progress in implementing the Strategic Plan, finalised in 1974, was thwarted by the combination of Lancashire's disquiet about the plan's apparent preoccupation with the Mersey Belt linking the two main cities, and by emphasis within the two conurbations on creating their own strategic plans in the wake of the creation, also in 1974, of the metropolitan county councils for Merseyside and Greater Manchester (Wannop, 1995).

As elsewhere in England, interest in regional planning in the North West, whether in relation to land use, economic development or, to a lesser extent, transport, faded dramatically after a short-lived upsurge in popularity in the 1960s (Wannop and Cherry, 1994), reaching its nadir in the 1980s (Baker *et al.*, 1999;

Breheny, 1991; Roberts and Lloyd, 1999, 2000). The implacable opposition of central government at the time to regional planning, combined with the predominant focus of land use and economic development policymaking on inner urban areas, meant that the great bulk of the 1974 Strategic Plan remained unimplemented, and many of its provisions obsolescent. The first signs of tentative revival came between 1987 and 1989 as the then regional office of the Department of the Environment (DoE), in consultation with four of the North West counties, many of its districts and other public agencies, oversaw the preparation of modest statements (as opposed to a plan or strategy) of planning guidance (Wannop, 1995). The focus was predominantly on the two metropolitan counties, abolished in 1986, rather than on the North West region. The metropolitan counties had statutory structure plans which had been in place since 1980 (Merseyside) and 1986 (Greater Manchester), and the regional DoE, responding to concerns within each conurbation, was concerned that there should continue to be some form of up-to-date strategic guidance, however modest in its scope, for the unitary development plans which were to emerge in the period until the late 1990s. The resultant statements, published for Merseyside in 1988 and for Greater Manchester a year later, were widely castigated as insipid, reflecting both the limited extent of central government ambition in relation to plan-making, and the need to develop an approach that was sufficiently vapid and uncontroversial to avoid rekindling the tensions which continued to underlie intra-regional relations (Williams and Baker, 2006). Yet while the focus of this renewed interest in strategic planning was at the scale of the conurbation rather than the wider North West region, the commitment amongst a range of actors to develop some form of supra-district guidance was significant in signalling the initial stirrings in a broader reawakening of interest in strategic planning.

Alongside the more general revival of interest in regionalism from the late 1980s and early 1990s, other issues, replacing or extending those on which earlier planning efforts had focused, helped to stimulate the gradual resurgence of regional land-use planning. One centred on the emerging debate about how best to accommodate the dramatic growth forecast for household numbers. Although a much more pressing issue in the context of London and the South East than in the North West (Robson *et al.*, 1996), this was an issue which preoccupied the planning profession from around the mid-1990s and which prompted most to conclude that, if inter-district wrangling was to be avoided and NIMBYism resisted, housing land allocations could only be decided at a regional level. Similarly, the allocation of industrial land was also a pointed issue in the North West, and one that again warranted a regional approach in order to reconcile longstanding inter-district arguments, especially about the location of major sites for inward investment and the apparent bias – in the eyes of the two main cities and their immediate neighbours – towards greenfield sites at the expense of brownfield ones (Wilks-Heeg *et al.*, 1999).

Other issues to the fore in the late 1980s and early 1990s reinforced this intra-regional desire for a North West-based approach. Transport infrastructure – and the need to lobby more effectively for improvements to it – was also an issue that was seen as meriting a regional approach (RTPI North West Branch, 1990). This applied particularly to concerns about the consequences for economic performance posed by what was seen as the inappropriately restricted capacity of the main north–south (the M6 to Birmingham) and east–west (the M62 to Leeds) motorway axes. It also applied to the woefully deficient intercity rail network, and to the upgrade of the West Coast Mainline linking the region with London, which government, in the early 1990s, was resisting. And there was also vigorous debate about policy for airports, centring on the rival claims of Manchester and Liverpool about where best to accommodate planned growth in passenger numbers, resulting, as Burch and Holliday (1993: 39) note, in delay to the publication of the regional economic strategy of 1993.

All of these factors contributed to a powerful case that new efforts ought to be expended to update regional planning policies that had last been agreed nearly twenty years previously. This went a step beyond DoE concern, expressed through its efforts to develop statements of guidance for the former metropolitan counties, to ensure a minimum degree of harmony across sub-regional structure and development plans. It signified a desire to build a more proactive, pro-growth strategy addressed not just to statutory land-use plans, but to much broader economic development concerns. Again, much of the argument was premised on the view that the creation of regional structures would help to enhance the capacity for effective lobbying of central government for increased resources (as, indeed, had been part of the rationale for the Strategic Plan for the North West, nearly twenty years previously (Wannop, 1995: 142)).

The first significant attempt to produce a region-wide strategic land-use planning document came with the publication, in 1993, of the North West Regional Association's *Greener Growth* strategy (North West Regional Association, 1993b). This was a cautious, unassuming document, which barely reflected the more ambitious, North West-focused agenda that was beginning to emerge. Its muted tone betrayed a deliberate attempt on the part of the Association to sacrifice specificity and ambition in favour of a pedestrian content that would avoid unduly antagonising competing interests within the region. Much of the content, in common with attempts to develop regional guidance elsewhere in the early 1990s (Baker, 1998), simply collated a series of lowest-common-denominator principles from the region's constituent development plans, and restated the fundaments of national planning policy. But there were also some newer concerns, reflecting the very different political and economic circumstances two decades on and responding, in particular, to the emerging pro-growth agenda of the private sector, as expressed through the Business Leadership Team, and shared by some within the region's local authorities.

These concerns within the North West were reinforced by pressures from central government. Part of this reiterated the still growing concern with the fraught issue of housing land allocations in the South East and in other areas of high demand. But part also involved a pragmatic attempt to coordinate the plan-led system supported by government in the early 1990s by formalising and institutionalising the system of regional planning that had emerged across much of England. The rationale within government was that quasi-statutory guidance would help ensure a minimum degree of coordination across development plans, without upsetting the plan-led system. Amongst some within the regions, however, the view was often that guidance should take on a more ambitious remit and begin genuinely to impact upon the content of development plans.

This tension – between government's minimalist vision of regional guidance as a modest, technocratic attempt to provide basic coordination across development plans, and the more ambitious remit favoured by some within the regions –

was evident as the North West's first formal Regional Planning Guidance document was published in 1996 (GO-NW, 1996). Its tone was overwhelmingly guarded, consciously evading some of the region's most sensitive issues. In contrast even to the tentativeness of the earlier *Greener Growth*, it asserted that green belt boundaries need not be considered in the immediate future. Citing the lack of robust data available at the time, it deferred firm decision-making about housing land allocations, arriving at the politically less contentious conclusion that the status quo could remain in the meantime. Likewise, a decision about the location of major sites for inward investment was, in effect, postponed, pending the commission of a further study.

There was recognition within the region that the initial RPG, partly by design but also as a result of central government strictures, was insufficiently aspirational and lacking in detail. This criticism grew in force as elite actors within the region began to lobby for a greater degree of 'vision', and (from 1997 in particular) as central government began actively to encourage the development of more ambitious, expansive and comprehensive RPG. Parallel encouragement from government for enhanced stakeholder involvement in the RPG process was expected to help further in reorienting RPG in this way (Baker *et al.*, 2003).

These pressures were evident in the subsequent iterations of the North West's RPG, published initially in 2003 (ODPM, 2003) and then, as a draft partial review in 2004, focusing largely on updating environmental provisions (NWRA, 2004). The inception of the North West Regional Assembly in 1999 immediately gave regional planning greater standing than before. Equally, the involvement of a variety of stakeholders – culminating in a formal examination in public in 2001 – meant that the resultant RPG benefited from an increased degree of specificity. The adopted 2003 RPG document not only detailed a fuller range of themes, but also revolved around a more sophisticated geography of categorised sub-regions. Reflecting a commitment to greater cross-sectoral breadth and an increased sensitivity to intra-regional geographies, it set its proposals within an overall 'spatial development framework' which emphasised the need to stimulate growth in the cores of most of the main town and city centres within a 'North West Metropolitan Area'. The clear definition of areas on which regeneration efforts should focus, and the concomitant identification of those towns (such as Warrington) in which development should be constrained, represented a departure from the geographical non-specificity of previous RPG documents.

The increased detail in the 2003 RPG was also evident in the prominence of a range of issues – not least housing land supply and strategic employment sites – about which earlier regional planning documents had remained coy. It addressed longstanding conflicts over housing land release by siding firmly with 'metropolitan concentration through urban regeneration': 'securing an urban renaissance' by focusing planned housing provision within the main urban areas of Manchester, Salford and Liverpool and encouraging densification and the recycling of brownfield land, in line with national policy in the wake of the Rogers Report (Urban Task Force, 1999) and the 2000 Urban White Paper (DETR, 2000). But it also deferred to market considerations and continued intra-regional politicking by identifying variable targets for the re-use of land and buildings in 'sustainable locations', ranging from 90 per cent of all development in Liverpool and Manchester, to 50 and 55 per cent in Cumbria and Cheshire respectively' (ODPM, 2003: 59).

Regional planning policy and the changing geography of growth in the North West

The succession of regional strategic land-use plans, from the 1974 Strategic Plan to the 2003 RPG, share many common themes. But these often conceal some subtle but instructive changes in emphasis. One relates to the lasting focus on the Mersey Belt between Liverpool and Manchester. The 1974 plan, reflecting the longstanding planning desire for containment, but also triggered by more recent anxiety about the diminishing socio-economic prospects of the area, suggested that new development be concentrated along this east–west axis. As Rodgers (1980: 302) notes, this was viewed by some in much more ambitious, national terms, as creating the potential for 'an urban unit large enough almost to rival London ... [a] northern equipoise to the overweening growth of the metropolis that, alone, can restore some balance of urban and economic power in Britain'. This is a theme that has grown in importance, forming a significant part of the 2003 RPG, which contained copious references to 'polycentricity' and the ostensible scope for Manchester and Liverpool to work collaboratively with each other, and with the other major provincial cities in the north of England. Similar arguments were beginning by 2004 to inform the preparation of the first Regional Spatial Strategy (RSS), which all regions are required to produce as part of a radical overhaul of the planning system instituted via the Planning and Compulsory Purchase Act 2004.

Beyond land-use planning, other regionally based strategy has also emphasised the supposedly under-exploited potential of a Liverpool–Manchester, Mersey Belt development corridor. But in contrast to the thrust of the 1974 plan, the rationale for these more recent efforts has revolved around international territorial competition, and the widely held view that the region's competitiveness could best be enhanced by encouraging the two cities to pool their efforts and avoid, where possible, competing with each other. This philosophy, which unsurprisingly has come in the main from the region rather than the cities, was evident in the Liverpool–Manchester Vision, developed for the NWDA, with the public support of the leaderships of the two city councils (Regeneris *et al.*, 2001). It was also evident in the later Mersey Belt study, commissioned by the NWDA to help lobby for a more imaginative sub-regional geography for the 2003 RPG (DTZ Pieda Consulting, 2002). But whereas the 1974 plan cast the strategy for the Mersey Belt unambiguously in terms of the need to offset population and employment loss from the two principal cities (Wannop, 1995: 144) and contain growth within the main existing area of urbanisation in the region, the intercity growth corridor envisaged by the later strategies was more multifarious, drawing inspiration from emerging neo-liberal orthodoxies about local economic development. The Mersey Belt Study, for instance, tried to develop ideas originally suggested in the 1999 RES, arguing that the corridor ought to be bifurcated, comprising a northern 'regeneration' zone along the M62 motorway (the 'metropolitan axis') and, more significantly, a 'growth' corridor along the M56 motorway through Cheshire, linking Chester in the west and Macclesfield in the east (the 'southern crescent').

Neither of these NWDA-commissioned studies proved by themselves especially significant in their impact. Despite its high profile, the 'vision', lacking wholehearted support from the two cities (and from Manchester in particular), never gained any real momentum and, within barely three years of its publication,

in effect was shelved by the NWDA (Deas, 2005). The Mersey Belt Study's proposed 'southern crescent' was also abandoned (though the thinking that informed
it remained prominent within the NWDA) after the 2003 RPG opted to formalise
a more conservative and less obviously growth-oriented geography. However, some
of the ideas which underlay the vision and the Mersey Belt Study helped to inform
a subsequent proposal from government, produced in partnership with the RDAs
and the Core Cities group, to develop a more extensive polycentric corridor, the
Northern Way, stretching from the west to the east coast of England, with its main
trans-Pennine branch running along the line of the M62 motorway axis (ODPM,
2004b). Again, the mechanisms through which this latest corridor was to be created were vague. Viewed in the context of its 14 million population, the £100
million 'growth fund' allocated in 2004 to help initial developmental efforts could
be viewed as derisory when measured against the strategy's '£29 billion challenge'
of narrowing the interregional disparity in output (Northern Way Steering Group,
2004). Nevertheless, the Northern Way did signify the continuing interest in
developing territorially based interventions which sat incongruously with, and
sometimes cross-cut, regional boundaries (echoing the geography of planning
efforts elsewhere (While *et al.*, 2004)). It also re-emphasised the related issue of
how best to construct regional strategy in a context within which the two main
cities increasingly viewed their relationship in terms of wider narratives about
inter-urban competition. And it provided further evidence of the subordination of
intra-regional (but not interregional) redistribution to growth concerns. These are
themes that, by 2004, were already informing the arrangements for RSS. Within
the North West, the deadline for preparation of the first draft RSS was set for
2005, specifically with the aim of harmonising plan-making timetables across the
three regions covered by the Northern Way. The shift in the focus of regional
strategy towards a more pronounced growth agenda, and its changing geography,
continued apace.

Regional planning and territorial conflict in the North West

Alongside strategic land-use planning at the regional level, the period since the
early 1990s has also seen the emergence of a succession of economic development
institutions and initiatives based around territories delimited in different ways. The
result is a variegated geography of policies and institutions, of which strategic land-
use planning comprises only one part. The remainder of this chapter charts the
complex territorial politics that have emerged in the North West in and around
spatial planning, looking first at the interactions between institutions at the *regional*
scale, and then at *intra-regional* relationships.

Regional institutional interactions and the tensions underlying the 'North West project'

The parallel growth of strategic planning and other policy with a focus on the
North West from the late-1980s and early-1990s onwards has meant some significant upheavals in the relations between regionally based institutions. As we
have seen, the thrust of spatial planning provisions has changed to a modest
degree, moving away from conservative and cautious land-use planning strategies
underlain by a commitment to both economic growth and promoting a degree of

territorial equity; and towards bolder strategy, extending beyond land-use concerns, viewing the region and its cities in an international context. Reflecting this shift and the related attempt to engineer inter-agency consensus about regional priorities, strategic land-use planning provisions in the North West accord more than before with the preoccupations of the NWDA's Regional Economic Strategy (RES). However, there remains an important tension between the two. The commitment in the 2003 RPG to 'greater economic competitiveness and growth' can be readily reconciled with the broader thrust of economic development policy in the North West and beyond; but its allied commitment to 'social progress' signifies a residual commitment to a more broadly based approach. Likewise, RPG in general retains much more of an alertness to environmental sustainability than (arguably) is the case for the RES (Counsell and Haughton, 2002). The focus of RPG's thematic priorities on 'prudent management of the region's environmental ... assets' and 'environmental improvement' contrasts with the much more limited priority to environmental protection and enhancement typically afforded by RDAs (Gibbs, 2000). The suggestion is of an outlook within the NWRA that is rather more expansive than the NWDA's narrower focus on securing economic growth.

These tensions continue a more general and longstanding difficulty in connecting economic development and land-use planning concerns at the regional level (see Baker *et al.*, 1999; Counsell and Haughton, 2003; see also Counsell and Haughton in this volume; Roberts and Lloyd, 1999, 2000). But this type of conflict has been particularly marked in the North West over many years. The North West Study of 1965 was prepared, at the behest of central government, as a regional economic planning document, and the subsequent strategies of 1966 and 1968 were produced under the auspices of the Regional Economic Planning Council, rather than the statutory planning authorities. And while there was a more concerted attempt to link narrowly defined land-use planning with other elements of spatial strategy such as transport and economic development in the 1974 Strategic Plan for the North West (Wannop, 1995), such efforts have tended to be the exception, as the contemporary experience with RPG and the RES in the North West in part testifies.

This dissonance in policy emphasis between different regional strategies is revealing. It might be interpreted, perhaps somewhat crudely, as symptomatic of a more general underlying tension between the conservative outlook of the NWRA and the more straightforwardly neo-liberal worldview that characterises the NWDA. This in turn relates to their contrasting composition and approach. On one hand, the NWDA is governed by a board of directors nominally drawn from the private sector, certainly upholds a business-oriented pro-growth agenda, and is always keen to present its modus operandi as a buccaneering one of entrepreneurial governmentality which contrasts with the passé bureaucratic inefficiency said to typify the public sector. The NWRA, on the other hand, is dominated by Labour local authority councillors, bases its decision-making on the deliberations of an elaborate network of committees, and operates to an administrative model in line with that of conventional (new) public management.

These contrasts are manifested in a wider relationship that has sometimes been one of mutual mistrust and an, often, uneasy coexistence. Part of this can be explained by inevitable inter-agency rivalries, particularly in a context in which both are trying to implant themselves as established features of the institutional landscape. But part relates to more deeply rooted ideological differences – between the narrow(er) economic development concerns espoused in the Regional Economic Strategy, and the broader, more cautious agenda articulated through RPG. This was certainly evident in the lead-up to approval of the 2003 RPG, as the NWDA, with more circumspect support from GO-NW, lobbied vigorously at the examination in public for a much stronger pro-growth emphasis, and a con-comitantly reduced focus on environmental sustainability (Counsell and Haughton, 2002). Conflict around the degree of emphasis on growth was evident in particular in the protracted disagreement about major employment sites. The 2003 RPG, despite intensive lobbying by the NWDA, restricted the number of 'regional investment sites' to 11, emphasising that the identification of these and any future areas should complement sustainability goals, contribute to 'urban renaissance' in the main conurbations and use recycled land where possible (ODPM, 2003: 46–7). By contrast, the NWDA continued, following an initial decision in 2001 in the lead-up to the RPG examination in public, to promote an additional 14 'strategic regional sites', basing their selection on the rather nar-rower objective of 'boost[ing] business growth opportunities', with apparently little concern for the distributional and environmental issues that were informing the NWRA standpoint (NWDA, 2003a: 8). This was linked in particular to concern that only one of the 11 RPG strategic sites was located in the 'southern crescent', limiting the scope for the development of 'knowledge-based' industries (2003a: 11). A subsequent policy statement on the Mersey Belt reiterated the NWDA's desire for more numerous strategic employment sites located in greenfield sites, committing the Agency to a further site in Macclesfield, again within the area for-merly covered by the now-shelved 'southern crescent' (NWDA, 2003b).

It is clear, then, that despite the considerable organic support that has emerged for some form of strengthened regional governance, the North West's regionalists do not constitute a homogenous grouping. This is a conclusion that is reinforced by the schisms within the private sector which, alongside inter-institutional fric-tions of the sort outlined, have also affected the shape of the unfolding regional institutional superstructure. In general, representatives of business in the North West have been supportive of the NWDA but lukewarm (at best) about the NWRA. Echoing its national policy, the regional CBI opposes a directly elected assembly, but supports the NWDA. Like the Federation of Small Business and Institute of Directors, the regional CBI is not one of the formal group of 'economic and social partners' which makes up the Assembly. The North West Chambers of Commerce in 2004 provided four Assembly members, but had previously threat-ened withdrawal because of its concerns about an NWRA perspective perceived as insufficiently attuned to business interests (Roberts, 2002). By contrast, the North West Business Leadership Team, also providing four Assembly members, has since its inception as a regional cheerleader in the 1980s been markedly more consistent in its support for the Assembly. Just as there is tension between regionally based institutions, there is far from being any unanimity of view amongst business repre-sentative groups at the regional scale.

There are, then, some clear areas of friction amongst regionally based bodies, and this applies both to the North West's principal regional agencies and to the var-ious private sector umbrella organisations. This might be interpreted as embodying more deep-rooted ideological tensions. Yet while the NWDA, in relative terms, may have devoted limited priority to tackling social exclusion and contributing to sustainability, its caricature as growth-fixated development 'enabler' ignores some signs that it has become progressively less preoccupied with a narrow, circumscribed view of economic competitiveness. Indeed, there is evidence that government, through its Public Service Agreement targets, is keen to encourage a broadening of the approach adopted across all the RDAs. For the North West, while economic targets continue to be central to the remit, they are defined in relative terms, implying at least some commitment to reducing interregional disparity. Thus, for instance, GVA per capita is to increase from 87 per cent of the national figure in 2003, to 89 per cent by 2006. At a rhetorical level, alongside the predictable clichés about building a 'world class' ultra-competitive region, the recognition of wider issues is also evidenced in the RES for 2003 by a narrative which talks of 'linking opportunity and need' and focusing upon 'equality and diversity' (NWDA, 2003c).

There is also the possibility that relations between the two principal regionally based bodies would change significantly in the unlikely event that a directly elected Assembly were ever to materialise. Proposals in 2004 in the Regional Assemblies Bill suggested that the Assembly role would move beyond the current informal scrutiny, inheriting the direct control (including key decisions about budget setting and board appointments) previously exercised by the Department of Trade and Industry (ODPM, 2004c). Reporting to what would almost certainly be a Labour-controlled regionally based body, rather than to Whitehall, could effect a significant shift in the agenda of agencies like the NWDA and greatly reduce, if not eliminate, the inter-institutional frictions that have been evident to date. But with the prospect of a directly elected assembly disappearing rapidly by the end of 2004, the potential for inter-institutional conflict at the regional scale remains a potent one.

Intra-regional tension, identity-building and the limitations to North West regionalism

The continued existence of conflicting agendas amongst regional institutions provides further evidence that the reshaping of the state's territorial form is one characterised by conflict and uncertainty. In the context of the North West, this has been compounded by protracted intra-regional discord. Some of this has related to a complex politics of territorial identity that has long generated academic interest. Fawcett (1919: 125), writing about a (greater) Lancashire area that roughly approximated to the contemporary North West, saw great potential for a degree of devolution on the back of popular regional identity: 'local patriotism is already strong; and it has developed a public opinion which makes the province quite as ripe for self-government as Scotland or Wales'. More recently, however, it has been clear that the faltering development of regional governance in the North West is in part attributable to a lack of any clear or cohesive sense of regional identity. Despite hopes to the contrary (see, for example, Burch and Holliday, 1993), institution-building in the period since the early 1990s appears not to have fostered any clearly discernible North West identity, in part explaining the government's decision in 2004 to delay (or, more cynically, abandon) plans to hold a referendum for an elected Regional Assembly.

This continuing absence of a clear North West identity has been in spite of efforts directly to foster the development of regional consciousness. A Constitutional Convention, drawing on the exemplar of Scotland in the 1990s and the subsequent experience of the North East of England, was established, under the chairmanship of the Bishop of Liverpool, to build a cross-party, cross-sector consensual identity amongst key opinion-formers (Deas and Giordano, 2003). Operating in a style that deliberately contrasts with the supposed staidness of the Constitutional Convention, but with broadly similar aims, another attempt to build regional identity has come through The Necessary Group. With the Mancunian music magnate Tony Wilson as its public face, this is an apparently genuine attempt to develop regional consciousness, largely through a series of high-profile, frequently tongue-in-cheek stunts, the most notable of which include a campaign to adopt a subverted flag of St George as the North West's emblem, and a 'declaration of devolution' pinned on strategically positioned lamp posts across London (The Necessary Group, 2004).

While it is easy to dismiss these efforts as rhetorical bluster of interest to few outside a narrow coterie of the most vehement pro- and anti-regionalists, the very fact that consciousness-raising is deemed necessary is significant in that it confirms the potency of the disparate, competing intra-regional territorial loyalties present in the North West, and the consequent scale of the challenge that confronts region-building. Reflecting this, efforts to develop regional governance have often generated little enthusiasm, and some have met with active resistance stemming from a variety of alternative sub-regional territorial configurations (Deas and Ward, 2000, 2002). The geographical focus of strategic land-use planning over time is illustrative of these kinds of territorial rivalry, principally relating to institutional identities, but underscored in some instances by corresponding sub-regional popular loyalties. As Williams and Baker (2006) note, the focus of strategic planning efforts has tended historically not to be on the North West as a whole, but on its conurbations – initially through voluntaristic efforts dating back to the 1920s, then via structure plans for the metropolitan councils adopted in the 1980s, and finally through strategic guidance documents published at the end of that decade. This has been echoed in more recent times, even in planning documents with a North West frame of reference. RPG, for example, acknowledges the continuing presence – or ascendancy – of policy elites based around more disaggregated city-based territories by giving explicit recognition to named sub-regions, by building some structures around them, and by directing some policies towards them.

The acknowledgement of sub-regions in these ways can be viewed simply as a pragmatic recognition of the existence of a more meaningful politico-institutional geography, which is widely accepted both institutionally and popularly and which undermines (sometimes deliberately) policy efforts focused on the North West region. Equally, it is also possible to view this as illustrative of the functional realities of a space economy that relates both to long-established territories already recognised as formal political units, and economic spaces of more recent origin which lack the same kind of associated political or institutional infrastructure. Conflict with regional institutions has been evident in relation to both these types of area. A significant portion of the opposition to region-building has stemmed from the region's two cities, and from Manchester in particular (Deas and Ward, 2002; Deas, 2005). In this instance, regional–local struggle has revolved around a series of issues, including the distribution of strategic sites for economic development, or the effectiveness of external promotional and image-building efforts (Hebbert and Deas, 2000).

From the late 1990s, these tensions have intensified as a result of a combination of factors. One has been the increased feelings of vulnerability at the city level with the growth of regional institutions, linked to central government embarking on a programme of regionally focused constitutional change and economic development activity. While such fears have lessened in the wake of the North East referendum of 2004, a second and more enduring concern has centred on continuing frustration about jurisdictional under-bounding of city government, accentuated by the consensus amongst policy actors that this has been injurious to their ability to 'compete' with cities elsewhere. There has also, thirdly, been growing concern about continuing (or widening) territorial socio-economic disparity, linked to perceptions about the reduced commitment to redistributive 'spatial Keynesian' regional policy, and to the related intensification of London-centric decision-making in line with hardening central government views about the need to promote the capital and accommodate growth therein in the interest of 'UK plc'. All of these concerns informed the Core Cities group, established to lobby central government on behalf of the major provincial English cities. Central to its agenda was a desire to persuade central government to amend decision-making with a view to enhancing the competitiveness of the cities, drawing on the supposed prototype of provincial cities elsewhere in Europe (see Parkinson *et al.*, 2004). All of this was premised on a clear view about territorial governance, albeit expressed often less than explicitly, that economic development efforts are pursued more effectively on a spatially and institutionally more flexible basis, at the scale of the city or city-region rather than the 'artificial' region. Although the presence of the RDAs within the Core Cities group meant that it could not be interpreted simplistically as anti-regionalist, its inception did denote a concern, on the part of both the RDAs and the cities, that regionalism ought principally to be about central government resourcing of provincial economic development efforts, rather than about the more wide-ranging, Celtic-style devolution, linked to enhanced democratic accountability and popular regional identity, which characterised the more expansive conception of regionalism common amongst many of the most prominent adherents of elected Regional Assemblies.

In addition to disquiet emanating from the cities, the development of the regionalism has also been constrained by the limited levels of support from the counties. Although comfortable with the semi-formal regionalism of the mid-1990s, the counties' atavistic response to proposals for an elected regional assembly (and with it the likely abolition of the upper tier of local government) was unsurprisingly unsympathetic. This was reflected in Lancashire's decision in 2003 to withdraw funding from the NWRA, and reports that Cheshire was considering following suit (Tomaney *et al.*, 2003).

Alongside opposition based on the view that cities or city-regions represent a more meaningful economic territory around which to create new structures of governance, the 'regional project' has also faced more recent, and less intense, challenge from a second type of area. As While *et al.* (2004) note in relation to Cambridge, the emergence of new economic spaces – expressed as 'growth corridors', neo-Marshallian nodes or 'clusters' and so on – has generated the rise of new regulatory structures which reflect both the territorial reach of these areas and the growth-oriented agendas of their constituent business elites. Their emergence has been something that some regionally based bodies, notably the

NWDA, have been keen actively to cultivate. But this raises the possibility, if the experience of other areas in the UK is to be applied to the North West, that further regional–local struggles could emerge in relation to these areas, particularly if agencies with a regional remit are required to pursue a less rigidly pro-growth agenda that is more in tune with the collective interest of the region, but jars against the specific economic development needs and wants of these new economic spaces. This is a conflict that has already been evident, in a limited way, in the North West in relation to the NWDA's proposal for the 'southern crescent' growth corridor, which generated opposition from many of the local authorities whose territories it traversed, as well as most of the Greater Manchester and Merseyside authorities (Counsell and Haughton, 2002). Following examination in public, it was eventually omitted from the adopted 2003 RPG, on the grounds that development in the area covered by the 'southern crescent' ought to be restrained and the pressures re-channelled towards economically less buoyant parts of the region (and the core urban 'North West Metropolitan Area' in particular).

This complex set of interrelationships in and around regionally-based planning (and public policy more generally) and an array of sub-regional territories is something which future RSS is already proposing to address. In light of the replacement of structure and local plans with local development frameworks, part of the role of RSS will be to manage planning issues on a sub-regional basis. This could involve the development of strategy for conurbations straddling several local authority districts, or in relation to emerging functional areas that cross-cut existing boundaries (such as North Wales/Chester, south Cheshire and the Potteries, or corridors along major roads such as the A69 through Cumbria, or the M62 and M6 motorways bisecting the whole region). It would, of course, be possible to interpret this as part of an attempt to promulgate a narrowly growth-oriented, 'picking winners' agenda. However, the recognition of sub-regions also involves other types of 'lagging' area, such as coastal areas, which implies a continuing commitment to address territorial inequality (even if this is conceived in the straightforwardly neo-liberal terms of boosting growth and encouraging structural adjustments in line with assumptions about the main ingredients for place competitiveness). The key point, though, is that the continued emergence of these myriad sub-regional spaces, including both those which are widely recognised as well as those more nascent, means that the future territorial structure of the state is one of uncertain shape, its precise delineation subject, inter alia, to a variety of struggles around the nature and form of economic development, land use and other strategy.

Conclusion: the contested territorial configuration of the state and the lessons from North West England

It is clear that over several decades, the process of regional institutional development has been a halting one, inhibited by numerous tensions around the North West's geography and the substantive emphasis of regionally based policy. These tensions relate to two particular factors.

The first centres on the way in which regional actors have responded to national policy changes. Reflecting a reoriented national framework for regionalism, conflict within the North West has been evident in a shift from a loose consensus built upon what was essentially a circumscribed regionalism, and towards contestation emerging in response to the bolder and more formal regional governance propounded by central government from 1997 (and which may have ended with the North East referendum in 2004). The consensus that took root in the first half of the 1990s revolved around a modest regionalism that was based on a broad but shallow form of flexible, semi-formal cooperation, with a shared desire to procure more resources as the key objective which bound together a range of interests. This was an unthreatening regionalism, licensed by the state. It was a regionalism to which business (at least through its representative organisations) could be a willing signatory. And it was a regionalism to which the overwhelming majority of local state actors could subscribe: the counties occupied a more prominent position in the regional coalition than had been the case in previous region-building exercises; and the cities (especially Manchester) could pursue a business-first growth agenda largely untrammelled by regional interference, allowing them to liaise with the region on a selective, opportunistic basis.

At that point, it is possible to argue that neighbouring regions such as Yorkshire and Humberside looked enviously to the North West as an example of a region that had overcome a long history of fragmentation and was beginning to work cohesively (While, 2001). A decade on and such a view looks less prescient. The experience of the North West suggests that the increasing formalisation of regional structures has meant that part of the compromise brokered around the more limited regionalism of the 1990s has begun to unravel. In particular, proposals for an elected regional assembly have been the catalyst for the re-emergence of some marked interregional tensions. The period following the formalisation in 2003 of abortive plans for a referendum to canvass support for an elected assembly has seen the mobilisation of actors relaxed about the restrained and restricted regionalism of the 1990s, but threatened by the more formal and better-resourced structures proposed. Fractures have opened, as we have seen, amongst business representative groupings, some of them at ease with the pro-growth, business-oriented perspective of the NWDA, but fearful of the challenge to this outlook that an elected assembly, with formal scrutiny and budgetary powers, might well exercise. Alongside this, perhaps the most potent challenge has come from within the Labour Party, with a powerful and wide-ranging coalition of Members of Parliament articulating a case against elected regional government, partly on the grounds of the perceived threat to a city- and local authority-based view of sub-national governance.

Part of the case of both business and politicians has centred on concerns that the advent of an elected assembly might disrupt the delicate balance of regional and sub-regional relations. The shared emphasis of policy at the regional scale (especially that emanating from the NWDA) and the city(-region) level on promoting growth and development has helped to minimise this kind of conflict. With the possibility that regional institutions might have a wider remit that included less exclusively growth-oriented statutory planning powers, there was an obvious risk of upsetting the precarious balance between region and city-based actors.

Second, alongside (and partly reinforcing) this territorial conflict around conceptions of deep(er) and shallow(er) regionalism, there have been significant tensions in the relationship between land-use planning and other forms of economic development strategy. One of these relates to the regulatory emphasis which land-use planning strategy is obliged to adopt, in contrast to other, looser forms of regional strategy. Regional land-use planning, in this sense, has to internalise the contradictions of national policy guidance informed by different branches of the state, including the sometimes different priorities of the Department of Trade and Industry in respect of RDAs, and the Office of the Deputy Prime Minister in relation to broader issues of regionalism. The re-emergence of seemingly dormant fissures within the region is partly a reflection of external changes which have gradually shifted regional land-use planning away from the often uncontentious banalities of the initial plans of the 1990s, and towards statutory Regional Spatial Strategies which genuinely impact upon a broader range of policy areas and 'tread on the toes' of sub-regions in a meaningful way for the first time.

The complexities and contradictions of regional institution-building reflect national policy based on what is in essence a competitive regionalism. This institutionalises a marked dissonance within attempts to build regional agencies and identity. Tension between the growth goals at the heart of the RDA agenda and the regulatory ones allotted to the assemblies is one aspect of this. Conflict between, on one hand, the city-focused emphasis of the urban renaissance and neighbourhood renewal agendas and, on the other, the geographically less constricted focus of efforts to develop clusters and new industrial spaces is another reflection of contradictory national policy. And interregional competition, as RDAs each seek to boost growth, is another aspect of this multi-stranded competitive regionalism. For actors within the North West, this means a whole range of 'scalar' institutional manoeuvres, resulting in an intricate, conflict-ridden territorial politics that is likely to endure.

References

Amin, A., Massey, D. and Thrift, N. (2003) *Decentering the Nation: A Radical Approach to Regional Inequality*, London: Catalyst.

Baker, M. (1998) 'Planning for the English regions: a review of the Secretary of State's Regional Planning Guidance', *Planning Practice and Research* 13(2): 153–69.

Baker, M., Deas, I. and Wong, C. (1999) 'Obscure ritual or administrative luxury? Integrating regional strategic planning and economic development', *Environment and Planning B* 26(5): 763–82.

Baker, M., Roberts, P. and Shaw, S. (2003) *Stakeholder Involvement in Regional Planning*, London: Town and Country Planning Association, 17 Carlton House Terrace, SW1Y 5AS.

Barlow, M. (1995) 'Greater Manchester: conurbation complexity and local government structure', *Political Geography* 14(4): 379–400.

Breheny, M. (1991) 'The renaissance of strategic planning?', *Environment and Planning B* 18(2): 233–49.

Brenner, N. (2004) *New State Spaces: Urban Governance and the Rescaling of Statehood*, Oxford: Oxford University Press.

Bridges, T., Edwards, D., Lloyd, G., Mawson, J., Roberts, P. and Tunnell, C. (2001) *Evaluation of RDA Strategies and Action Plans*, London: Department of Transport, Local Government and the Regions. Available: http://www.dti.gov.uk/rda/reports/dundee uni.pdf (accessed 28 September 2003).

Bristow, R. (1987) 'The North West', in P. Damesick and P. Woods (eds) *Regional Problems, Problem Regions and Public Policy in the UK*, Oxford: Clarendon Press.

Burch, M. and Holliday, I. (1993) 'Institutional emergence: the case of North West England', *Regional Policy and Politics* 3(2): 29–50.

Cabinet Office / DTLR [Department for Transport, Local Government and the Regions] (2002) *Your Region, Your Choice: Revitalising the English Regions*, Cm 5511, London: The Stationery Office.

Counsell, D. and Haughton, G. (2002) *Sustainable development in Regional Planning Guidance, regional report 6: the North West*, Hull: Department of Geography, University of Hull, Cottingham Road, HU6 7RX. Available: http://www.hull.ac.uk/geog/PDF/nwest.pdf (accessed 8 October 2004).

Counsell, D. and Haughton, G. (2003) 'Regional planning tensions: planning for economic growth and sustainable development in two contrasting English regions', *Environment and Planning C* 21(2), 225–39.

Deas, I. (2005) 'Reinventing the metropolitan region: experiences of scalar conflict in Manchester', in A. Harding (ed.) *Rescaling the State in Europe and North America*, Oxford: Blackwell.

Deas, I. and Giordano, B. (2003) 'Regions, city-regions, identity and institution-building: contemporary experiences of the 'scalar turn' in Italy and England', *Journal of Urban Affairs* 25(2): 225–46.

Deas, I. and Ward, K. (2000) 'From the new localism to the new regionalism? The implications of RDAs for city-regional relations', *Political Geography* 19(3): 273–92.

Deas, I. and Ward, K. (2002) 'Metropolitan manoeuvres: making Greater Manchester', in J. Peck and K. Ward (eds.) *City of Revolution: Restructuring Manchester*, Manchester: Manchester University Press.

DETR (Department of the Environment, Transport and Regions) (2000) *Our Towns and Cities: The Future – Delivering an Urban Renaissance*, Cm 4911, London: The Stationery Office.

DETR/MAFF (Department of the Environment, Transport and Regions; Ministry of Agriculture, Fisheries and Food) (2000) *Our Countryside: The Future – A Fair Deal for Rural England*, Cm 4909, London: The Stationery Office.

DTI (Department of Trade and Industry) (1998) *Our Competitive Future: Building the Knowledge Driven Economy*, Cm 4176, London: The Stationery Office.

DTZ Pieda Consulting (2002) *Mersey Belt Study*, Warrington: NWDA, Renaissance House, Centre Park, WA1 1XB. Available: http://www.nwda-cms.net/Document Uploads/MERSEYBELTSTUDY.pdf (accessed 1 October 2004).

Fawcett, C. (1919) *Provinces of England: A Study of some Geographical Aspects of Devolution*, London: Williams and Norgate.

Gibbs, D. (2000) 'Ecological modernisation, regional economic development and regional development agencies', *Geoforum* 31(1): 9–19.

GO-NW (Government Office North West) (1996) *Regional Planning Guidance for North West England*, RPG 13, London: Department of the Environment.

Harding, A., Deas, I., Evans, R. and Wilks-Heeg, S. (2004) 'Reinventing cities in a restructuring region? The rhetoric and reality of renaissance in Liverpool and Manchester', in M. Boddy and M. Parkinson (eds) *City Matters: Competitiveness, Cohesion and Urban Governance*, Bristol: The Policy Press.

Haughton, G. and Counsell, D. (2004) 'Regions and sustainable development: regional planning matters', *The Geographical Journal* 170(2): 135–46.

Hebbert, M. and Deas, I. (2000) 'Greater Manchester – up and going?', *Policy and Politics* 28(1): 79–92.

Lefevre, C. (1998) 'Metropolitan government and governance in western countries: a critical revew', *Intenational Journal of Urban and Regional Research*, 22(1): 9–25.

Mawson, J. and Spencer, K. (1997) 'Origins and operation of the Government Offices for the English regions', in J. Bradbury and J. Mawson (eds) *British Regionalism and Devolution*, London: Jessica Kingsley.

Nathan, M., Roberts, P., Ward, M. and Garside, R. with NCVO and Southernwood, K. (1999) *Strategies for Success? A First Assessment of Regional Development Agencies' Draft Regional Economic Strategies*, Manchester: Centre for Local Economic Strategies (CLES), Express Networks, 1 George Leigh Street, M4 5DL.

The Necessary Group (2004) *A Flag for the North West*, Available: http://www. itsnecessary.co.uk/page.asp?content=pressrelease (accessed 27 September 2004).

Northern Way Steering Group (2004) *Moving Forward: The Northern Way – First Growth Strategy Report*, Newcastle: Northern Way Steering Group. Available: http://www.the northernway.co.uk/documents_report.html (accessed 1 October 2004).

North West Regional Association (1993a) *A Regional Strategy for the North West*, Wigan: NWRA, North West Assembly House, Dorning Street, WN1 1HJ.

North West Regional Association (1993b) *Greener Growth: Draft Advice on Regional Planning Guidance for North West England*, Wigan: NWRA, North West Assembly House, Dorning Street, WN1 1HJ.

NWDA (North West Development Agency) (2000) *England's North West: A Strategy Towards 2020*, Warrington: NWDA, Renaissance House, Centre Park, WA1 1XB. Available: http://www.nwda-cms.net/DocumentUploads/RegionalStrategy.pdf (accessed 29 September 2004).

NWDA (North West Development Agency) (2003a) *Strategic Regional Sites: First Monitoring Report*, Warrington: NWDA, Renaissance House, Centre Park, WA1 1XB. Available: http://www.nwda-cms.net/DocumentUploads/pub_STRAT_REG_SITES.pdf (accessed 6 October 2004).

NWDA (North West Development Agency) (2003b) *Mersey Belt Study: Policy Statement*, Warrington: NWDA, Renaissance House, Centre Park, WA1 1XB. Available: http://www.nwda-cms.net/DocumentUploads/MerseyBeltStudyPolicyStatement.pdf (accessed 8 October 2004).

NWDA (North West Development Agency) (2003c) *Regional Economic Strategy 2003*, Warrington: NWDA, Renaissance House, Centre Park, WA1 1XB. Available: http://www.nwda-cms.net/DocumentUploads/RES2003.pdf (accessed 29 September 2004).

NWRA (North West Regional Assembly) (2004) *Partial Review of Regional Planning Guidance for the North West (RPG13)*, Wigan: NWRA, North West Assembly House, Dorning Street, WN1 1HJ.

ODPM (Office of the Deputy Prime Minister) (2003) *Regional Planning Guidance for the North West*, RPG 13, London: The Stationery Office. Available: http://www.go-nw.gov. uk/planning/rpg13.html (accessed 28 September 2004).

ODPM (Office of the Deputy Prime Minister) (2004a) *A New Opportunity for the North West*, London: ODPM. Available: http://www.nwra.gov.uk/documents/801078186349. pdf (accessed 20 January 2006).

ODPM (Office of the Deputy Prime Minister) (2004b) *Making it Happen: The Northern Way*, London: ODPM, Available: http://www.odpm.gov.uk/embedded_object.asp?id= 1139974 (accessed 20 Jauary 2006).

ODPM (Office of the Deputy Prime Minister) (2004c) *Draft Regional Assemblies Bill*, July 2004, Cm 6285, London: HMSO. Available: http://www.archive2.official-docu-ments.co.uk/document/cm62/6285/6285.pdf (accessed 20 January 2006).

Parkinson, M., Hutchins, M., Simmie, J., Clark, G. and Verdonk, H. (2004) *Competitive European Cities: Where do the Core Cities Stand?* London: Office of the Deputy Prime Minister, London ODPM http://www.odpm.gov.uk/embedded_object.asp?id=1127441 (accessed 20 January 2006).

Regeneris Consulting, SURF and DTZ Pieda Consulting (2001) *Liverpool Manchester Vision: Strategy and Action Plan*, Altrincham: Regeneris Consulting, Downs Court, Cheshire, WA14 2QD. Available: http://www.nwda-cms.net/DocumentUploads/LivMan.pdf (accessed 1 October 2004).

Roberts. P. (2002) 'Chambers threaten assembly walkout', *Manchester Evening News*, 19th April. Available: http://www.manchesteronline.co.uk/business/general/s/3/3009_chambers_threaten_assembly_walkout.html (accessed 29 September 2004).

Roberts, P. and Lloyd, G. (1999) 'Institutional aspects of regional planning, management, and development: models and lessons from the English experience', *Environment and Planning B* 26(4): 517–31.

Roberts, P. and Lloyd, G. (2000) 'Regional development agencies in England: New strategic regional planning issues?', *Regional Studies* 34(1): 75–9.

Robson, B. (2001) 'Slim pickings for the cities of the north', *Town and Country Planning* 70: 126–8.

Robson, B., Deas, I. and Lawson, N. (1996) 'Housing need and provision, and the economic prospects of the North West', in M. Breheny and P. Hall (eds) *The People – Where Will They Go?* London: Town and Country Planning Association.

Rodgers, B. (1980) 'The North West and North Wales', in G. Manners, D. Keeble., B. Rodgers and K. Warren (eds) *Regional Development in Britain*, second edition, Chichester: Wiley.

RTPI (Royal Town Planning Institute) North West Branch (1990) *North West 2010: The Pressing Case for Strategic Planning*, Sale: RTPI North West Branch, Friars Court, Sibson Road, Cheshire, M33 7SF.

Senior, D. (1965) 'The city region as an administrative unit', *Political Quarterly* 36(1): 82–91.

Tickell, A., Peck, J. and Dicken, P. (1995) 'The fragmented region: business, the state and economic development in the North West of England', in M. Rhodes (ed.) *The Regions and the New Europe*, Manchester: Manchester University Press.

Tomaney, J., Hetherington, P. and Pinkney, E. (2003) *Nations and Regions: The Dynamics of Devolution*, Report no. 11, Newcastle: Centre for Urban and Regional Development Studies, University of Newcastle upon Tyne, NE1 7RU. Available: http://www.ucl.ac.uk/constitution-unit/monrep/er/regions_june_2003.pdf (accessed 29 September 2004).

Urban Task Force (1999) *Towards an Urban Renaissance*, final report of the Urban Task Force chaired by Lord Rogers of Riverside, London: E&F Spon.

Valler, D., Wood, A., Atkinson, I., Betteley, D., Phelps, N., Raco, M. and Shirlow, P. (2004) 'Business representation in the UK regions: mapping institutional change', *Progress in Planning* 61(2): 75–135.

Wannop, U. (1995) *The Regional Imperative: Regional Planning and Governance in Britain, Europe and the United States*, London: Jessica Kingsley.

Wannop, U. and Cherry, G. (1994) 'The development of regional planning in the UK', *Planning Perspectives* 9(1): 29–60.

While, A. (2001) *Regional Partnerships and Economic Development in England*, unpublished PhD thesis, Leeds Metropolitan University.

While, A., Jonas, A. and Gibbs, D. (2004) 'Unblocking the city? Growth pressures, collective provision, and the search for new spaces of governance in Greater Cambridge, England', *Environment and Planning A* 36(2): 279–304.

Wilks-Heeg, S., Deas, I. and Harding, A. (1999) *Does Local Governance Matter to City Competitiveness? The Case of Inward Investment in Merseyside and Greater Manchester*, Working Paper 2, ESRC Cities programme, Manchester: Centre for Urban Policy Studies, University of Manchester, M13 9PL.

Williams, G. and Baker, M. (2006) 'Strategic planning and regional development in North West England', in H. Dimitrou and R. Thompson (eds) *Strategic and Regional Planning in the UK*, London: Spon.

7 Advancing together in Yorkshire and Humberside?

Dave Counsell and Graham Haughton

Mutant planning: an introduction to the contemporary dynamics of planning

'Advancing Together' is the title of the high level policy framework for the Yorkshire and Humber region, produced on behalf of regional partners by the Regional Assembly. It provides a common vision and a set of key objectives that are intended to inform a range of other regional and local strategies. *Advancing Together* was first published in 1998 and it was reissued in a revised form in 2004. In both cases, the framework was endorsed by all of the key institutions in the region, a process of institutional collaboration which reflected the importance given to integrated policy making. This in turn reflects the political concern nationally to promote 'joined-up thinking', one of the core political ideas of post-1997 'New Labour' administrations.

In analysing the shift towards integrated plan making, we want to examine in tandem how the scales and the scope of planning are being reworked. At a theoretical level this links to debates about rescaling (see Goodwin, Jones and Jones in this volume) and also what we refer to as the 'politics of scope' (Allmendinger and Haughton, 2004). We are concerned with seeing both policy scale and policy scope as political and social constructs that are highly contested. The politics of scale in planning requires analysis of how powers and responsibilities are being continuously reworked in non-linear, non-hierarchical ways across various 'governance lines', which run across neighbourhood, local, sub-regional, regional, nation and supra-national levels. For instance in the case of English planning, the powers of county level planning have recently been reduced as part of a shift towards giving regional planning frameworks a statutory status. In similar vein, all levels of English planning are now expected to pay attention to the European Spatial Development Perspective (ESDP). Powers are therefore being reworked in particular ways across the scales of planning as they relate to England.

The powers of planning are being dramatically rescaled at the formal level. But at the more informal level too planning is being rescaled, as new planning tiers are created; for instance, Regional Spatial Strategies can now involve the creation of new sub-regional studies as part of their processes. The Sustainable Communities Action Plan (ODPM, 2003) announced four growth areas for the south of England, three of which involve new planning sub-regions to run alongside existing 'local' or 'regional' systems of planning. The same document created nine housing renewal 'Pathfinder' areas, again running outside of, yet alongside the formal boundaries of statutory local planning frameworks and Regional

Spatial Strategies. As if that were not complex enough, we have seen the creation of the Northern Way and Midland Way, new meta-regional strategies that span existing regional boundaries, with the Northern Way covering Yorkshire and the Humber, the North West and the North East (ODPM, 2004).

Running alongside this shift towards multi-scalar governance, there has been the move towards integrative planning, not least in response to the ESDP (Counsell *et al.*, 2003; see Healey's contribution to this volume). It is these moves away from narrow sectoral models of policy formation towards more holistic, trans-sectoral policy making that are what we refer to when we speak of the changing scope of policy. These processes are not unproblematic, however. As Box 7.1 intimates, there are a variety of ways in which horizontal policy integration can be understood and pursued. At its very basic level, this policy integration imperative is about improving the links between land use planning strategies at various scales and the growing number of other, potentially related strategies, notably those for economic development, environmental and resource management, and transport.

Box 7.1 Dimensions of vertical and horizontal policy integration

Vertical integration linking policy actors
1   With UK national and European policy statements, in planning and related functional areas, such as environmental protection area designations
2   Linking regional strategies for planning etc. with e.g. county council, local authority and neighbourhood plans and strategies
3   Linking to strategies at 'new' flexible policy scales, such as the new sub-regional economic strategies and spatial planning studies, and new meta-regional strategies such as the Northern Way.

Horizontal integration of policy domains
1   Between places
   a   Administrative areas within an administration
   b   Adjoining regions/nations e.g. border issues
2   Between policy themes/sectors
   a   Departments within an administration
   b   Other public bodies within the administrative area, for instance linking local plans to wider strategies of a local strategic partnership or municipal authority.
   c   Other strategies and themes
3   Aiming for integration across the social, economic and environmental components of sustainable development
4   Building better integration between public, private, voluntary and community sectors
5   Linking the provision of infrastructure, services and investment, for instance involving transport planning, water management, industrial sites and spatial planning.

Our argument here is that we are witnessing a parallel and interconnected process to the rescaling of planning, involving major changes in the definition of planning in terms of its scope, and also its intended relationship with other sectoral strategies. In effect, different governments have had different views on what constitutes the legitimate scope of planning and the extent to which it should inform and be informed by, for instance, social policy. In the last ten years in particular, English planning has been reinterpreted once again, broadening out from the predominantly land use focus imposed on it during the 1980s toward becoming a much more holistic policy arena. Planning has been positioned as a key sector in the government's pursuit of sustainable development, adopted in recent years as the formal goal for the planning system. This is not an apolitical act, however, as the government has adopted an 'integrative' definition of sustainable development which emphasises that it should not be seen as an environment-led agenda, but rather an agenda for integrating economic, social and environmental issues. Planning is therefore being used to pursue a particular political viewpoint, where integration has a particular political purpose.

We would argue that it is important to examine as interrelated processes *both* the scope of territorial structures and their scale – that is, geographical coverage. Referring to Box 7.1, we can see that the distinction between horizontal and vertical policy integration is actually a false one. It is a useful heuristic device at one level, but actually policy is always being reshaped and reformulated in ways that involve changing powers and responsibilities simultaneously across vertical and horizontal dimensions. In addition, our approach to the issue of policy integration is to bring to the fore that choices over policy scales and policy scope are inherently politicised, where powerful alliances are formed to shape policies in ways which best serve the interests of particular groups (Haughton and Counsell, 2004).

These are perhaps the defining characteristics of the emergent era of fluid and increasingly diverse planning regimes across the UK space. Planning powers are reworked extensively and intensively across policy scales whilst new ways are being opened up to ensure that land use plans and other sectoral strategies are, if not quite mutually constituted, then at least mutually informed; it is precisely these issues that we refer to when we talk of mutant planning. There are new ways of undertaking planning that are emerging with amazing speed, involving planning powers being strengthened at some scales, reduced at others, reworked at all scales, and new scales of planning inserted, with sub-regional and metaregional strategies. The old rigid hierarchical systems of formal planning have been reshaped but nevertheless provide the skeleton of the formal planning system, but they are now overlain with a range of other informal and formal ways of working, involving links to other policy sectors and to new scales of planning with less rigidly defined statutory roles. These related strategies often work to different spatial and temporal horizons, creating an ever-more complex web of links, which can work to broaden and enable what it is to be a planner, or induce narrowness and conformity. The processes will work out differently in different areas. Those practices that work will be evaluated, judged and presumably backed by new resources and directives by the state. Those new practices that fail to meet expectations can expect to be allowed to wither or to be closed down. This is the era of mutant planning, reshaping what it means to be a planner anywhere in the UK but also, crucially, in different ways in different parts of the post-devolution

UK space, where how planning is constituted and performed in Wales will differ from Scotland, and what happens in London may well differ from what happens in Yorkshire and the Humber.

Using these insights, in this chapter we explore whether and how the main policy actors in the Yorkshire and Humber region are advancing together towards integrated regional policy. The chapter draws on data from a three year long study of regional planning guidance in all eight English regions carried out between 2000 and 2002 (see Haughton and Counsell, 2004). This is supplemented by the results of a study undertaken for regional bodies in Yorkshire and Humber on integrating regional strategies (Yorkshire Futures, 2003), an evaluation of the EU Objective 2 Programme in the region (by Leeds Metropolitan University and University of Hull, 2003) and by two additional interviews undertaken in 2004. In total, 45 people involved with regional policy related issues in Yorkshire and the Humber were interviewed over the period 2000–2004.

Regional planning in Yorkshire and the Humber: historical perspectives

Regional policy in the UK in the three decades following the second world war focused on addressing the north–south divide using a variety of policy tools, applied with varying degrees of enthusiasm and levels of financial backing. In essence, the dominant approach was one of 'carrot and stick', providing incentives for employment to move towards the less prosperous areas, such as Yorkshire and Humber, plus planning regulations to constrain growth in the more prosperous parts of the country. With major losses of traditional manufacturing jobs occurring, particularly in the northern industrial towns, the political imperative was to stem job losses and reduce outward population movement by bringing in new jobs, particularly through attracting in large factories and, to a lesser extent, offices. The other key aspect of most northern regional plans of the period was to address the problem of poor quality housing stock, seeking to upgrade, demolish and rebuild as appropriate.

The history of regional planning in Yorkshire and the Humber has been particularly well documented by Pearce (1989) and Roberts (1994). According to Roberts (1994: 214), up to the mid-1960s planning at the sub-regional and regional level created a 'legacy of parish-pump spatial policy'which provided a powerful force for 'policy inertia'. The creation of the regional economic planning council signalled a shift towards a more regional strategic process, notably with the 1966 document *A Review of Yorkshire and Humberside* (Yorkshire and Humberside Regional Economic Planning Council, 1966). Whilst not a strategy in itself, Roberts argues that it had a stronger spatial development dimension than the actual 1970 strategy (see below). Interestingly, both the 1966 Review and the 1969 report *Humberside – A Feasibility Study* (Central Unit for Environmental Planning, 1969) focused on the potential of the Humber estuary to be developed as a major maritime industrial area. This emphasis on the Humber inevitably caused friction with the political authorities in the more urbanised western parts of the region, which wanted an emphasis on addressing their legacy of problems involving unemployment and poor quality housing stock.

The first formal regional strategy for the Yorkshire and Humber region was published in 1970 (Yorkshire and Humberside Regional Economic Planning Council, 1970). Six broad planning objectives were identified, the first of which was the creation of new job opportunities, particularly in services and science-based industries.

Spatial development issues were addressed in the strategy, albeit in a mainly descriptive way, linked to infrastructure investment and providing land for industrial development. By and large the document avoided major strategic decisions on preferences for how development should be guided. So whilst the broader spatial strategy was to focus development on existing cities and towns, the strategy was keen to promote the potential of smaller towns too. As part of the overall strategy, detailed descriptions of sub-areas within the region were provided along with suggested goals in guiding future development.

The regional strategy was reviewed in 1975 (YHEDC, 1975) when the broad objectives of the earlier strategy were confirmed and three priorities were identified for implementation:

1   To encourage industrial investment,
2   To complete the basic communications infrastructure, and
3   To tackle environmental issues.

Environmental issues were defined as: housing, water supply and sewage disposal, air pollution, health, education and training, and river pollution.

Looking back at the 1970 strategy, it is interesting to see how it manages to be both more 'integrative' in its concerns in some respects than contemporary regional strategies, but less holistic in others. For instance, the 1970 strategy shows a much more clearly articulated concern with aspects of health and educational infrastructure than its counterparts 25–35 years later. It also has a clearer spatial dimension to it, in terms of its understanding of the sub-regions. In terms of the policy process, it involved a wide range of partners from local authorities, businesses, universities, colleges, nationalised companies and utilities (rail, electricity, mining, docks, steel), and trade unions, with professional support coming from central government officials. Alternatively, it lacks many of the strengths of the 2001 Regional Planning Guidance for Yorkshire and the Humber, such as the enormous amount of consultation, its use of targets and its attention to a much wider range of environmental issues, such as biodiversity, and flood risk. The 1970 strategy was also very much the key regional document, which as we will see below compares with the contemporary proliferation of regional strategies.

By 1979, the political mood nationally had changed and one of the early acts of the new Conservative government was to dismantle the Economic Planning Councils, with regional policy residualised in the process to little more than a centralised function of allocating grants to mobile firms. After a decade of little or no strategic regional thinking, there was growing concern about the lack of regional cohesion amongst the main players (Leigh and Green, 1990). However, in 1991 the Yorkshire and Humberside Regional Association published a detailed overview and outlook of the region's economy and infrastructure on behalf of local authorities. Reflecting the political climate of the times perhaps, it did not aspire to providing a strategic framework, but it did have a strong European dimension.

The European Commission came to the fore in 1993, promoting and indeed paying for the first regional strategy since 1970. Concerned that its Regional Structural Funds were being allocated in a strategic vacuum, the Commission required most eligible UK regions to develop across the 'social partners' an agreed regional strategy. In Yorkshire and the Humber the work involved a major overview by a consortium of regional universities and a firm of consultants. This work informed the resulting strategy. The strategy group was chaired by a representative of the regional Trades Union Council, with representatives from the various regional offices of central government, local authorities, and the regional groupings for universities and for Training and Enterprise Councils (Yorkshire and Humberside Partnership n.d., c.1993). The strategic element of this document covered just over six pages, focused mainly on economic development issues, with some attention to transport protecting sensitive environmental areas and protecting and developing the region's historic and cultural, scenic and wildlife heritage. It lacked a clear spatial development framework and had little substantive to say about improving the regional environment.

Following the revived fortunes of regional planning in the early 1990s, the first round of Regional Planning Guidance (RPG) for Yorkshire and Humberside was issued in 1996 (Department of Environment, 1996). This first version of RPG covered a wider range of issues than the previous regional strategies. The overall aim of regional planning policy was the pursuit of sustainable development, reflecting the emerging discourse of sustainability and the growing policy attention to environmental concerns. Nevertheless, within this broad aim two of the four objectives in the RPG still emphasised economic development (ibid.):

1   Promoting economic prosperity and achievement of a competitive position;
2   Conserving and where possible enhancing the region's environment;
3   Facilitating processes of industrial adjustment, economic diversification and urban and rural regeneration and renewal; and
4   Making best use of the available resources and encouraging efficient use of energy.

In contrast to the 1970 strategy, social issues such as health, and education and training had been relegated in importance, but environmental issues were given rather more prominence. The 1996 RPG was produced with advice from local planning authorities, but with final control resting with the relevant central government department. In essence, it was a document for the region more than a document from the region. In common with other RPGs of this period, it adopted a lowest common denominator approach, avoided making hard decisions, tended towards a bland repetition of government advice, and lacked a distinctive spatial strategy.

This was the state of regional planning policy in Yorkshire and Humber when New Labour came to power in 1997. After its fading fortunes throughout the 1980s, by the early 1990s the main regional planning actors were starting to come together, albeit hesitantly (Green and Leigh, 1990), assisted by the creation of the Government Offices for the Regions in 1994 and the focus of the European Commission on the regional scale for its economic support programmes.

Advancing Together: an integrated approach post-1997?

The election of the Labour Government in 1997 saw a surge of interest in regional governance issues. As a consequence, the institutional architecture of regional governance has also expanded rapidly since 1997 with a range of new duties and responsibilities devolved to the regional scale. In the absence of political devolution, though, the English regional architecture appears somewhat disjointed, with separate institutions responsible for producing different sectoral strategies emphasising the need for more integrated approaches. The Regional Development Agencies, established in 1998, were given the role of preparing Regional Economic Strategies, the first round of which were prepared in the period 1999/2000. Regional Chambers, which increasingly became known as (unelected) Regional Assemblies, were encouraged in each region to provide a link to local democratic systems, involving representatives from local authorities and a wide range of other key regional stakeholders. The Chambers quickly took on key scrutiny roles, not least in relation to the RDAs, but also in most cases becoming the Regional Planning Bodies, responsible for producing draft Regional Planning Guidance (RPG).

In 1998, the government introduced new arrangements for the production of Regional Planning Guidance. This led to a second round of RPG preparation over the following 5–7 years, with a requirement to include summary Regional Transport Strategies within the documents. The Planning and Compulsory Purchase Act 2004 replaced the RPG system with Regional Spatial Strategies that are intended to be both more comprehensive and possess statutory status. Responding to calls for clarity over how various strategies interrelated, the government issued advice on how to produce Regional Sustainable Development Frameworks (RSDFs), leading to England-wide coverage during 2000/2001.

Other strategies produced by a range of partnerships across the region in this period include: Housing, Culture, Employment and Skills, Environmental Enhancement, Health, Sports, Innovation and International Trade, together with various regional action plans for tourism and sustainable communities; these are in addition to strategies prepared for funding programmes, such as the EU Objective 1 and 2 programmes.

In Yorkshire and the Humber, work on both RPG and RES was initiated in 1999 within the overall framework provided by its overarching strategy *Advancing Together*, first published the previous year (it was subsequently revised and reissued in 2004). The role of *Advancing Together* as a framework for policy integration is reflected in its use in the titles of the other key regional strategies. Talking about the role of *Advancing Together*, a senior regional planner we interviewed stated:

> What we have is the *Advancing Together* objectives and the RSDF ... the one providing the over-arching objectives and the other a check-list to be considered at every level. If nothing else, as I've said, it means that we are not all sitting round asking what our objectives are – they are agreed now and set out so the focus of attention is on how can we ensure that these are brought about. So that's where the RES starts from and it's where the other strategies start from.
>
> (Interview, 2004)

The Regional Development Agency Yorkshire Forward, published its draft Regional Economic Strategy (RES), *Advancing Together: towards a world class economy*, in mid 1999 and, after a period of consultation, the final version was endorsed by central government early in 2000. The draft RPG, *Advancing Together: towards a spatial strategy*, was published by the Regional Assembly for consultation in July 1999 and for submission to public examination in October 1999. It was not finally approved until October 2001.

Since 1999, there has been a requirement that RES and RPG should be subject to a process of sustainability appraisal during which their core strategies and policies are appraised against a set of sustainability objectives. As we have already noted, the New Labour government sees sustainable development as a concept that brings together social, economic, and environmental objectives in an integrated way, involving a search for policy solutions that benefit all three dimensions – known as 'win–win–win' solutions. This is not without its critics. The Campaign for Rural England (CPRE) for instance is very critical of the government's preference for sustainability appraisal rather than strategic environmental assessment, arguing that in practice whilst ministers talk about balancing economic, social and environmental objectives, 'traditionally this has meant that the environment almost always loses out' (Hamblin, 2004: 13)

Nonetheless, in England sustainability appraisal is the recommended method for achieving integration between objectives. Central government advice (DETR 2000a and 2000b) suggests that strategies are appraised against the objectives agreed to in each region's Sustainable Development Framework. Whilst this creates scope for some regional sensitivity, in practice all RSDFs necessarily reflect the government's view of a 'balanced' interpretation of sustainable development. Unusually for this round of regional strategy documents in England, sustainability appraisals of the initial Yorkshire and Humber RES and RPG were undertaken in parallel, using the same appraisal criteria and the same consultants.

Although starting on a parallel timescale in 1999, the Yorkshire and Humber RES and RPG subsequently diverged (see Table 7.1), largely because of the different regulatory requirements for the two strategies. RESs are produced internally by RDAs, in this case Yorkshire Forward, which retain ownership despite the strategies having to be submitted to central government for endorsement. The preparation of RESs is relatively quick, the whole process taking less than a year to complete. RPG by contrast is subject to a lengthy process of public examination, followed by the Secretary of State taking ownership from the Regional Planning Body and issuing 'proposed changes' for consultation before publishing the final report; in the case of Yorkshire and Humber, this occurred in October 2001 through the Government Office for Yorkshire and Humberside.

The timescales of RPG and RES did begin to move out of synchronisation as they developed at different paces. This divergence has continued during subsequent revisions to the agreed strategies. RESs are reviewed on a three-yearly basis, with the first review initiated in 2002 and completed in 2003. A partial review of RPG was begun in 2003 and was not completed until 2005. In the meantime, the Regional Assembly is already well advanced in its work on producing a Regional Spatial Strategy under the new arrangements introduced in 2004.

Table 7.1 Diverging timescale for preparing RPG and RES

	Draft	Public examination	Strategy published	Draft review	Public examination	Strategy published
RPG	1999	2000	2001	2003	2004	2005
RES	1999	N.A.	2000	2002	N.A.	2003

Despite these diverging timescales most stakeholders who were interviewed in a study of strategy integration in the region (Yorkshire Futures, 2003) felt that the 2003 version of the RES was better integrated with RPG than the earlier version, although some tensions still remained. Perhaps inevitably most of the tensions were between economic development policies in the RES and environmental protection policies in RPG. Environmental organisations in the region have consistently pointed out all iterations of regional strategy preparation and highlight that the promotion of development on greenfield sites next to motorway junctions, associated with the RES, conflict with environment and transport policies in RPG:

> So, there are some critical things to deal with [in preparing the RSS] … things like the greenfield allocations next to motorway junctions … and also the approach that is being taken to the Humber Trade Zone in view of flood risk and other environmental impacts.
> (Interview: Government Agency representative, 2003)

Whilst recognising that there were tensions, another interviewee suggested that these emerged more strongly down-stream when RES policies were being implemented:

> I think there are bound to be tensions … its difficult to reconcile those isn't it … I don't think there are necessarily tensions with the words in the RES … but when those are translated into investment programmes and particular projects … that's when the tensions come out.
> (Interview: Government Office representative 2004)

Tensions such as these between the RES and some environmental policies in RPG are perhaps aggravated because many of the RES policies are not expressed in spatial terms. For example, tensions over the Humber Trade Zone (HTZ), a policy aimed at optimising the economic benefits to be gained from the Humber Ports, remain in part because the RES does not clearly identify the boundaries of the HTZ (Leeds Metropolitan University and University of Hull, 2003). In the absence of this spatial detail, those debating the impacts of development in the Humber Estuary on biodiversity and flooding can only speculate about conflicts which might occur, sometimes resulting in tensions which could have been avoided had more detail been available at the time.

Looking next at vertical integration between local, regional and national policies, recent research in the region has identified tensions between the different scales (Counsell and Haughton, 2002; Yorkshire Futures, 2003). One particular tension concerns the Sustainable Communities Plan (ODPM, 2003), an ODPM (Office of the Deputy Prime Minister) initiative to address shortages of housing land in the South East and low demand for housing in the North. There is a view amongst some regional stakeholders that this whole policy agenda is being imposed on regional planning bodies from outside normal regional planning processes. In Yorkshire and the Humber, tensions have arisen in particular over proposals for the Northern Way, a concept linking together the three northern regions and focusing on eight city-regions, which has arisen as part of the Sustainable Communities Plan work. The idea was launched by central government (ODPM, 2004) at a stage when the preparation of the new RSSs was well advanced, without any detail being provided about what it might entail for spatial development in the region, and the task of pursuing it was given to the three northern RDAs. Their report, *Moving Forward: the Northern Way – first growth strategy* (2004) was widely criticised for being prepared without widespread consultation and for its strong economic focus:

> There are always going to be tensions between the national and the regional, just as the local. The current tension is the Northern Way … that's an RDA/ODPM driven initiative … it's not an RSS driven initiative … we are probably going to feel as though we have been handed something … that 'thou shall incorporate'. So there are a number of policy areas such as this where we feel that things are imposed … .
>
> (Interview: senior regional planner, 2004)

Another interviewee noted the problems of joining up policy at the regional level when nationally there were diverse agendas:

> Well yes it's difficult … you can't argue it doesn't exist … You can talk about joined-up government, but there are such huge agendas and such a huge range of people it's inevitable there are differences. [Referring to the discussion on climate change at the RPG Partial Review EIP] On the one hand you had the government arguing that it wants to meet climate change targets [for reducing CO_2] and on the other what is the Transport Strategy doing about this … Of the three main areas producing greenhouse gas, transport is the one that is growing … and how does that fit?
>
> (Interview: Government Office representative, 2004)

Our recent work has revealed rather fewer tensions between the new regional scale policies and local policies, though undoubtedly these exist. It probably helps that political representatives of local authorities sit as members of the Regional Assembly, which may mean that controversial proposals are side-lined early on as the emphasis is on region building, involving a tendency to focus on developing areas of agreement rather than flushing out areas of disagreement.

Advancing Together: towards a sustainable region

As we noted earlier, RSDFs are intended to bring about greater coherence in approaches towards sustainable development in regional strategies by establishing a common vision for the region, providing common sets of objectives to address sustainable development, and common indicators to measure progress. Yorkshire and Humber's first RSDF, *Advancing Together: towards a sustainable region*, was adopted by the Regional Assembly in November 2000, and published in February 2001 (Regional Assembly for Yorkshire and the Humber, 2001). The vision statement contained in the RSDF is shared with the RES and RPG. Fifteen aims for sustainable development in Yorkshire and the Humber are identified in the RSDF. In addition, four cross-cutting themes were included in the revised RSDF, published in 2003 (YHA, 2003): social equity across all sectors; a partnership and participative approach; geographic adaptation to the needs of rural and urban communities; and creativity, innovation and the appropriate use of technology.

Worryingly, in view of its intended integrative role, interviews undertaken in early 2003 suggested that there was some confusion among regional policy makers in Yorkshire and the Humber about the status and purpose of the region's RSDF (Yorkshire Futures, 2003). Whilst staff from the Regional Assembly and RDA who had been involved with using the RSDF were clear about its important role in policy appraisal, other stakeholders, for example some of those preparing the Regional Housing Strategy, were largely unaware of its purpose. At the time these interviews were being undertaken, the RSDF was being reviewed and a new glossy version has now been issued, incorporating a guide to using the framework in sustainability appraisal. The role that the RSDF has played in achieving inte-grated policy will be more apparent when considering its use in sustainability appraisal of the RES and RPG.

Advancing Together through sustainability appraisal

Recent research (Counsell and Haughton, 2002) suggested that of all the English regions, the approach to sustainability appraisal in Yorkshire and the Humber most closely approximated to that recommended in the 'good practice guide' published by the government (DETR, 2000a and 2004b). Sustainability appraisal in many regions was criticised for being an 'add-on' at the end of the process but, in Yorkshire and the Humber, it began at the 'strategic options' stage of draft RPG and was continued through the other major stages in the process. It was also regarded as having pursued a more inclusive approach than elsewhere. Finally, the Yorkshire and Humber region was distinctive by virtue of setting in train a parallel appraisal for both RPG and RES, using the same set of objectives and the same consultants.

The first pubished sustainability appraisal of draft RPG was undertaken by the consultants ECOTEC at the time of the 'consultation draft' in July 1999 (ECOTEC 1999a). The draft Regional Economic Strategy was also published at this same time by Yorkshire Forward and ECOTEC was commissioned to undertake a sustainabil-ity appraisal of this using the same methodology as for RPG (ECOTEC 1999b). This 'twin-tracking' of RPG and RES appraisals was unique in the english regions, and resulted in a local gvernment officer commenting:

> I think we had a much better understanding of the RES because the two things were being appraised at the same time.
>
> (Interview, 2000: 14)

The sustainability appraisal for the 1999 version of the RES concluded that it was generally compatible with RPG, but on environmental protection issues, reflecting concerns of environmental bodies expressed earlier, it suggested that:

> The thrust of the RES in 'getting the best out of physical assets and conserving environmental assets' may conflict with the stronger conservationist stance of the RPG natural resource use policies.
>
> (ECOTEC, 1999b: 24)

Despite the more co-ordinated and inclusive approach to Sustainability Appraisal in Yorkshire and the Humber, our research (Counsell and Haughton, 2002) sug-gested that it had had only a marginal effect on the policy content of RPG, limited in effect to policies on sustainable development itself. More particularly it appeared to have had little influence on the major policy debates on housing and employment land. A government agency official commented at the time:

> I saw no significant change to RPG before and after the appraisal ... Well, perhaps no change is too strong, but there was no material change. The big issues were sidestepped.
>
> (Interview, 2000)

This research nevertheless pointed to evident benefits in the Yorkshire and the Humber approach:

> The integrated approach to sustainability appraisal provided a thread of continuity through the RPG process and assisted in integrating policies in RPG and RES – at the very least these separate regional strategies have been appraised on the same basis and against the same criteria. It has also served to flag up conflicts between policies, but has been less successful in helping to resolve these conflicts.
>
> (Counsell and Haughton, 2002: 35)

Interestingly, the conclusions of the most recent sustainability appraisal of the draft revised RPG (ENTEC, 2003) suggest that there is now a good synergy between the RES and the revised RPG. This document notes how some key RPG themes such as climate change, renewable energy and waste management were being referred to in the RES, and how its initiative on 'opportunities for a low carbon economy is an excellent example of how the RES can contribute to RPG objectives' (2003: 29).

Achieving integration: are the barriers being overcome?

It was always unreasonable to expect that the so-called silo mentality that characterised economic development and spatial planning (Vigar *et al.*, 2000, Haughton and Counsell, 2004) would suddenly melt away as regional institutions began to address policy integration. Indeed, to some degree, these barriers are likely to have increased with the expanding regional policy agenda and the growing number of organisations focusing on the regional scale.

We have already flagged up some key potential barriers to achieving better policy integration: different regulatory requirements; different timescales (both policy horizons and timescales for preparing and reviewing strategies); and different knowledge, expertise and professional cultures (Owens and Cowell, 2002). For example, on policy horizons, RPG covers a 20 year period, currently from 1996 to 2016 whilst the current RES covers the ten years from 2003 to 2012. Recent work in Yorkshire and the Humber has exposed considerable differences in the knowledge bases and professional cultures of those involved in the preparation of regional strategies (Yorkshire Futures, 2003). On the issue of sustainable development, for example, it has been suggested that planners involved in preparing RPG were much more aware and comfortable with addressing this concept than those involved with the initial RES, with these differing approaches creating the potential for tension (Benneworth, 1999; Gibbs and Jonas, 2001). Differences are also apparent between the professional cultures of economic development staff and planners and housing professionals (Yorkshire Futures, 2003). More worryingly, the strategies examined in the Yorkshire and Humber research appeared sometimes to be based on different data sets and assumptions resulting in an absence of shared common opinion on some key issues.

The messages from recent interviews are mixed on whether these barriers are being overcome. Certainly, in 2000 those we spoke to were encountering considerable problems, particularly around the issue of sustainable development, but three or four years later these seemed to have started to diminish. Some fairly typical comments in 2000 included:

> I think that over time they will diverge [RPG and RES]. I think it very difficult to reconcile some of the objectives in the RES with the overall approach in RPG.
>
> (Interview: pro-development lobbyist, 2000)

> At the outset it was my perception that the RDA was reluctant about sustainable development – here we go again, it's just an environmental constituency trying to put the brakes on sustainable development.
>
> (Interview: local planner, 2000)

But just a few years later, a different tone had emerged:

> Our response [to the 2003 RES] was fully taken into account by the RDA the final RES document was signed off by the Assembly. So we do feel they are compatible.
>
> (Interview: regional planner, 2003)

> The different RPG and RES processes led to some diversion between the strategies ... But if you look at the most recent RES completed last year ... there was much more agreement ... and I think that reflects the agreement in the region ... and within Yorkshire Forward ... to actually carry that through. So they were prepared to change their RES to bring it into line with RPG as it had ended up being issued.
>
> (Interview: senior regional planner, 2004)

On this evidence, some of the barriers do appear to be being addressed as regional bodies develop capacity and improve procedures for strategy preparation. This appears to support speculation by Owens and Cowell (2002: 66) that better integration is likely to arise gradually from 'a process of learning within and between coalitions'. One of our interviewees in 2004 explained that one of the reasons for the current more positive relationship between RPG and RES is that the organisational arrangements for preparing RPG/RSS have been devised to include those responsible for the other principle regional strategies:

> Basically what we have is the full assembly with fifteen leaders (of councils) and twenty-two strategic and economic partners, and we also have a number of 'observer members' which include Yorkshire Forward (YF) and Government Office (GO), so in that sense, at full assembly level there is a link there to who we see as our main regional partners in terms of getting the joining up ...

between the membership of the assembly and the observer members, all of those responsible for drawing up those [main] strategies are all covered.
(Interview: senior regional planner, 2004)

Returning to the concerns identified in previous research (Counsell and Haughton, 2002; Yorkshire Futures, 2003), about the tensions between strategies, it is notable that the regional bodies have now commissioned work on the spatial implications of the RES which should help resolve issues such as the impact of the Humber Trade Zone on the environment. There was some consensus amongst people interviewed in the region that there is a need for greater clarity about policies such as the HTZ. The mid-term review of the Objective 2 Programme for the region reached a similar conclusion (Leeds Metropolitan University and University of Hull, 2003). Alternatively, it is widely acknowledged that consensus will not always be possible:

At the end of the day there are very different perspectives on the world ... and you are not going to get agreement on anything ... and there are going to be some pinch-points where it is very clear that you haven't got agreements.
(Interview: Government Office representative, 2004)

Perhaps what has changed since the earlier attempts at preparing regional strategies is that the main regional institutions are sharing knowledge information and language to a much greater extent, as they develop their capacity for joint working. Data and research, for example, are increasingly being collected and commissioned to standardised formats, managed by Yorkshire Futures, the regional observatory.

Conclusions

In drawing together our conclusions on regional policy integration in Yorkshire and Humber, it is useful to summarise the interaction between policy scales and sectors in the form of a chart (see Figure 7.1).

This clearly illustrates some of the complexity to be overcome to achieve policy integration. We say some, because this is a very partial picture looking at only five policy streams: The EU structural programmes; RES preparation; Sustainable Communities Plan; spatial planning; and transport planning. On the other hand, it excludes the policy streams relating to sustainable development, culture, housing, health, and education. The picture is further complicated by the fact that the policy flow is not a simple cascade from top to bottom. Most policy processes at the regional scale inform and are informed by policy processes at the national and local scales. For example, whilst RPG is adopted by central government, it is prepared by the regional planning body (the Regional Assembly in Yorkshire) and is informed by work on local plans. Similarly, the RES is prepared at the regional scale despite having to be endorsed by central government and have its spatial implications tested in RPG. So rather than flowing in one direction, the policy process ebbs and flows between the different scales and sectors.

What the experience of Yorkshire and the Humber demonstrates is how planning is being both rescaled and its scope reworked in complex, non-linear ways. In an earlier chapter in this book, Goodwin *et al.* argued that we should

Figure 7.1 Interaction between policy scales and sectors: Yorkshire and Humber

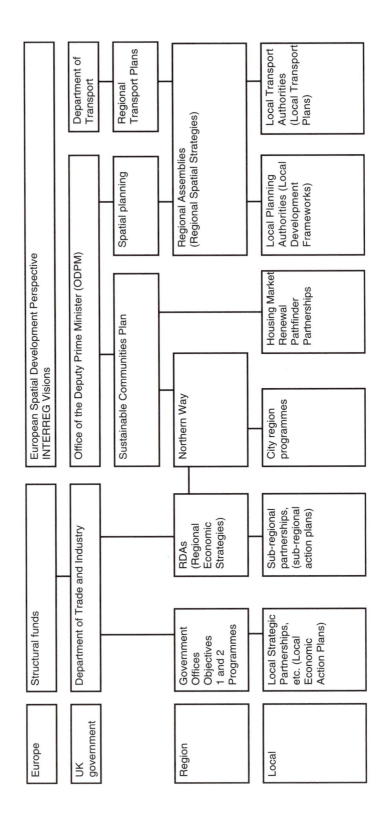

focus not so much on the so-called 'hollowing out of the nation state' as the processes of 'filling in'. They argue convincingly about the need to see how the new institutions are peopled. Our work goes further in suggesting that we need to see how the new territorial structures are being required to adopt and adapt to a range of new policy techniques, such as sustainability appraisal, which enforce powerful disciplinary effects on how actors work both within and across their individual sectors. In particular, our work helps to illustrate how new regional planning and economic development strategies have been introduced for the region, initially in rather poorly connected ways. But under pressure from central government to show greater evidence of policy integration, policed through the new policy techniques of Regional Sustainable Development Frameworks and Sustainability Appraisals, the processes of revising the RES and the RPG have begun to be better integrated in recent years.

In short, what we saw in the late 1990s was how the new regional governance systems were being used very cannily by powerful alliances to pursue particular sets of interests in the differing arenas of planning and economic development. Since then, the process of 'joining up' policy has begun to subtly, yet quite distinctively, rewrite the rules for those seeking to shape the new territorial structures at the regional level. This has forced those seeking to shape and engage with the new territorial systems to acknowledge – or to confront more explicitly – the legitimacy of alliances with alternative knowledges, techniques and cultures, of how planning and economic development might be practised at regional and local scales. These processes are inherently multi-scalar and cross-sectoral in nature, involving a wide ranging series of experiments, conflicts, challenges over institutional responsibilities, techniques and approaches, working in complex iterations which involve tangled networks and alliances operating across governance scales and across traditional sectoral boundaries.

Acknowledgements

This article draws on work funded by ESRC grants R000238368 and R000230756, and by Yorkshire Futures.

References

Allmendinger, P. and Haughton, G. (2004) 'The fluid scales and scope of spatial planning', (mimeo, available from g.f.haughton@hull.ac.uk).

Benneworth, P. (1999) 'Sustainable development, regional economic strategies and the RDAs', *Regions: the newsletter of the Regional Studies Association*, 222: 10–19.

Central Unit for Environmental Planning (1969) *Humberside – A Feasibility Study*. London: HMSO.

Counsell, D. and Haughton, G. (2002) *Sustainability Appraisal of Regional Planning Guidance: final report*, London: ODPM, http://www.odpm.gov.uk/index.asp?id=1145500.

Counsell, D., Haughton, G.F., Allmendinger, P. and Vigar, G. (2003) 'New directions in UK strategic planning: from development plans to spatial development strategies', *Town and Country Planning*, 72: 15–19.

Department of the Environment (1996) *RPG12: Regional Planning Guidance for Yorkshire and Humberside*. London: HMSO.

DETR (2000a) *Guidance on preparing Regional Sustainable Development Frameworks*, London: HMSO.

DETR (2000b) *Good Practice Guide on Sustainability Appraisal of Regional Planning Guidance*, London: HMSO.

ECOTEC Research and Consulting (1999a) *Sustainability Appraisal of Draft RPG for Yorkshire and the Humber*, Birmingham: ECOTEC.

ECOTEC Research and Consulting (1999b) *Sustainability Appraisal of Regional Economic Strategy for Yorkshire and the Humber*, Birmingham: ECOTEC.

ECOTEC Research and Consulting (2001) *Sustainability Appraisal of the Revised RPG for Yorkshire and the Humber*, Birmingham: ECOTEC.

ENTEC UK Limited (2003) *Regional Planning Guidance for Yorkshire and Humber (RPG12) Consultation Draft: Sustainability Appraisal*, Leamington Spa: ENTEC.

Gibbs, D. and Jonas, A. (2001) 'Rescaling and regional governance: the English Regional Development Agencies and the environment', *Environment and Planning C: Government and Policy*, 19: 269–88.

Green, H. and Leigh, C. (1990) 'The strategic imperative: the role of agencies in the region's development', *Regional Review*, 1.0: 2–3.

Hamblin, P. (2004) 'What's the damage?' *The Guardian, Guardian Society*, 12–13.

Haughton, G. and Counsell, D. (2004) *Regions, Spatial Strategies and Sustainable Development*, London: Routledge.

Leeds Metropolitan University and University of Hull (2003) *Mid-term Evaluation of the Yorkshire and Humber Objective 2 Programme*, Leeds: LMU.

ODPM (2003) *Sustainable Communities Plan*, London: The Stationery Office.

ODPM (2004) *Making it happen: the Northern Way*. http://www.odpm. gov.uk/index.asp?id=11139962 (accessed 19 January 2006).

Owens, S. and Cowell, R. (2002) *Land and Limits: Interpreting Sustainability in the Planning Process*, London: Routledge.

Pearce, D. (1989) 'The Yorkshire and Humberside Regional Planning Council 1965–79', in P. Gardside and M. Hebbert (eds) *British Regionalism 1900–2000*, London: Mansell, 129–41.

Regional Assembly for Yorkshire and Humber (2004) *Advancing Together Towards a Sustainable Region: The Regional Sustainable Framework for Yorkshire and Humberside*, Wakefield: RAYH.

Roberts, P. (1994) 'West Yorkshire in a regional context: retrospect and prospect', in G. Haughton and D. Whitney (eds) *Reinventing a Region: Restructuring in West Yorkshire*, Aldershot: Avebury, 211–31.

Vigar, G. Healey, P. Hull, A. and Davoudi, S. (2000) *Planning, Governance and Spatial Strategy in Britain*, Basingstoke: Macmillan Press.

Yorkshire and Humber Assembly (YHA) (2003) *Building the Benefits: Yorkshire and Humber Regional Sustainable Development Framework Update 2003–2005*, Wakefield: YHA

Yorkshire and Humberside Partnership (n.d., *c*.1993) *Yorkshire and Humberside Regional Strategy: A Partnership for Europe*, Barnsley: Yorkshire and Humberside Partnership.

Yorkshire and Humberside Economic Development Council (1966) *A Review of Yorkshire and Humberside*, London: HMSO.

Yorkshire and Humberside Economic Development Council (1975) *Yorkshire and Humberside Regional Strategy Review 1975: the next ten years*, London: HMSO.

Yorkshire Futures (2003) *Links and Consistencies between Regional Policies*, http://www.york-shirefutures.com/siteassets/documents/YorkshireFutures/8/B/8B00D354-7CCF-44E9-9 91D-0C115025D9ED/final%20report%204.pdf (accessed 19 January 2006).

8 Spatial governance in contested territory

The case of Northern/North of Ireland

Geraint Ellis and William J.V. Neill

Introduction

Northern Ireland was born out of, and continues to suffer from, the consequences of a contested re-territorialisation (or perhaps more appropriately, a *de*-territorialisation) of the UK state. In this context, the devolution project in Northern Ireland and its consequences for spatial governance have been, and will continue to be, driven by a very different set of political objectives from those experienced in Britain. It would be wrong, however, to see Northern Ireland's tortuous path towards devolution (see Figure 8.1) as being profoundly exotic to New Labour's regional project in Britain. Indeed, like other UK regions, Northern Ireland has experienced a new phase of governance since 1997, and its experience of devolution dating back to the 1920s touches many of the policy debates now facing regional governance across Britain. Utmost amongst these issues, have been those related to contested territory and cultural identity. Although experienced far more acutely than in Britain, Northern Ireland may offer some important lessons on how spatial planning should respond to such tensions. Indeed, in a region where even self-ascription is hotly disputed,[1] spatial governance continues to reflect territorial struggle, with the 'peace process' preceding and following the Good Friday Agreement in 1998, representing its most recent phase.

This chapter focuses on these issues by reviewing the history of regional spatial policy in post-partition North of Ireland, with particular emphasis given to policy debates of the last ten years. The argument is made that underlying ethnic tensions, coupled with overly technocratic approaches to planning, have not only shaped the way the region is framed in the territorial and political imagination, but have also thwarted open debate on serious matters of spatial governance (including those related to infrastructure and service provision, economic development policy and island-wide management of natural resources), resulting in a discourse of planning that too often resorts to fudge.

The beginning

The establishment of Northern Ireland by the Government of Ireland Act in 1920 was a process of re/de-territorialisation of the UK state, which satisfied neither Unionists nor Nationalists.[2] Unionists, whose identity and sense of divine territorial right ('God and Ulster') had been shaped by the 'heroic' siege of Derry ('No Surrender'), victory of William of Orange over Catholic King James in 1690 and the sacrifice of a generation in the form of the 36th Ulster Division[3] in the trenches

of the Somme, saw the creation of a Northern Ireland Parliament as a partially successful challenge to an ascendant cultural Irish nationalism from which they felt
alienated (Stewart, 1989). For Nationalists, the very existence of Northern Ireland
was seen as an artificial remnant of an incomplete process of de-colonialisation,
marking the tail end of centuries of British domination of Ireland. Indeed, from a
Nationalist perspective, the Northern Ireland state is often characterised as being
embedded in, and a symptom of, a social relation of subordination (O'Dowd *et al.*,
1980). The understanding of spatial governance amid such contestation cannot
thus be divorced from a deeper appreciation of two cultural identities[4] in conflict
where the meaning of place is constitutive of identity itself.

The symbolic language of territory and identity

The meanings that people attach to place are now recognised as a key concern of
any sensitive approach to spatial governance (Healey, 2001:278; Hillier, 2001:71;
Neill, 2004). Both cultural traditions in Northern Ireland draw emotional suste-
nance from contrasting place visions of territory, which, while becoming gradually
outdated in an increasingly multi-cultural Britain and Ireland, nevertheless,
remain powerful. On the one hand there is a Nationalist cultural tradition that
focuses on a Gaelic vision of Ireland based on pre-Christian folklore, the potato
famine and the 1916 Easter Uprising, as articulated through the pen of W.B. Yeats,
Christy Moore and even the musical *Riverdance*. Although northern Nationalists
have forged their own distinctive post-partition identity, this is a cultural image
that has some resonance (O'Connor, 1993). On the other hand, Unionists have,

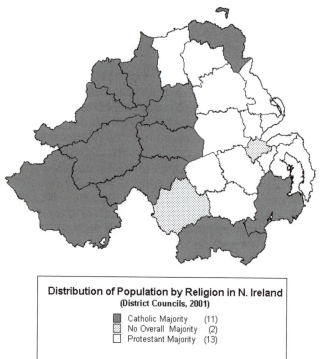

Distribution of Population by Religion in N. Ireland
(District Councils, 2001)

■ Catholic Majority   (11)
▨ No Overall Majority   (2)
☐ Protestant Majority   (13)

Figure 8.1 Northern Ireland district council areas showing predominant religious group

post-partition, expressed a sense of exclusion and alienation towards Irish Nationalism and culture, seeing the South as Gaelic, aligned with the Catholic church (and by implication controlled from Rome) and harbouring territorial claims over the North. This view was summed up in 2002 by David Trimble, leader of the Ulster Unionist party, when he referred to the Republic of Ireland as 'a sectarian, mono-ethnic, mono-cultural state' (*Belfast Telegraph*, 2002).

Yet Unionism itself has never convincingly established its own cultural image as one having legitimate territorial claim. While the Unionist tale is a complex one with roots in centuries of contact and mixing on an archipelago of islands on the edge of Europe, to most of the world, particularly following the coverage of Drumcree or the harrowing pictures from the school blockade in Ardoyne, their cultural image is one of intransigent, even bigoted, settlers who have refused to integrate. Although Republicans have accused Ulster-Scots as being a 'rediscovered identity' as a reaction to the greater cultural fluency of Nationalism, it is one that is gradually being presented with increasing self-confidence as evidenced in the musical *On Eagles Wings*.[5] However, one can argue that as the cultural tradition that has had dominant political control of Northern Ireland for most of the twentieth century, the 'corrosive siege mentality' (Elliott, 2000) of the Unionist psyche can be linked to the weak production of a territorial imagination for the region. This has been overly influenced by notions of the greatness of a British past (Graham, 1994) now forced into re-evaluation by a multicultural Britain (Parekh, 2000) and where 'territory' is appropriated with intensity through the spatial practice of marching. The cultural identity of Unionism, dominant until the 1970s at least, forms one of the ethno-tectonic plates over which spatial management of Northern Ireland has been confronted.

Spatial governance: the early years

Northern Ireland's urban areas have long faced intractable problems arising from ethnic segregation where more than 60 per cent of the population live in areas that have more than 80 per cent of one religion, with Catholic communities tending to be more deprived (O'Reilly and Stevenson, 1998). At a regional level there is a danger of oversimplifying the complex ethnic geography, which exhibits the reality of many enclave communities living in the territory of 'the other' (Murtagh, 2003), yet one bold demographic generalisation stands out. Crude statistical analyses show that since its creation, the east of Northern Ireland has remained predominantly Protestant and the west predominantly Catholic. The River Bann dividing East and West also runs through the mind. Recent population data demonstrates that roughly two thirds of the population west of the Bann is Catholic and over 70 per cent east of the Bann is Protestant. The latter figure rises to 75 per cent if Catholic west Belfast is excluded from the calculation (Compton, 1995). In general demographic terms there are, therefore, two Ulsters, leading some more extreme views to even make a case for a repartition (Kennedy, 1986). While such drastic re/de-territorialisation is not a likely prospect, ethnic demographic asymmetry does cast a shadow over many spatial governance concerns in the North. In the early years of Northern Ireland's existence, it could be argued that the economic viability of the Province depended upon the considerable concentration of Unionist industrial capital in the Belfast area, integrated as it was with the economies of Liverpool and Glasgow, part of the workshop of the world which made the British Empire possible

(O'Dowd *et al.*, 1980: 30) and relying on the West for agricultural produce. In a laissez-faire approach to regional policy the pre-Second World War economic geography of Northern Ireland was 'lopsided'. In 1935 over 60 per cent of all jobs in manufacturing employment were in the Belfast County Borough alone (Wiener, 1980: 29), while the rural west was characterised by intense rural poverty.

Spatial modernisation in the 1960s

In the late 1940s, with almost 80 per cent of the insured workforce concentrated within a 30-mile radius of Belfast, the NI economy still remained heavily specialised around its nineteenth century industrial base of linen, shipbuilding and engineering (O'Dowd *et al.*, 1980: 30–31). The devolved government's post-war policy to modernise the region's economic base was strongly supported by reformist Unionism in the 1960s faced with the pressures of international competition in Northern Ireland's traditional markets. Yet it is against the perceived shortcomings of the economic and spatial modernisation plans of this era that all subsequent spatial policy in Northern Ireland has been judged. The new spatial vision that emerged in the 1960s reflected the Anglo-modernist planning paradigm of the time and drew heavily on the work of 'mainland' planning consultants (Matthew 1963; Wilson, 1965; Travers Morgan, 1969; Building Design Partnership, 1969). The strategy can be summarised around a few key points:

- A development 'stopline' around Belfast to contain the outward sprawl of the city, a policy which has been largely successful (although housing development has continued unabated in the countryside beyond, see below).
- A major programme of transportation investment following the principles of the Buchanan report in Britain (Buchanan, 1963) for a car-based urban future. This involved the building of motorways focused on revamping access to and within the city of Belfast, which again has been largely successful with Northern Ireland having 2.2 times the per capita road length of Britain (DoE NI, 1999).
- Most controversially, the demagnetising of Belfast, assisted by the 'stopline', would be complemented by the designation of other growth centres aimed at more widely spreading international investment. Bolstered by the growth-pole theory of the time and concerned with the service and economic efficiencies of settlement concentration, most designated development nodes, however, remained east of the Bann.

Although presented as the result of the cold economic calculus of the technical reports, Nationalist opinion did criticise this policy approach for underplaying spatial equity – a fact reinforced by other aspects of Unionist spatial governance of the time. For example, the major new growth centre, the new city of Craigavon, was named after the first Prime Minister of Northern Ireland who famously referred to Stormont Buildings in Belfast as 'a Protestant Parliament for a Protestant people'. Furthermore the region's new university was controversially located near the predominantly Protestant town of Coleraine rather than in the second city of Derry/Londonderry which has a Catholic majority, while the motorway network snaked towards the major Unionist towns of Ballymena, Portadown and Enniskillen, but left the majority of the Derry–Belfast link as single carriageway. All these decisions compounded Nationalist feeling that reformist Unionism, whatever battles it had to fight with defenders of the status quo was 'fine sentiment but little substance' (Osborne and Singleton, 1982: 177) and contributed to priming the conflict that was to dominate the region for the next three decades.

Territorial governance under direct rule

The outbreak of 'The Troubles' in the late 1960s and subsequent suspension of the devolved government in Northern Ireland in 1972 was to be accompanied over the next 25 years by Westminster-determined planning policy that struggled to cope with the socio-spatial expression of ethnic identity. Abstract codifications and statistical classifications of space in the 1960s planning documents were to quickly pale before the brute reality of the hastily prepared Orange and Green maps of sectarian space familiar at every 'security force' base in the Province. The hard edge of ethnic spatial management was evident in the all too visible presence of CCTV surveillance, watchtowers, listening antennae, fortified police stations and 'peace' walls. In the military management of ethnic conflict, the suspicion that urban planning decisions involving roads and housing layouts, not to mention ethnic interface or 'peace' wall planning, involved direct, if subtle, military and security considerations, still lingers (Cowan, 1982; Hillyard, 1983; Dawson, 1984). The foremost ethnographic mapper of Northern Ireland's divisions considers that 'such considerations must have impinged' on spatial policy of the time (Boal, 1990: 10). Strategic spatial management during these years of Westminster rule also had a softer side. Foremost in tackling some of the background causes of conflict was the vigorous approach to social housing provision started in the 1970s and pursued with increased zeal in the 1980s by the Northern Ireland Housing Executive (NIHE) even when such programmes were being decimated on the Thatcherite 'mainland' (Brett, 1986). The marshalling of development as a feel-good factor and the promotion of spaces of consumption as 'oases of normality' (NIIS, 1978) was to achieve prominence in the 1980s (Neill, 1993).

Without outright acknowledgement, the basic foundations of strategic planning were subject to radical change during the 1970s, as the *Regional Physical Development Strategy for Northern Ireland* (DoE NI, 1975) put forward for public discussion no less than six spatial options with relative degrees of laissez-faire, intervention, agglomeration and dispersal. A 'district towns' strategy of relative dispersal was eventually adopted as official policy. Acknowledging 'the peculiar circumstances of Northern Ireland' (1975: 32) the principal aim of the planning strategy would be 'to distribute population, development and activities of all kinds in such a way that irrespective of place of residence, everyone will enjoy at least reasonable access to a high level of employment opportunities and to services and facilities of all kinds' (1975: 25–6). In 1978, a relaxation of policy with respect to development in the open countryside added a greater leaning towards settlement diffusion and away from the growth centre thinking of the 1960s. Although this policy draws on a range of issues related to rural development, in the political context of the 1970s this cannot be disengaged from issues of 'territorial equity' which was to become, in fact, the spatial orthodoxy or 'planning doctrine' for Northern Ireland.

While 'territorial equity' is indeed a laudable aim, it has been applied blindly in such a crude and technocratic way that it has left some of the crucial questions over spatial governance unchallenged and more awkwardly, even unasked. To understand why this is so, it is useful to consider the institutional and political context of planning in the region over the last 30 years. It should be acknowledged that Northern Ireland's planners were faced with the arduous task of maintaining a normal regulatory planning function in extremely difficult circumstances. Not only was the legitimacy of any state intervention questioned by a large proportion of the Nationalist community, but 'The Troubles' created a society polarised along sectarian lines, which resulted in specific planning related issues of social and economic deprivation, distorted land markets, blighted space and the need for duplication of many urban services (Boal, 1996). The planning system responded to such difficulties by becoming heavily technocratic and relying on the notion of a unitary 'public interest' as its guiding light. The formal neutrality of planning in Northern Ireland has been widely acknowledged (Benvenisti, 1983; Boal, 1996; Bollens, 1999) and clearly illustrated by Blackman's (1991) analysis of the adoption of the Belfast Urban Area Plan. The culture of technocratic neutrality adopted by Northern Ireland's planners since 1972 was a tactic to promote stability and overcome sectarian discrimination, but has also been perceived as aloof and insensitive to the particularity of place. Bollens has gone so far as to suggest that 'neutrality is associated with unequal outcomes, poor public perception, and ineffective upliftment of Belfast's economically deprived' (1999: 268).

The sole, centralised planning authority, the Northern Ireland Planning Service has been established as a 'Next Steps Agency' within the region's Department of the Environment, staffed by planners who have only a remote relationship with the communities and politicians they serve and who are constantly caught in the tension between being independently minded professionals and small cogs in the civil service bureaucracy. While the centralisation of planning responsibilities has been geared to eradicating sectarianism in policy making the institutional arrangements have, ironically, made it much more difficult to develop a more progressive, culturally inclusive approach to planning (Ellis, 2001). It has been suggested therefore that it is not so much sectarian issues that have dogged successful planning in Northern Ireland but professional dominance, the pretence of a questionable public interest, a narrow technical definition of planning's role and a lack of participation (Ellis, 2000). Indeed, planning in Northern Ireland has even been described as being 'autistic' (Ellis, 2005) in that it can sometimes be seen as being 'morbidly self-absorbed and out of contact with reality'.[6] No one would deny that it has not been easy over the past 30 years to fulfil normal planning functions in Northern Ireland, but while there *may* have once been a case for isolating planning away from the most contentious arenas of political conflict to facilitate basic land use regulation, it seems an inappropriate response to the current political climate of reconciliation.

Regional spatial strategy since the Good Friday Agreement

The Good Friday Agreement of 10 April 1998 has formed the basis for all governance in Northern Ireland and represents a significant re-/de-territorialisation of relations between the UK and Republic of Ireland nation states. Following a referendum, the Irish government has removed the strong territorial claim on the North from its constitution and the British government has revised the 1920 Government of Ireland Act, which instituted partition. More plural forms of Irish/British political association have ensued, most significantly a North–South Ministerial Council on the island of Ireland and a British–Irish Council (Council of the Isles) bringing together no less than eight governmental entities from the British Isles (Westminster government, Republic of Ireland, Scotland, Wales, Northern Ireland, Jersey, Guernsey and the Isle of Man). One commentator optimistically sums up the new constitutional arrangements thus:

> With the ratification of the 1998 Belfast Agreement, both sovereign governments signed away their exclusivist sovereignty claims over Northern Ireland – and came of age. This signalled, I believe, the cessation of the long constitutional battle over the territory of Ulster: that contentious piece of land conjoining and separating the islands of Britain and Ireland for so long. The Siamese twins can now, one hopes, learn to live in real peace, accepting that their adversarial offspring in Northern Ireland may at last be British or Irish or both.
>
> (Kearney, 2003: 28)

More pessimistically, another political analyst criticises the 'consociational' basis of the Belfast Agreement arrangements, novel in comparative politics (Lijphart, 1977) in that it puts an explicit recognition of cultural difference on a constitutional basis. This carries the possibility that despite far reaching equality legislation in Northern Ireland (Ellis, 2000), cultural differences will remain sharp and more integrationist and assimilationist outcomes impeded (Wilson, 2003). While at the time of writing (August 2005) devolved government in Northern Ireland remains suspended in the face of the recent announcement by the IRA that their war is, in effect, over, it remains likely that spatial planning in the Province will continue to proceed within the basic framework of principles embedded in the Good Friday Agreement. Indeed the North's current statutory regional spatial plan (Regional Development Strategy for Northern Ireland 2025, or RDS), considered in the remainder of this chapter, carries the distinction of quasi-constitutional status being explicitly mentioned in the Good Friday Agreement itself, with the British Government pledging to make rapid progress with:

> a new regional development strategy for Northern Ireland, for consideration in due course by the Assembly, tackling the problems of a divided society and social cohesion in urban, rural and border areas, protecting and enhancing the environment, producing new approaches to transport issues, strengthening the physical infrastructure of the region, developing the advantages and resources of rural areas and rejuvenating major urban centres;
>
> (Belfast Agreement, 1998: 19)

The process leading up to the production of the RDS has been documented elsewhere (see Neill and Gordon, 2001; Murray and Greer, 2000; McEldowney and Sterrett, 2000). In essence, the final plan published in late 2001 (DRD, 2001) and endorsed by the Assembly before suspension does not stray far from the doctrinal principles endorsing relative economic and population dispersal embodied in Northern Ireland's previous 1975–1995 Regional Physical Development Strategy. On this occasion, with the watchwords of 'balanced development', alignment is sought with the vanguard of European spatial planning thinking with the importation of the abstract spatial generalities of 'hubs, gateways and corridors' (ESDP, 1999) all servicing a happy 'family of settlements'.

While it is a notable achievement to produce a spatial plan in the context of ongoing ethnic strife, in our view the success of the plan has been unduly exaggerated (e.g. Morrison, 2000; McEldowney *et al.*, 2000, Royal Commission for Environmental Pollution, 2002), particularly given the current lack of evidence of how successful its implementation will be. While a full judgement of the RDS can only reasonably be made with a longer perspective of hindsight, a critical evaluation has been made of how the plan was developed (Neill and Gordon, 2001) and we offer additional reservations here related to how the strategy relates to the broader issues of spatial governance in the region. These concerns include:

- The aspirational rhetoric of the plan;
- A truncation of discussion on possible spatial futures;
- An uncritical acceptance of the nature of rural development;
- Lack of consideration of cross-border governance issues.

Aspirational rhetoric

The preparation of the RDS rode a wave of post-Agreement optimism that has increasingly looked naive in the context of faltering devolution, stalled decommissioning of paramilitary arms, occasional flaring of acute sectarian violence and ongoing lack of agreement on the nature of power sharing. At the time, the future of Northern Ireland did indeed seem as if it was entering a new period of brotherly love, but can now be seen as a step that has taken the region only halfway to paradise. From this context came a strategy that has been described as a 'society-wide hug' (Greer and Murray, 2003: 293) or a 'cotton wool approach to spatial ethnic management' (Neill, 2004: 202) while others have defended the metaphors and vocabulary of the RDS as helping mobilise an alternative spatial imagery that could help displace the geography of old sectarian politics (Healey, 2004). Here Graham correctly points out that 'one of the primary cultural factors demarcating Northern Ireland from its southern neighbour is that it has never evolved that sense of an invented landscape which might help unify its population in a shared communal sense of identity' (Graham, 2003: 265–6). The jury remains out on whether European gateways and corridors will provide that unifying catalyst.

The rhetoric of the RDS may also be critically evaluated from more functional perspectives. In attempting to 'go beyond land use planning' the strategy attempted not just to shape physical development, but also influence the spatial expressions of other forms of environmental policy and socio-economic development. However, aspiration appears to already be cracking under some of its first tests. In the wake of the RDS, the Department of the Environment embarked on an ambitious programme of Area Plans, through which the RDS objectives should have been reflected given its statutory status. However, even before the first of these plans reaches its public enquiry stage the Planning Service has issued blanket rebuttals to objections which are based on RDS policy (such as those related to securing affordable housing) as simply being outside their statutory remit and therefore not to be expressed in a land use plan. On economic issues also, the RDS's vision of 'balanced development' across the region remains disconnected from the prioritisation of public expenditure commitments, leading one local think tank to criticise it for steering clear of controversy rather than tackling key economic questions (NIEC, 1999: 7). A recent research study points to a reason:

the public finance culture of Northern Ireland has to change. There has been a financially irresponsible culture, in the sense that the UK Exchequer would insulate the residents of Northern Ireland from the cost of internal conflict and the resulting inefficiency in both microeconomic and macroeconomic terms. Even when – perhaps especially when – there are generous fiscal equalisation arrangements, there has to be a credible budget constraint.

(Heald, 2003: 59)

Spatial futures?

As already indicated, the previous regional strategy for Northern Ireland did at least try to articulate some alternative spatial scenarios. No such thinking informed the public consultation process associated with the RDS – despite the intervening technological revolution, paradigm shifts in terms of debates over sustainable development and the relationship between development and transport infrastructure. Thus the strategy sought to apply a neutral concept of territorial equity, with little or no debate of the relative merits of settlement concentration and dispersal (e.g. Williams, 2000), the potential for public transport-led development patterns (Transport 2000 NI, 1999) nor how the roll-out of telecommunications infrastructure could inform spatial trade-offs and choices (Blair, 2004).

The contentious nature of 'the rural'

The failure in the development of the strategy to have an informed debate over the fundamental nature of social needs and the distribution of economic development in the region partly stems from an unwillingness to address a number of contentious issues related to the economic, environmental and cultural dimensions of rural development. When the RDS was being drawn up, the agricultural sector dominated by small, family-run farms was beginning to face major uncertainties with collapsing farm incomes, major readjustment in the face of CAP reform and specifically in Northern Ireland potential major environmental restrictions on agricultural practices arising from enforcement of the EU's Water Framework Directive and end to structural funds in 2006. The RDS failure to seriously grapple with these issues has been subject to an extensive critique, particularly on the links between its land use planning objectives and the needs of rural development (Greer and Murray, 2003).

Significantly, the RDS did not challenge the prevailing policy that there would be 'a presumption in favour of planning permission for single new dwellings in the countryside, outside of Greenbelts and Countryside Policy Areas' (DoE NI, 1993: 43). Although this is now being tentatively reviewed in the form of an 'issues paper' for a new Planning Policy Statement, it has not come soon enough for the rural landscape to have been subject to an extraordinary level of dispersed housing, with severe consequences for water quality, car use, viability of many public services and a range of other sustainability objectives. Figures recently released by the Department of Regional Development show that over 5,600 single new dwellings in the Northern Ireland countryside were awarded permission in 2002/3 – nearly *treble* that of England, Scotland and Wales combined (DRD, 2004). While these figures are damning enough, the government's policy response, including that of the RDS, has been described as 'an objective lesson in obfuscation' (National Trust Northern Ireland Planning Commission, 2004, para. 3.5.7). This criticism continues:

The cumulative effect of the discussion of the single dwellings issue in planning documents is a wringing of hands about the unfortunate adverse effects accompanied but no real action to address the issue. Policy is aimed at ducking and weaving around the personal interests of the beneficiaries of current policies. The manner of the wording and implementation of policy is to guarantee that constraint on development is minimal now and, by placing in its own way a quagmire of impediments to change, is clearly intended to remain so. Indeed it is remarkable that so much policy text should be devoted to doing so little. This is bureaucracy with hardly any benefit and rules without any regulation. This is simply not tolerable in 2004.

(2004, para. 3.5.9)

While in Britain there has been some reflection on the relative environmental and economic benefits of urban agglomeration and rural development, including those related to the cross-subsidy towards rural services such as postal delivery and telephone use (Travers, 2000), such issues continue to be taboo in Northern Ireland. One reason why this is so is because Northern Irish rural society continues to be a repository of political power and moreover cultural imagination for both sides of the sectarian divide.

For Unionists, particularly those supporting the more extreme Democratic Unionist Party of Ian Paisley, agricultural policies are seen as being 'very important' (Irwin, 2002) and with dwindling farm incomes, the territorial holdings of the (protestant) farming class have looked increasingly vulnerable. Because of this, the party took up the chair of the agricultural policy committee under devolution. Indeed, so important are rural issues to maintaining the broad-based support for the DUP that one of its key policy papers specifically addresses 'Rural reform and rejuvenation', which describes Ian Paisley as 'the farmer's friend' (DUP, 2003: 10) and incredulously advocates even further relaxation of planning controls in the countryside.

This is not, however, just a Unionist issue, as Nationalism is also imbued with strong rural narratives (Graham, 2003: 269), but for very different reasons. As Greer and Murray point out, 'the public discourse on rural disadvantage as conveyed by its spatial representation has become a powerful metaphor of key socio-economic differentials which have built up over time between the competing traditions in Northern Ireland' (Greer and Murray, 2003: 17). Gerry Adams, the leader of Sinn Fein, now the North's largest nationalist party, argues that the skewing of resources, infrastructure and job investment to the more deprived areas 'west of the Bann and along the border' should be put on a statutory basis (Adams, 2001), which echoes the call for 'rural-proofing' contained in the RDS. A Sinn Fein spokesman has recently associated attempts to restrict single dwellings in the countryside under the rationale of sustainable development as an attempt to kill off a rural way of life that has existed for over 800 years (*Belfast Telegraph*, 2004). Furthermore, those motivated by concerns over rural deprivation also contribute to the 'obfuscation' identified by the Northern Ireland Planning Commission by suggesting that concerted planning efforts aimed towards the Belfast Metropolitan Area Plan are 'relegating the rural to a state of limbo' (Greer and Murray, 2003: 297) while accusing the RDS as offering 'almost primeval antipathy to new single homes in the countryside' (2002: 298).

Whilst spatial equity must remain a policy goal in contexts of ethnic division (Neill and Schwedler, 2001: 209), the fact that fundamental issues related to sustainability, service efficiency and economic development performance are suppressed for fear of exposing tender questions of ethnic geography, illustrates how some questions are too hot to handle in the still tender post-conflict society of Northern Ireland.

Cross border governance

A further dimension where the RDS has failed to live up to its expectations is in facing up to the realities of cross-border governance. Indeed, unique as it is within the UK as having an international land border and sharing a broad ecological niche with a neighbouring sovereign state, the regional strategy makes only the feeblest attempt to place Northern Ireland in its island context. There are, of course, intense Unionist sensitivities about the territorial relationship with the Republic of Ireland, but as with other aspects of the strategy, any engagement with such issues is therefore shunned to achieve the woolly consensus needed for a cross-party buy-in. This has been done even in a context where not only is there some history of cross-border co-operation on a range of environmental matters (see Buick, 2002) but also where all the major parties have signed up to the need for increased co-operation on 'strategies and activities which would contribute to a coherent all-island approach to the achievement of sustainable development' (North–South Ministerial Council, 2000). Indeed, in the only UK region without its own sustainable development strategy, the RDS is seen by some as the next best thing. The opportunities for a more progressive all-island approach to sustainable development have been highlighted by Ellis *et al.* (2004) and include not only resource management issues where the RDS claims some competence (e.g. habitat conservation, waste management, transport, coastal management and tourism) but also issues that are unique within the UK, including cross-border distortions to economic and environmental regulation and a duty to recognise trans-boundary participation rights in environmental decision making (Macrory and Turner, 2002).

Conclusions

The Good Friday Agreement in 1998 marked a critical milestone in the history of Ireland, unlocking an impasse that had plunged Northern Ireland into three decades of bitter ethnic conflict. As most now acknowledge, the Agreement was only the start of another chapter in the resolution of a centuries old territorial dispute, yet unlike the preceding phases, has taken place largely in the absence of violence. The Agreement was also novel in that for the first time, spatial planning was acknowledged as having a crucial and constitutionally recognised role in preparing the region for what was hoped to be an enduring peace. In the optimistic culmination of the 1998 peace talks, the Agreement pledged to deliver a regional development strategy that would, inter alia, tackle 'the problems of a divided society and social cohesion'. Notwithstanding the difficulties of preparing the RDS in this context, we should not patronise the fact that a strategy did actually emerge, nor deceive ourselves that it has the competence to really address these issues. A critical cause of this is the historic and contemporary context of a sectarian society, but also the unwillingness of planners to raise sensitive issues of territory and cultural identity. By pursuing an obsessive line of technocratic neutrality, Northern Ireland's planners do not catalyse a radical regional vision, but bring forth the greyness of the lowest common denominator.

Appendix

1922	Partition of the island of Ireland, establishing the Irish Free State and first Northern Irish Parliament.
1968	Civil Rights movement followed by an upsurge in sectarian violence, ushering what was to become known as 'The Troubles'.
1972	Amidst continuing sectarian conflict, Direct Rule imposed by Edward Heath, and establishes post of Northern Ireland Secretary.
1974	Power sharing NI executive established in January only to be brought down by strike by Protestant workers in March.
1985	Anglo–Irish Agreement establishes number of cross-border initiatives and strongly opposed by Unionist community.
1994	Republican and Loyalist paramilitary groups announce ceasefire.
1995	IRA ceasefire called off, followed by bomb attacks at Canary Wharf and Manchester.
1997	IRA calls another ceasefire.
1998	Good Friday Agreement followed by its overwhelming endorsement in referenda either side of Irish Border. The peace deal includes paramilitary commitments on decommissioning, removal of territorial claims of the North from the Republic of Ireland's constitution and devolution to a consociational executive and assembly. Assembly first meets on July 1st and elects David Trimble as First Minister, Seamus Mallon as Deputy First Minister.
1999	Devolution to power sharing NI Executive in December.
2000	Continuing dispute over IRA decommissioning results in suspension of NI Executive and Assembly and imposition of direct rule once again in February. Progress on decommissioning results in devolution again to NI Assembly and Executive in May.
2001	Power sharing government suspended for one day to buy time for further negotiations centred on decommissioning. Negotiations fail and devolution suspended again in September, only to be re-established in October.
2002	Continuing Unionist dissatisfaction with IRA activity results in the fourth suspension of devolution and return to Direct rule, pending further negotiations.
2003	NI Assembly elections return Sinn Fein and Ian Paisley's Democratic Unionist Party (DUP) as largest parties, DUP refuses to enter power-sharing without disbanding of IRA.
2004	In September political talks at Leeds Castle come close but fail to produce breakthrough deal. Northern Bank robbery in Belfast in December and ongoing concerns regarding criminality and IRA disbanding continues to intensify and remains a key blockage to a lasting settlement.

Figure 8A.1 Northern Ireland devolution: a synopsis

Notes

1   'Northern Ireland', 'North of Ireland', 'the Province', 'Ulster', 'the Six Counties', are various attempts, with differing degrees of subtlety, to claim territory by naming. A variety of these terms will be applied in this chapter, as will the abbreviation 'NI'.

2   Both Unionism and Nationalism span a wide range of political opinion and both terms are used here to represent a broad spectrum of beliefs. More extreme versions of these ideologies are sometimes referred to as Republicanism and Loyalism respectively.

3   This was also known as the Ulster Volunteer Force (UVF), from which a present Loyalist paramilitary group takes its name.

4   Even here, however, there is contestation, with Unionist identity seen in some Republican perspectives as inauthentic, maintained by a continued colonial prop (Adams, 1986).

5   This is a musical about the Ulster-Scots frontiersmen who contributed to the founding of the USA, which toured the States in May 2004 before a run in Belfast.

6   This is the description of the range of language and communication disorders associated with autism, as noted in the *Oxford English Dictionary*.

References

Adams, G. (1986) *The Politics of Irish Freedom*, Dingle, Co. Kerry: Brandon Books.

Adams, G. (2001) 'The Equality Agenda', *Belfast Telegraph*, January 17.

Belfast Agreement (1998) Belfast, p. 19.

Belfast Telegraph (2002) 'Trimble in new swipe at Republic', March 14.

Belfast Telegraph (2004) 'Plan to kill off countryside', September 27.

Benvenisti, M. (1983) *Jerusalem: Study of a Polarised Community*, Jerusalem: West Bank Data Project.

Blackman, T. (1991) *Planning Belfast*, Aldershot: Avebury.

Blair, N. (2004) 'Telecommunication infrastructure in N. Ireland. Unpublished PhD thesis', Queen's University Belfast, School of Environmental Planning.

Boal, F.W. (1990) 'Belfast: hindsight or foresight – planning in an unstable environment.' Chapter 1 in P. Doherty (ed.) *Geographical Perspectives on the [Belfast Region*, Geographical Society of Ireland, Special Publications 5: 4–14.

Boal, F.W. (1996) 'Integration and division: sharing and segregation in Belfast', *Planning Practice and Research*, 11(2): 151–8.

Bollens, S.A. (1999) *Urban Peace-Building in Divided Societies: Belfast and Johannesburg*, Colorado: Westview Press.

Brett, C.E.B. (1986) *Housing a Divided Community*, Dublin: Institute of Public Administration.

Buchanan, C. (1963) *Traffic in Towns*, London: HMSO.

Buick, J. (2002) *Crossing the Border: A Regional Approach to Environmental Management*, Report 2002.2, Lund: International Institute for Industrial Environmental Economics.

Building Design Partnership (1969) Belfast Urban Area Plan, Belfast: BDP.

Compton, P. (1995) Demographic Review Northern Ireland, Northern Ireland Economic Council, Belfast.

Cowan, C. (1982) 'Belfast's hidden planners', *Town and Country Planning*, 51(6): 163–7.

Dawson, G. (1984) 'Defensive planning in Belfast', *Irish Geography*, 17: 22–41.

Democratic Unionist Party (2003) *Rural Reform and Rejuvenation: Policy Paper*, Belfast: DUP.

Department of the Environment (NI) (1975) *Regional Pysical Development Strategy, 1975–1995*, Belfast: DoENI.

Department of the Environment (NI) (1993) *A Planning Strategy for Rural Northern Ireland*, Belfast: DoENI.

Department of the Environment (NI) (1999) *Transport Statistics, 1998–9*, Belfast: DoENI.

Department of Regional Development (DRD) (2001) *Shaping Our Future: Regional Development Strategy for N. Ireland*, Belfast: DRD.

Department of Regional Development (DRD) (2004) *Sustainable Development In The Countryside (PPS14): Issues Paper*, Belfast: DRD.

Elliott, M. (2000) *The Catholics of Ulster*, London: Penguin Press.

Ellis, G. (2000) 'Addressing inequality: planning in Northern Ireland', *International Planning Studies*, 5(3): 345–64.

Ellis. G. (2001) 'The difference context makes: planning and ethnic minorities in Northern Ireland', *European Planning Studies*, 9(3): 339–57.

Ellis, G. (2005) 'City of the black stuff: Belfast and the autism of planning', unpublished paper.

Ellis, G., Motherway, B., William J.V., Neill, W.J.V. and Hand, U. (2004) *Towards a Green Isle? Local sustainable development on the Island of Ireland*, Armagh: Centre for Cross Border Studies.

European Spatial Development Perspective (1999) European Commission, Brussels.

Graham, B.J. (1994) 'Heritage conservation and revisionist nationalism in Ireland', Chapter 8 in G.J. Ashworth and P.J. Larkham (eds), *Building a New Heritage: Tourism, Culture and Identity in the New Europe*, London: Routledge.

Graham, B.J. (2003) 'Interpreting the rural in Northern Ireland: place, heritage and history', Chapter 12 in J. Greer and M. Murray (eds) *Rural Planning and Development in Northern Ireland*, Dublin: Institute of Public Administration.

Greer, J. and Murray, M. (2003) 'Rethinking rural planning and development in N. Ireland', Chapter 13 in J. Greer and M. Murray (eds) *Interpreting the Rural in N. Ireland*, Dublin: Institute of Public Administration.

Heald, D. (2003) *Funding the N. Ireland Assembly: Assessing the options*, Belfast: Northern Ireland Economic Council, Research Monograph, 10.

Healey, P. (2001) 'Towards a more place-focused planning system in Britain', Chapter 12 in A. Mandanipour, P. Healey and A. Hull (eds) *The Governance of Place: Space and Planning Processes*, Aldershot: Ashgate.

Healey, P. (2004) 'The treatment of space and place in the new strategic spatial planning in Europe', *International Journal of Urban and Regional Research*, 28(1): 45–67.

Hillier, J. (2001) 'Imagined value: the poetics and politics of place', Chapter 4 in A. Mandanipour, P. Healey and A. Hull (eds) *The Governance of Place: Space and Planning Processes*, Aldershot: Ashgate.

Hillyard, P. (1983) 'Law and order', in J. Darby (ed.) *Northern Ireland: The Background to the Conflict*, Belfast: Appletree Press.

Irwin, C. (2002) *The People's Peace Process in Northern Ireland*, London: Palgrave.

Kearney, R. (2003) 'Ireland and Britain – towards a Council of the Isles', Chapter 1 in R. Savage Jr. (ed.) *Ireland in the New Century*, Dublin: Four Courts Press.

Kennedy, L. (1986) *Two Ulsters: A Case for Repartition*, Belfast: Queen's University Belfast Press.

Lijphart, A. (1977) *Democracy in Plural Societies*, New Haven: Yale University Press.

Macrory, R. and Turner, S. (2002) 'Participatory rights, transboundary environmental governance and EC law', *Common Market Law Review*, 39: 489–522.

Matthew, R.H. (1963) *Belfast: Regional Survey and Plan*, Belfast: HMSO.

McEldowney, M., Sterrett, K., Gaffikin, F. and Morrissey, M. (2000) *Shaping our Future: Public Voices – a new approach to public participation – the Regional Strategic Framework for Northern Ireland*. Paper presented at Planning Research 2000 conference, London School of Economics, March.

McEldowney, J.M. and Sterrett, K. (2001) 'Shaping a regional vision: the case of Northern Ireland', *Local Economy*, 16(1): 38–49.

Morrison, B. (2000) 'Staying ahead of the game on quality front', Special Irish Supplement, *Planning*, June 8–9.

Murray, M. and Greer, J. (2000) *The Northern Ireland Regional Strategic Framework and its Public Examination Process: Towards a New Model of Participatory Planning?* Belfast: Rural Innovation and Research Partnership.

Murtagh, B. (2003) 'Dealing with the consequences of a divided society in troubled communities', Chapter 7 in J. Greer and M. Murray (eds) *Rural Planning and Development in Northern Ireland*, Dublin: Institute of Public Administration.

National Trust Northern Ireland Planning Commission (2004) *A Sense of Place: Planning for the Future in Northern Ireland*, Belfast: National Trust.

Neill, W.J.V. (1993) 'Physical Planning and Image Enhancement: recent developments in Belfast', *International Journal of Urban and Regional Research*, 17(4): 595–609.

Neill, W.J.V. (2004) *Urban Planning and Cultural Identity*, London: Routledge.

Neill, W.J.V. and Gordon, M. (2001) 'Shaping our Future? The Regional Strategic Framework for N. Ireland', *Planning Theory and Practice*, 2(1): 31–52.

Neill, W.J.V. and Schwedler, H.-U. (2001) 'Planning with an ethic of cultural inclusion', Chapter 14 in W.J.V. Neill and H.-U. Schwedler (eds) *Urban Planning and Cultural Inclusion: Lessons from Belfast and Berlin*, London: Palgrave.

North–South Ministerial Council (2000) *Joint Communiqué*, Environment Sector, Interpoint, Belfast, June 28.

Northern Ireland Economic Council (1999) *A Response by the N. Ireland Economic Council to Shaping our Future*, Belfast: NIEC Advice and Comment Series.

Northern Ireland Information Service (1978) '9 point package to spell the rebirth of Belfast', Belfast: Northern Ireland Information Service.

O'Connor, F. (1993) *In Search of a State: Catholics in Northern Ireland*, Belfast: Blackstaff Press.

O'Dowd, L., Rolston, B. and Tomlinson, M. (1980) *Northern Ireland: Between Civil Rights and Civil War*, London: CSE Books.

O'Reilly, D. and Stevenson, M. (1998) 'The two communities in Northern Ireland: deprivation and ill health', *Journal of Public Health Medicine*, 20(2): 161–8.

Osborne, R. and Singleton, D. (1982) 'Political processes and behaviour', Chapter 7 in F.W. Boal, and N.J. Douglass (eds) *Integration and Division: Geographical Perspectives on the Northern Ireland Problem*, London: Academic Press.

Parekh, B. (2000) *The Future of Multi Ethnic Britain*, London: Profile Books.

Regional Physical Development Strategy for Northern Ireland (DoE NI, 1975), Belfast.

Royal Commission for Environmental Pollution (2002) *Twenty-third report: Environmental Planning* (Cm 5459), London: The Stationery Office.

Stewart, A.T.Q. (1989) *The Narrow Ground: Aspects of Ulster 1609–1969*, Belfast: Blackstaff Press.

Transport 2000 (NI) (1999) *Response to Shaping our Future*, T2000 NI, Belfast.

Travers, Morgan and Partners (1969) *Belfast Transportation Plan*, London.

Travers, T. (2000) 'Town and Country', *The Guardian*, December 12.

Wiener, R. (1980) 'The Rape and Plunder of the Shankill', Community Action: the Belfast Experience, Belfast: Farset Press.

Williams, K. (2000) *Achieving sustainable urban form*, London: E & FN Spon.

Wilson, R. (2003) *Northern Ireland: What's gone wrong? Working Paper 1*, Institute of Governance, Queen's University Belfast.

Wilson, T. (1965) *Northern Ireland Economic Plan*, Report to Government of Northern Ireland, Belfast.

9 Redefining 'the space that is Wales'[1]

Place, planning and the Wales Spatial Plan

Neil Harris and Alan Hooper

Introduction

The purpose of this chapter is to introduce readers to the preparation of the Wales Spatial Plan. The Plan is a recent exercise in devising a spatial planning instrument for Wales and a number of important changes in both the status and purpose of the plan have taken place in the four years in which the plan has been in preparation. In the following account, we provide a brief introduction to the establishment of the National Assembly for Wales as a new, devolved democratic institution. We highlight how the establishment of the Assembly, and the wider project of political devolution, have been important factors in setting the preconditions for the introduction of a national spatial planning framework. We then continue to identify some of the potential functions of the plan – which, during the process of writing the present chapter, has progressed from the stage of a consultation draft document to a final, approved version – and outline the contents of the plan. Following this, we set about reporting on how various constituencies have responded positively to the principle of a national spatial planning instrument, yet have found the details of the plan lacking in several respects. We address these issues within the context of the plan as a *potential* instrument in supporting a territorially based project of creating and reinforcing the spatial dimensions of Wales and even Welsh identity. In doing so, we highlight how recent sustainability appraisal of the plan suggests that important elements of this – including issues such as culture and language – were underdeveloped in the consultation draft and have not been developed significantly in the final version of the plan. The chapter argues that, far from realising its potential as a vehicle for developing spatial identity and promoting territorial cohesion – a feature of various other spatial planning instruments, including the European Spatial Development Perspective – the plan at present is best described as a practical mechanism for sectoral policy integration. The plan describes spatial planning as 'an important tool for reconciling [the] different policy and activity strands which impact upon our various geographic areas' (Welsh Assembly Government, 2004a: 4). So, while the plan does engage in the activity of 'place-making' – here understood as setting out the role and function of different places – the objective of policy integration dominates the plan. This in itself raises significant challenges, for the plan relates closely to the Assembly's strategies in respect of economic development and sustainable development. As a vehicle for policy integration, the plan incorporates a series of potential tensions between different policy areas, some of which will only surface and be capable of resolution as part of the implementation of the plan. Nevertheless, and despite the adoption of a less politicised approach to spatial planning, the Wales Spatial Plan cannot be described as a particularly technical exercise in spatial planning, one that is embedded in an extensive and detailed understanding of spatial patterns and trends. The approach that has been adopted in Wales to spatial planning lies somewhere *between* the political and technical.

Devolution and the establishment of the National Assembly for Wales

The recent history of devolution to Wales is a chequered one. In two separate referenda within the space of twenty years, the people of Wales have repeatedly failed to produce a resounding vote of support for the establishment of a devolved, democratically elected institution. The earlier referendum on the establishment of a Welsh Assembly in 1979 was defeated. Many regard this as unfortunate and an outcome that failed to recognise the political developments that would affect Wales for much of the next two decades. The period after the defeated referendum was marked by successive Conservative Governments at Westminster at odds with the traditional voting allegiances of the majority of Wales. The tensions between Westminster and Wales were repeatedly played out until the election of a Labour government in 1997. The New Labour Government moved quickly on its proposals for Welsh devolution that it made the subject of a referendum in 1997. The referendum produced a result in favour of establishing an Assembly by the very narrowest of margins and with a relatively low voter turnout. The outcome of the referendum has often been referred to as evidence of the continuingly cautious attitude of the Welsh population to devolved government. The Assembly has, as a consequence, adopted an agenda that deliberately aims to demonstrate that devolution can make a real difference to the development and delivery of public policy and public services within Wales.

The legislation providing for devolution in Wales was enacted in 1998 and allowed for the establishment of the National Assembly for Wales in 1999. The creation of the Assembly was regarded as democratising the existing Welsh Office machinery that had been established in 1964 as part of a ministerial portfolio at Westminster (Morgan and Mungham, 2000; Osmond, 1998). The Act transferred the various powers of the Secretary of State for Wales to the National Assembly for Wales in accordance with a model of executive devolution. The powers devolved to the Assembly relate primarily to secondary legislation and policy-making functions. It has no capacity to enact primary legislation and does not have tax-varying powers unlike the Scottish Parliament. This apparent constraint has usually been identified as a weakness of Welsh devolution, and certain important policy and decision-making spheres remain outside of the scope of the Assembly's immediate powers. The nature of the devolution settlement also leads occasionally to some confusion on the locus of power and responsibility in relation to particular issues. Nevertheless, various commentators have pointed out that, in both planning and other functions, secondary legislation and policy-making provide significant scope for manoeuvre and pursuing an alternative, Welsh agenda (Bosworth and Shellens, 1999). This agenda and programme of reform has often been presented as one in which a 'distinctively Welsh' dimension to policies can be developed. The Wales Spatial Plan is no different and states clearly that, 'Devolution has given us the opportunity to shape distinctively Welsh answers to Welsh questions' (Welsh Assembly Government, 2004a: 4). The Assembly has also added to the considerable institutional capacity in economic development, planning and other related policy areas that has existed for some time in Wales, including the Welsh Development Agency and a variety of other all-Wales organisations. The early devolution settlement has now been reviewed through the work of the Richard Commission (2004) that was established in 2002 to consider the powers and electoral arrangements of the National Assembly for Wales. It reported that the Assembly had been able to fulfil its various functions within its existing powers and capacities, but was increasingly pressing on the limits to those powers. The report has been interpreted by many as a strong case for an increase in the capacities of the Assembly, including in primary legislation and tax-variation. It is often remarked that devolution is a process and not an event, a view confirmed by the early consideration of the Assembly's powers.

Devolution and spatial planning

One of the most significant effects of the establishment of the Welsh Assembly Government as far as spatial planning is concerned is that it has politicised the Welsh territory, referred to in the consultation draft version of the Wales Spatial Plan as 'the "space that is Wales"' (WAG, 2003: 5). It has transformed a physical space into an explicitly political space. The approved version of the plan even hints at the practice of 'the management of resources *and territory*' (2004a: 4, emphasis added), a distinct sideways glance at the similar practice embedded in the European Spatial Development Perspective which promoted a European territory. This politicisation of space becomes particularly evident as Wales is not functionally coherent as a region[2] – its different parts are not well connected in terms of infrastructure and, in some ways, culture. This is especially the case between north and south Wales, significant parts of which are better connected functionally and indeed culturally to adjacent parts of England than to each other. This situation is popularly described as the tensions arising between Wales' north–south political axis and its east–west functional axes. The plan also contains a mapping of the various sub-regional responsibilities and arrangements within Wales, which further highlights the practical difficulties in defining and elaborating upon bounded spaces for the purposes of policy formation. This apparent disjuncture between Wales as a functional or physical space and Wales as a political arena helps to explain both some of the content of the Wales Spatial Plan and the debates that have helped to frame it. The attempt in the consultation draft plan to introduce zones and areas, albeit ones loosely defined, that cut across differing functional, political and cultural boundaries proved especially problematic and have since been deleted from the revised version. Nevertheless, the further politicisation of the Welsh territory has also been instrumental in helping the planning system in Wales to move beyond a situation in which it has been described as 'poised between a distinctive socio-political formation and the centralized legal and bureaucratic apparatus of the British state' (Tewdwr-Jones, 1997: 54), enabling it to move towards a form that is increasingly different from that it has historically shared with England.

Devolution to Wales has been coincident with academic and professional attempts to redefine the activity of planning to encompass a wider and more strategic series of activities. This has been defined as the activity of 'spatial planning', definitions of which are provided elsewhere in this edited volume (although see also Tewdwr-Jones, 2004; Harris and Hooper, 2004). Devolution and the establishment of the National Assembly for Wales set out some important preconditions for the emergence and development of spatial planning and the Assembly has been among the first of the devolved administrations to take forward a national spatial planning instrument. Indeed, the Assembly (2004a: 4) describes itself as 'one of the leading protagonists in the British Isles' within the field of spatial planning. First among these preconditions is the promotion of renewed thinking about Wales as a political rather than simply administrative entity, as a space that is addressed in a particular way as part of a political project rather than an administratively expedient exercise. This has led in some quarters to renewed emphasis on issues focused on internal accessibility, such as the extent to which different parts of Wales are more or less accessible to each other and, within the context of polycentric and balanced development, also to other areas such as Ireland, England and beyond. Second, the period immediately after the creation of the Assembly and since has witnessed an acceleration in the number of policy initiatives and strategy documents that is characteristic following the establishment of a new political institution. This has led to a profusion of different policies and precipitated the need for some joined-up thinking and policy integration. The activity of spatial planning is being promoted by a range of professional interests as a vehicle for integrating various policy areas on a territorial basis. Third, the Assembly has been keen to demonstrate as a new political institution that its policy agenda is somehow different and distinctive from earlier efforts under the Welsh Office, and that policies and strategies are not based on the same kind of material that is simply produced closer to home. Devolution has therefore been an important factor in cultivating the practice of spatial planning at the national level. The Assembly's initiative in spatial planning has been progressed since 2000 and, following publication of the draft Wales Spatial Plan in 2003, the approved version of the Plan appeared in November 2004.

The Wales Spatial Plan

The Assembly made a public commitment in its first strategic plan to the preparation of a national spatial planning framework. The purpose of the proposed framework was to establish a context for sustainable development and environmental quality. The framework was initially conceived of as a document closely related to the land-use planning system although its political salience has increased dramatically in the period since that first commitment. The framework has enjoyed an increased profile at Cabinet level in the Welsh Assembly Government and is now regarded, both within the Assembly itself and a number of its sponsored bodies, as one of the Assembly's most significant policy documents. The plan makes clear its own status and position within the hierarchy of Government documents – it 'forms one of the high-level strategic guidance "building blocks" of the Welsh Assembly Government' (2004a: 5). The profile now afforded to the Wales Spatial Plan has surprised and delighted even the most fervent advocates of spatial planning. The preparation of a national spatial planning framework was initiated as a voluntary exercise although the Assembly is now under a legislative duty to prepare a Wales Spatial Plan following the enactment of the Planning and Compulsory Purchase Act 2004. This recent legislative duty does not, however, give any special status to the plan in decision-making, other than the general weight that is usually attached to statutory documents. The plan therefore relies on mechanisms other than formal, statutory status for its implementation. The extensive consultation process and the engagement with stakeholders has become a characteristic feature of the plan, with the 'bottom-up' approach to preparation intended to foster support for the plan and facilitate its implementation.

The early commitment of the Assembly to prepare a national spatial planning framework was taken forward in a series of related research exercises commissioned by its Planning Division. Two early items of research confirmed the significance of spatial planning as part of the division's work. The first of these identified spatial planning as one of three recommended themes for inclusion in the Assembly's planning research programme. This is significant, for it has ensured that spatial planning has had research resources dedicated to it in order to take forward the preparation of the Wales Spatial Plan. The second started to address the spatial dimensions of some of the Assembly's corporate strategies. The concept of spatial planning was evidently becoming more firmly embedded in the Assembly's thinking. Consequently, further work of a comparative nature was commissioned to inform the Assembly's selection of an approach to or methodology for preparing its national spatial planning framework. The research identified a series of different potential functions of such a framework:

- To establish a spatial context for social, economic and environmental activity in Wales;
- To act as a strategic framework for investment, resource allocation and development decisions;
- To explain the differential impact of policies across Wales and address the compatibility of different sectoral policies; and
- To identify and express the character of different functional areas across Wales.

The Wales Spatial Plan reflects each of these functions to a differing degree. However, the emphasis in the published document is on the last two of these. The first of the functions recommended above was for the framework to establish a spatial context for social, economic and environmental activity in Wales. Its purpose would be to ensure that policy development had sufficient regard to European and cross-border issues with adjacent regions, and that policies did not become too domestic or insular. The plan, published in draft form and subsequently published in revised form since the conclusion of that research, has not emphasised this particular function and has rather neglected this dimension. The European context, for example, is dealt with principally by means of an appendix, and this is primary descriptive of European-scale documents. It does not, as one might expect that it should, translate various supra-national spatial policies into the Welsh context and 'position' Wales within Europe. It is also instructive that the plan contains no 'context' maps and relies principally on Wales-outline mapping. The second potential function of the framework is for it to act as a strategic framework for investment, resource allocation and development decisions. The plan in its present form is considered to be insufficiently detailed or connected to funding programmes for it to act in such a way. It does not presently provide sufficient confidence for the private sector in making investment decisions and similarly does not provide a robust context for major public investment decisions. Nevertheless, the plan is allocated a specific role in acting as a framework for future programmes as part of the European Union Structural Funds. The third potential function of the framework was outlined as its ability to explain the differential impact of Assembly policies across Wales and address the compatibility of different policies across different sectors. This does feature as one of the principal roles of the published Wales Spatial Plan and it is its potential as a policy integration tool that explains some of its profile across and within the Assembly. The fourth and final identified function is to identify and express the character of different functional areas across Wales. This is very clearly one of the functions of the published plan, identifiable by its inclusion of various geographic sub-areas of Wales for which it highlights a series of propositions. For example, the plan defines a vision for south east Wales – which it portrays as 'The Capital Network' – as 'An innovative skilled area offering a high quality of life – international yet distinctively Welsh'. It is an area that is to be integrated functionally and better networked to both raise international competitiveness and reduce disparities within the area. However, several key agencies operating throughout Wales maintain that the functional areas defined in the plan lack sufficient clarity to be accepted as genuine, functional areas that are interconnected in important and intricate ways. Noticeably absent from this list of four potential functions of the spatial plan is the possibility of it acting as an instrument to assist with cultivating a sense of national identity or enhancing territorial cohesion, a function which is elaborated upon more fully below and identified as being underdeveloped in the published version of the plan. Nevertheless, the plan does promote the theme of respecting distinctiveness, highlighting how 'A cohesive identity which sustains and celebrates what is distinctive about Wales ... is central to promoting Wales to the world' (WAG, 2004a: 33). Yet even here, identity is reduced simply to about being recognised in the wider world.

The Welsh Assembly Government published 'People, Places, Futures – The Wales Spatial Plan' as a consultation draft in 2003 against a background of high expectations. The consultation version was, according to the responsible minister, designed to act 'as a vehicle to get people thinking about the future of Wales'. The consultation draft was published a year after it had been intended to publish the first final version of the plan. This delay can be attributed to a number of factors, including the increased political profile that the plan enjoyed in its later preparation stages and the Assembly's open and inclusive approach to earlier consultation stages. The Assembly has prepared its Wales Spatial Plan in an open and inclusive manner that has permitted consultation and engagement at a number of different stages. This open and inclusive mode of working is characteristic of the Assembly as a whole and is commended in the Richard Commission's report. The consultation draft of the plan was very much a document designed to facilitate consultation and engagement and the final version of the plan was expected to be more detailed and robust as a consequence of consultation responses and further work undertaken by the Assembly. The final version that appeared in November 2004 has expanded on the consultation draft, although not significantly so, and it appears in a form and character that is very similar to the earlier draft. Therefore, a number of deficiencies of the plan have not, as yet, been addressed and remain valid, including those related to issues of culture and identity.

An outline of the Wales Spatial Plan

Spatial planning is defined simply in the Wales Spatial Plan as 'the considera-tion of what can and should happen where' (WAG, 2004a: 5). This simplistic definition obscures some of the complexities of what the document does in fact aim to achieve. Like many spatial planning instruments, the Wales Spatial Plan addresses a range of different policy sectors ranging from environmental protec-tion, waste, transport and economic development to health, service provision and multiple deprivation. The Wales Spatial Plan itself is a relatively short doc-ument at some 76 pages in length. It is an example of an objectives-led spatial planning instrument and its content is based around a vision, accompanied by a series of five guiding themes for which a number of objectives and actions are stated. Despite some perceptions to the contrary among politicians, the plan is not a 'national blueprint for Wales', but is rather a strategic plan. The spatial vision is also presented diagrammatically at an all-Wales level as a schematic map that identifies six areas with socio-economic hubs, international and inter-regional as well as regional links, and a network of key centres or settlements. The plan provides a summary account of various trends, related to each of the stated objectives, that is presented through selected data, some of which is rep-resented in the form of map-based figures. This does not provide an extensive analysis of available data and no formal analysis of datasets is publicly available. The data is simply sufficient to describe key trends and issues as they affect Wales and its different areas. The provision of an overall picture using key data results in there being little explanation as to why the plan defines the six geo-graphic sub-areas (or 'areas with socio-economic hubs') in the way that it does, or why this has been reduced from the eight geographic sub-areas identified in the consultation draft. Each of the identified areas has a particular vision to guide future collaborative work, together with an overall strategy, a series of propositions, and a twin set of actions comprising those at the local level and those at the national level. The plan is completed with a section on implemen-tation and monitoring and a small number of appendices covering demographics, the European context and existing arrangements for regional collaboration.

Responses to the Wales Spatial Plan

The activity of spatial planning is recognised as being a new one that is being defined partly through the various plans, strategies and programmes of those organ-isations charged with preparing them, either in the devolved administrations, the English regions or at local authority level. The consultation exercises surrounding the Wales Spatial Plan provide an interesting opportunity to identify the reactions of stakeholders and interest organisations to the introduction of a national spatial planning instrument and how it has been received. It is clear that the Assembly's initiative in taking forward a spatial planning exercise has been very widely wel-comed in professional and political circles and is regarded by many as innovative in its approach. Indeed, the Assembly reports that the plan is one of its most requested documents issued for public consultation. Spatial planning has evidently struck a chord both within the Assembly and with the wide range of public, private and vol-untary organisations throughout Wales. In particular, the Wales Spatial Plan has

Box 9.1 Outline of the Wales Spatial Plan's key elements

The National Framework

- Vision: 'We will sustain our communities by tackling the challenges presented by population and economic change; we will grow in ways which will increase our competitiveness while spreading prosperity to less well-off areas and reducing negative environmental impacts; we will enhance our natural and built environment for its own sake and for what it contributes to our well-being; *and we will sustain our distinctive identity.*' (emphasis added)
- Building Sustainable Communities
- Promoting a Sustainable Economy
- Valuing our Environment
- Achieving Sustainable Accessibility
- Respecting Distinctiveness
- Working With Our Immediate Neighbours

Areas of Wales and their visions

- *North West Wales – Eryri a Môn.* 'A high-quality natural and physical environment supporting a cultural and knowledge-based economy that will help the area to maintain its distinctive character, retain and attract back young people and sustain the Welsh language'.
- *North East Wales – Border and Coast.* 'An area harnessing the economic drivers on both sides of the border, reducing inequalities and improving the quality of its natural and physical assets.'
- *Central Wales.* 'High-quality living in smaller-scale settlements within a superb environment, providing dynamic models of rural sustainable development, moving all sectors to higher value added activities.'
- *South East – The Capital Network.* 'An innovative skilled area offering a high quality of life – international yet distinctively Welsh. It will compete internationally by increasing its global visibility through stronger links between the Valleys and the coast and with the UK and Europe, helping to spread prosperity within the area and benefiting other parts of Wales.'
- *Swansea Bay – Waterfront and Western Valleys.* 'An area of planned sustainable growth and environmental improvement, realising its potential, supported by integrated transport within the area and externally and spreading prosperity to support the revitalisation of West Wales.'
- *Pembrokeshire – The Haven.* 'Strong communities supported by a sustainable economy based on the area's unique environment, maritime access and tourism opportunities.'

been eagerly anticipated in land-use planning circles as a possible solution to the diminished strategic planning capacity that has existed in Wales since local government reorganisation in the mid-1990s (Harris and Tewdwr-Jones, 1995; Tewdwr-Jones, 1998). However, the Wales Spatial Plan is intended to have wide relevance across a significant number of different policy fields and the connection to the planning system has not been made explicit within the document. This is despite the statutory requirement upon the Assembly to produce the Wales Spatial

Plan being contained within recent planning legislation. While the plan states that it will be of relevance in the preparation of development plans at the local level and in individual planning decisions, it fails to elaborate upon its role in relation to the land-use planning system. It appears that the authors of the plan have been reluctant to spell out in detail how the Wales Spatial Plan will be relevant to different policy sectors. Indeed, to do so would risk constraining the potential of the plan in the various spheres to which it *could* be relevant, and so the linkages between the plan and different fields are to be negotiated rather than prescribed.

One of the recognised deficiencies of the consultation document was a lack of specificity on the role and status of the document, particularly with respect to implementation of its proposals and recommended actions (WAG, 2004b). Nevertheless, the vision, objectives and values expressed in the consultation document were widely supported by many of the respondents to the consultation exercise conducted for the Assembly. One of the most contentious elements of the consultation draft of the plan and that attracted many negative comments was the use of zones that identified different parts of Wales that faced common or similar spatial issues and challenges. The way in which the zones were presented in the consultation draft was potentially confusing and it quickly became clear that the concept of zones needed to be reconsidered in preparing the final version of the plan. This issue is related to another in the consultation responses on the definition of boundaries for geographic sub-areas referred to in the plan. The concern here appears to be that the defined areas are based on functional interrelationships and do not follow the existing administrative boundaries of, say, local government and various agencies that operate both at the all-Wales level and sub-regionally. There is concern in some quarters that this may hinder the effective implementation and delivery of the objectives contained within the Wales Spatial Plan. On the other hand, the elaboration in the plan of strategic issues and common difficulties that transcend local government boundaries may be read as a statement that the present administrative structures and boundaries of local government are poorly matched to the issues that need to be addressed. Consequently, the zones as they appeared in the draft version of the plan were the subject of extensive comments and have been simplified and modified in the approved version.

Sustainability appraisal of the Wales Spatial Plan

The Assembly has a legislative duty to promote sustainable development across all of its activities. It is not surprising therefore to find that sustainable development is central to the Wales Spatial Plan and this has been widely welcomed, particularly among environmental interest groups. Nevertheless, the initial intention that the plan should provide 'a context for sustainable development and environmental quality' has to some extent been departed from and added to. Some concerns have been expressed that certain elements of the plan in fact exhibit a distinct socio-economic focus and fail to give adequate expression to environmental issues. The consultation version of the plan has also been subjected to a sustainability appraisal (Forum for the Future, 2004). The formal report on the sustainability appraisal states that the plan 'has clearly demonstrated its *potential* to make a significant positive contribution to the wider vision for the spatial sustainable development of Wales' (p. 1, emphasis in original). The report's authors also commend the plan on its emphasis on sustainability issues, but continue to state that the potential of the plan in draft form is threatened by the failure to establish mechanisms for assess-ment and monitoring of the implementation of the plan. The approved version of the plan now has a limited section on monitoring and implementation, although this focuses primarily on establishing an infrastructure to support the plan and a programme of actions, such as the development of a series of indicators by which to monitor the future implementation of the plan. The capacity and ability to imple-ment the plan as yet remains unproven.

Identity, commonality and difference

The Wales Spatial Plan is an interesting and original attempt at devising a spatial planning instrument. It is of particular interest for its process of preparation that has been open, inclusive and focused on the engagement of a wide range of stake-holders. The question of whether the plan is an active component in the formation or deepening of a particular, Welsh national identity is an interesting one. If spatial plans are to be regarded as political exercises in national identity formation, then it is necessary in reading such plans to assess the mechanisms by which identity is established. Common or shared attributes or characteristics are one of the principal means by which a particular identity can be shaped. The Wales Spatial Plan, however, makes very limited use of mechanisms to create a common, 'Welsh' identity. This is despite the stated emphasis on establishing 'a cohesive identity which sustains and celebrates what is distinctive about Wales' (WAG, 2004a: 33). A close reading of the plan throws up little that could be read as a project in nation-building or identity formation. Yet a spatial plan, if it is to be accepted as a realistic instrument and widely accepted, also needs to recognise the differences that exist *between* areas. The emphasis in the Wales Spatial Plan is clearly on the different roles, functions and meanings attached to different parts of Wales. In the political foreword to the consultation draft of the Wales Spatial Plan, the relevant minister declared that despite Wales' small size 'it contains an amazing diversity: of people, of industry and of geography' (WAG, 2003: i). This diversity is not only recognised but also to be built upon and strengthened. The plan, it is stated, recognises the different needs and potential of different places, and rejects 'one size fits all' solutions (WAG, 2004a: 4).

The Wales Spatial Plan emphasises 'respecting distinctiveness' as one of its five key themes. The theme was expressed as one of two values that underpinned the content of the consultation draft. The consultation draft pointed clearly to a strong sense of place in Welsh communities founded on history and tradition and went on to proclaim that: 'Distinctiveness, sense of identity and pride in place are important elements of successful communities *and countries*' (WAG, 2003: 5, emphasis added).

The plan translated this value into a practical context by confirming that the Assembly will encourage the local determination of spatial priorities within the context of the national framework that it outlines. Consultation responses to the draft of the Wales Spatial Plan demonstrate strong and widespread support for the value (and subsequently theme) of respecting distinctiveness (WAG, 2004a: 7). This emphasis on diversity and difference is also promoted by the practices associ-ated with partnership working which has come to characterise policy development in Wales since the establishment of the Assembly. Partnership working has dis-placed the historically antagonistic and conflict-based working practices between local authorities and the former Welsh Office. The Assembly is unable, due to the circumstances that it finds itself in, to forge ahead with top-down approaches to policy implementation. The interdependencies that exist between the Assembly and the 22 unitary local authorities mean that partnership working is necessary in order to ensure policy effectiveness and efficient service delivery.

The debate initiated in the plenary session of the Assembly at which the full consultation draft of the Wales Spatial Plan was officially launched provides a fascinating insight into the potential of the plan to act as an explicitly political or cultural instrument in nation-building. Although some disappointment was expressed with the draft product, the initiative was widely welcomed across the different political parties represented in the Assembly. One Plaid Cymru Assembly Member was concerned that the plan had demonstrated but not realised its potential 'to develop and maintain a sustainable Welsh nation'. The language used here is interesting for the use of such explicit nation-building language is not evident in the document itself. Other Assembly Members participating in the debate also called for greater emphasis in the plan on improving transport connections between north and south Wales in order to bring the regions of Wales closer together. In the plenary session on publication of the final version of the Plan, similar sentiments were expressed:

> Much is made of east–west links, but Wales is, and needs to be, far more than a scenic backdrop to two main arteries of east–west communication.

> I am sure that everyone would agree that transport between north and south Wales is important. We must remember that there is a vast area between the A55 and the M4 [the east–west corridors in the north and south respectively]. This needs to be addressed.

This has parallels with similar transport and other infrastructure improvements at the level of the European Union that have the explicit purpose of assisting integration of the European territory. It is not clear, however, whether the tensions between internal accessibility and the preservation of cultures and lifestyles in particular areas of Wales have been fully thought through. Proposals such as that made in the plenary session are designed to have a similar purpose of integrating the Welsh territory, of ensuring that the physical space of Wales supports and aligns with the political project of devolution. For some, the document pulled in the opposite direction, with one nationalist party member expressing disappointment that the Wales Spatial Plan 'increasingly binds us to England'. For yet others, the attention given in the plan to the economic drivers outside of Wales with which the Welsh economy is highly intertwined was a welcome acknowledgement of the practical realities of policy implementation and the artificial importance of political and administrative boundaries.

The sustainability appraisal of the plan also comments critically that while the consultation version of the plan does address cultural and heritage assets in a limited way, it failed to have regard 'to promoting culture in other ways' (Forum for the Future, 2004: 2). The character of the consultation draft as a rather practically focused policy integration tool meant that it underplayed some important issues of cultural significance. This too was raised in the Assembly Members' plenary debate, with one member arguing that reference is needed in the plan 'to the factors that make Wales unique, which are its culture and its language'. Cultural diversity, in so far as it is acknowledged, is largely restricted to consideration of the Welsh language and its particular significance in certain geographic communities across Wales. It consequently highlighted particular geographic areas in which the significance of Welsh culture needs to be further emphasised with revision of the plan. Indeed, consideration of Welsh language issues and culture are identified as weaknesses in several of the area perspectives presented as part of the draft plan. The overall evaluation of the appraisal may be summarised as being that the draft Wales Spatial Plan did little to either address or promote the culturally distinct characteristics of the different parts of Wales, including language-based cultural issues. Cultural and identity issues are, then, ones that the Wales Spatial Plan in its consultation draft form did little to recognise or address. The sustainability appraisal of the plan demonstrates clearly that these are issues that the plan needs to address as part of building sustainable communities across Wales. Nevertheless, the commentary clearly suggests that culture and identity are issues that *can* be successfully promoted through spatial planning instruments, and that there is latent potential in such spatial planning approaches. Against the background of the appraisal and these debates, the final version of the plan does very little to address such criticisms. Issues of culture generally and language in particular remain underdeveloped, in spite of the headline emphasis in the plan on 'distinctiveness' and 'identity'. Consideration of the Welsh language appears as a key issue in only one of the area visions, being that for the North West of Wales where the language has demonstrated some evidence of decline and appears to be most threatened as an element of the fabric of local communities. In conclusion, a reading of the plan suggests that issues of culture, identity and language are important and significant issues, with the plan having a recognised potential in addressing such issues. Yet these key issues remain ill-defined and it is not clear how, or in what ways, the plan contributes in these fields.

Conclusions

The Wales Spatial Plan is an example of an objectives-led approach to spatial planning, as opposed to an approach that is founded on a more extensive and technical spatial analysis of the Welsh territory. Such approaches are potentially capable of being strong vehicles for the design and elaboration of particular visions and images of territorial identity. Nevertheless, it is not obvious in reading the plan that it is particularly concerned with issues of identity, nor does it elaborate well on such issues where they are alluded to. The Wales Spatial Plan can so far be characterised as an instrument to facilitate policy integration rather than as an explicitly political instrument designed to foster a particular and strong sense of Welsh identity. It is, first and foremost, a policy integration tool. The principal concerns of the plan centre on the integration of a range of sectoral issues in the economic, social and environmental spheres. However, the political significance and potential of the Wales Spatial Plan has been recognised and it is for this reason that the plan can be described as being located between the functional and the political. This reflects the development of the document from the initial conception of a national spatial planning framework related to the land-use planning system to a wide-ranging, corporate policy document with application across the Assembly's various functions and divisions. Its importance as a plan is in recognising the distinctive needs and differences between the different parts of Wales and not in creating or extending some sense of Welsh identity.

Notes

1 The consultation draft of the Wales Spatial Plan interestingly used the phrase 'the space that is Wales' to refer to the Welsh territory.
2 Debates on spatial planning in Wales take place within the constraints of a peculiar vocabulary that has failed to resolve some basic issues of nomenclature, including on such issues as 'national/nation' and 'regional/region'. This is a long-standing feature of Welsh life related to the history of Wales, but it is nevertheless brought into sharp relief when debating explicitly spatial issues at different scales. In this particular phrase, we use the term 'region' to describe Wales as a whole, whereas others may describe Wales as a nation. The term 'territory' is largely avoided in practical spheres, although it is occasionally brought into use when relating spatial planning issues to European debates and contexts, and does now also feature incidentally in the Wales Spatial Plan (Welsh Assembly Government, 2004a: 4).

References

Bosworth, J. and Shellens, T. (1999) 'How the Welsh Assembly will affect planning', *Journal of Planning and Environment Law*, March, pp. 219–24.

Commission on the Powers and Electoral Arrangements of the National Assembly for Wales (2004) *Report of the Richard Commission*, Cardiff Bay: Publications Centre, National Assembly for Wales.

Forum for the Future (2004) *Sustainability appraisal of 'People, Places, Futures: the Wales Spatial Plan': Final Report*, London: Forum for the Future.

Harris, N. and Hooper, A. (2004) 'Rediscovering the 'spatial' in public policy and planning: an examination of the spatial content of sectoral policy documents', *Planning Theory and Practice*, 5(2): 147–69.

Harris, N. and Tewdwr-Jones, M. (1995) 'The implications for planning of local government reorganisation in Wales: purpose, process and practice', *Environment and Planning C: Government and Policy* 13(1): 47–66.

Morgan, K. and Mungham, G. (2000) *Redesigning Democracy: The Making of the Welsh Assembly*, Bridgend: Seren.

Osmond, J. (ed.) (1998) *The National Assembly Agenda: A Handbook for the First Four Years*, Cardiff: Institute of Welsh Affairs.

Tewdwr-Jones, M. (1997) 'Land-use planning in Wales: the conflict between state centrality and territorial nationalism' in R. Macdonald and H. Thomas (eds) *Nationality and Planning in Scotland and Wales*, Cardiff: University of Wales Press, pp. 54–76.

Tewdwr-Jones, M. (1998) 'Strategic Planning' in J. Osmond (ed.) *The National Assembly: A Handbook For The First Four Years*, Cardiff: Institute of Welsh Affairs, pp. 152–62.

Tewdwr-Jones, M. (2004) 'Spatial planning: principles, practices and cultures', *Journal of Planning and Environment Law*, May, pp. 560–9.

Welsh Assembly Government (2003) *People, Places, Futures: The Wales Spatial Plan. Consultation Draft*, Cardiff: Welsh Assembly Government.

Welsh Assembly Government (2004a) *People, Places, Futures: The Wales Spatial Plan*, Cardiff: Welsh Assembly Government.

Welsh Assembly Government (2004b) *People, Places, Futures: The Wales Spatial Plan. Consultation Report*, Cardiff: Welsh Assembly Government.

10 Escaping policy gravity

The scope for distinctiveness in Scottish spatial planning

Philip Allmendinger

Introduction

Devolution is not simply a UK phenomenon but part of a global trend towards more flexible, entrepreneurial and accountable forms of governance (Keating, 1998; Loughlin, 2000). There has been a great deal of debate in recent years concerning the changing nature of the state and the ways in which it is being 'hollowed out' by a range of factors which in turn are causing a restructuring of scale, governance and functions. According to such views the notion of 'hollowing-out' involves three interrelated processes: the *de-statisation* of the political system, most clearly reflected in the shift from government to governance; the *internationalisation* of policy communities and networks; and the *denationalisation* of the state, in which state political and economic capacities are being reconfigured territorially and functionally along a series of spatial levels – sub-national, national, supra-national and trans-local (see the chapter by Goodwin *et al.* in this volume).

The state has been reconfigured to a role as site of power dynamics, rather than an arbiter of power, subject to a plethora of competing pressures and demands. This can, and does, lead to a range of competing and contradictory regulatory and policy tendencies (Haughton and Counsell, 2004). But states are not passive recipients of such pressures. Instead, they select strategies and seek to build coalitions for support and, critically, implementation. Such a view allows for resistance to policy by those outside such coalitions: 'Various groups can draw on selective values, ethical codes and scientific knowledges to seek to counter dominant tendencies' (Haughton and Counsell, 2004: 40).

Devolution may be regarded by some as a democratic revolution (Barnett, 1997) but the outcomes may not necessarily, or even ordinarily, amount to a variance in policy between devolved administrations. As Curtice (2003: 15) points out, a clear majority, 57 per cent, believe that the Scottish Parliament has made no difference to the way that Scotland is governed. Distinctiveness, and the scope for policy divergence, is bounded by a range of cultural, political, institutional, professional as well as supra-national governmental and legal factors. Within the space carved out for autonomy centrifugal and centripetal forces are simultaneously pushing *for* and *against* change. A further dimension has been added by Storper (1997) who makes a distinction between *institutional* change and *organisational* change (see also Mitchell, 2001 for a discussion of this distinction in relation to Scotland). The latter concerns concrete administrative and political bodies while the former has more to do with the rules, habits and customs that bind individuals and groups of individuals. With regard to spatial planning, there have been identifiable organisational changes post devolution (Allmendinger, 2002) though it is far from clear whether and to what extent there have been changes in the practices or institutions of planning post devolution (Allmendinger, 2001a; 2001b). Despite some minor differences and nomenclature since devolution spatial planning across the UK has begun to continue its convergence around 'key themes'. Significantly, and contrary to expectations (e.g. Tewdwr-Jones and Lloyd, 1997; Boyack, 1997; Hayton, 1997), changes to spatial planning in England have been more radical and groundbreaking than in Scotland since 1999. If planning in Scotland was once distinct, radical and groundbreaking (Wannop, 1980), it is no longer. There are a number of possible institutional and organisational explanations for this situation that are at the heart of assessments of the impact of devolution and spatial planning (Allmendinger, 2001a; 2001b; 2002). What is clear is that, in the view of Scots, the Parliament itself has a decreasing significance (Table 10.1).

One dimension that has been largely overlooked in the assessment of policy change post devolution is the significance of resistance within the understanding of the state as an arena for dynamic, contested and conflicting policy pressures. Here, resistance does not necessarily imply 'blocking' change but providing powerful and influential policy, ideological and administrative trajectories whose 'gravity' makes policy diversion difficult. As North (1990) argues, the structuring effects of particular socio-spatial arrangements are frequently powerful constraints upon emerging or new organisational forms and practices. Such path dependency can operate in a variety of ways though largely reflects highly asymmetrical power relations (see also Swyngedouw, 1997).

Like other areas of public policy and the state generally, spatial planning is subject to the same forces that are compelling change and simultaneously resisting it. Similarly, planning is not a neutral recipient of change nor is it separate from the struggle itself. It may well find itself subject to mêlées over the future of the state but it is both a contributor *to* and reflection *of* such struggles. Johnson (1993), for example, sees professions such as planning as devices through which the means and ends of government are constructed – they identify societal problems and provide solutions. Such relations are themselves dynamic: the 1990s witnessed a shift in the relationship between the local and national state and the planning profession (Hill, 2000) particularly following widespread disquiet about misconduct and inappropriate behaviour on the part of elected officials (Nolan Committee, 1997).

Table 10.1 Significance of Scottish Parliament on Scottish affairs

Which of these has most influence over the way Scotland is run:	1999	2000	2001	2003
	%	%	%	%
The Scottish Parliament	41	13	15	17
The UK government at Westminster	39	66	66	64
Local councils in Scotland	8	10	9	7
The European Union	5	4	7	5

Source: Curtice (2004: 16)

The questions that this chapter seeks to address, therefore, are have the trajectories of change in spatial planning shifted as a result of devolution and what factors have driven and constrained change? The broad argument of this chapter is that despite expectations of change the Scottish state is caught in a 'pincer movement' of, at one level, global and UK trajectories of change largely beyond its control and, at another level, 'policy gravity' and resistance to the limited scope for change. Regardless of devolution, the trajectories of change in planning policy north and south of the border are being driven by common concerns and objectives that derive from a range of influences. On the whole, the trajectories of change in spatial planning have not altered and are still converging.

A related hypothesis that cannot be tested in this chapter is that the changing nature of the role of the state towards a more strategic 'steering' function and as a recipient and arbiter of dynamic, contested and powerful interests leads to ambiguous national spatial guidance thereby placing a greater role on planners at the local level to choose particular courses and priorities. Such a role is itself framed by common doctrines and orthodoxies of practice that have also influenced the boundaries of policy options at the national level.

The remainder of this chapter highlights and compares the raised expectations for change in Scottish planning with what has actually occurred since 1998 as well as identifying some common key themes that have underpinned policy trajectories both north and south of the border. The chapter then briefly explores these key themes before drawing some general conclusions.

The scope for and impact of change in Scottish spatial planning

Prior to 1999 the Scottish Office was subject to similar national initiatives to those in England and Wales (McGarvey, 2002). The extent to which there was scope for discretion and autonomy is subject to two broad schools of thought. Some, such as Paterson (1994) and Kellas (1989), perceive a large degree of autonomy for Scotland before devolution. Others such as Keating and Midwinter (1981) and Griffiths (1995) see policy making in the UK as more centralised and give greater weight to the role of the Westminster Parliament in setting agendas and developing policy. The majority view seems to be that the pre-devolution system tended to policy convergence, albeit with differences among Scotland, Wales and Northern Ireland (Midwinter, Keating and Mitchell, 1991; Connolly and Loughlin, 1990; Griffiths, 1995). There were certainly expectations of change to spatial planning concerning both output and process. According to one former planning minister 'there is a hope that the Parliament will be able to operate in a more modern way than Westminster, with more of an emphasis on openly involving people, community, business and civic groups' (Boyack, 1997: 308). Nevertheless, I will be focusing here on policy output.

It is still unclear whether devolution is affording opportunities for policy output divergence and distinctiveness. Examples of both can be highlighted; nevertheless, while devolution invites new bodies to develop their own policy styles and priorities pressures to converge will persist (Allmendinger, 2001a; 2002). Although there is evidence of policy divergence between Scotland and the rest of the UK since devolution concerning, for example, tuition fees, care for the elderly, etc., there is also emerging evidence of policy continuity or even convergence (e.g., Mooney and Johnstone, 2000; Taylor and Sim, 2000; Mitchell, 2004; Fawcett, 2004).

Policy output in spatial planning reflected the general trends towards convergence. Despite the crude claim that 'Scottish planning has much to teach the English' (Hague, 1990: 287), both distinctiveness and dependence in planning policy existed side by side prior to devolution with coordination *and* separation characterising policy making on planning north and south of the border (Allmendinger, 2002). This is, perhaps, more the case in planning than other areas of policy. Multilevelled policy coordination and governance – spatial planning – requires both national (UK) and supra-national coordination to be effective in policy fields such as environmental regulation and transport infrastructure. As a result the differences are increasingly cosmetic (Rowan-Robinson, 1997) particularly following the increased role of the European Commission in dictating environmental regulation, the impact of pan-national environmental treaty obligations, local government reorganisation and its impact on regional planning and the commodification and bypassing of planning generally under the Thatcher and Major years. A more realistic perspective is given by Hayton who states that Scotland 'has a similar but subtly different planning system (to England)' (1996: 78).

In contrast to these expectations of change, planning in Scotland post devolution, like the rest of the UK, has been framed by a range of key themes that have arisen from a plethora of sources. Two dimensions of these themes are worth highlighting. The first is continuities. There is a long history to some themes that goes back beyond 1998 and devolution. There are also strong continuities or similarities between the themes across England and Scotland. The second dimension is the conflicting nature of some of the themes and the scope for interpretation by professionals such as planners to reconcile competing policy objectives. The principal themes comprise:

- **Economic competitiveness.** Since 1979 there has been a strong theme in planning concerning the need to contribute to economic growth. In the view of Thornley, 'The purpose is one which has its primary aim in aiding the market' (1993, p. 143). Perceived as a 'supply side constraint' by the New Right planning was similarly castigated by New Labour ministers as an impediment to growth: 'It is clear, and as I am well aware from my time at the Department of Trade and Industry, that the planning system plays a big part in determining business opportunities' (Byers, 2001). While the Office of the Deputy Prime Minister (ODPM) has been less anti-planning than its Conservative predecessors the pressure for a deregulation has come from other central government departments (Brown, 1998; HM Treasury, 1998; McKinsey, 1998) as well as business interests. In Scotland there have similarly been strong moves to embed the sceptical and largely anti-devolution business community within policy generally (Raco, 2003) and planning in particular: 'the slowness and the negative nature of the system, which is seen as getting in the way of business competitiveness' (Scottish Office, 1999: 1). This constant chiding, backed up by central policy guidance, has led to a culture of defensiveness on the part of planners and an unwillingness to challenge new development that is justified through employment and economic growth. Planning practice in all parts of the UK is now firmly market supportive.
- **Sustainability and sustainable places.** The notion of sustainability is one that permeates planning policy north and south of the border and has been a

theme of government policy since the early 1990s. Despite its vagueness as a concept and the lack of guidance about the implications for policy on the ground planning has been reoriented towards sustainability in terms of policy outputs and processes and as a profession: 'Planning must offer a means to mediate consciously between these (environmental, social and economic) competing objectives' (RTPI, 2001:3). This vagueness has led Haughton and Counsell to point out that: 'With its three interrelated themes of economic growth, social equity and environmental responsibilities, sustainable development has come to legitimate processes of global competitiveness and expansion, by neutralising opponents who would argue that growth is necessarily social polarising and environmentally damaging' (2004: 32). In England Planning Policy Statement 1 (ODPM, 2004) claims that sustainable development is the core principle that underpins planning while Scottish Planning Policy 1 (Scottish Executive, 2002a) states that the planning system should ensure that development and changes in land use occur in suitable locations and are sustainable. After the anti-planning attacks of the 1980s the concept of sustainability has provided planning with a *raison d'être* that is vague enough to justify and deliver considerable discretion and power to planners.

- **Community involvement and participation.** Since the late 1960s there has been a shift away from the idea of planning as a bureaucratic, scientific enterprise towards a more inclusive and democratic enterprise (Healey, 1997). While much of this movement has been driven by a growing distrust of officialdom in general, a greater desire to get involved on the part of the public and an increasing awareness of environmental issues there has also been a good deal of self interest that has persuaded planners to allow the public greater involvement in the process. Many of the problems of towns and cities that planning promised to address were, by the early 1970s, clear and still very much present. Further, many of the solutions in the form of high-rise housing and urban motorways were clearly exacerbating many of the problems. One solution was to claim that planning was not a technical or scientific process but a political one that required a wider input and, therefore, blame. Planning has always had strong advocates for greater involvement though the profession and the state have been wary of devolving too much decision making to those outside the planner–applicant axis. The Scottish Executive's coolness to the introduction of a third party right of appeal (Scottish Executive, 2004) to planning decisions despite widespread calls for its introduction typifies the reluctance to open up the system to wider views. Both Westminster and Holyrood have proposed changes to planning that reduce as well as selectively extend public involvement.

- **From physical to social welfare planning.** Unlike its US counterpart planning as a profession in the UK has not traditionally had an ethical agenda. Members of the profession are required not to discriminate but also not to be advocates or actively work towards a social agenda of, for example, eliminating poverty. However, a key theme of Labour administrations both north and south of the border has been to use planning as a tool towards wider social ends including inclusion. As the former Minister for Social Justice in the Scottish Executive has put it, 'I see positive benefits from the wider planning approach (which) recognises that access to existing jobs, promoting new business development in urban areas, creating transport links, focusing new development where it will be served by public transport, and sustaining and enhancing town centres all have a role to play' (Baillie, 2004: 4). This broadening of planning from the narrow market supportive confines of the 1980s and early 1990s has received support from a core of planners who had tried to use planning towards redistributive ends though had been restricted by political, institutional and professional constraints (Macdonald and Thomas, 1997). Such views exist side by side and through time with a variety of interpretations of the role of the planner. Both Westminster and Holyrood have begun to reorient and re-emphasise social welfare dimensions of planning thereby echoing a strong, though by no means unanimous, undercurrent in planning thinking.

- **'Space and Place'.** The 'repackaging' of planning by the RTPI over the last few years has sought to provide a focus upon the multi-tiered spatial planning and detailed place making through an emphasis upon design: 'globalisation tendencies have served to re-emphasise the importance of place making activities at the local and regional scale, as policy makers seek to build positive images in order to attract and retain both mobile investment and consumer spending within localities' (Haughton and Counsell, 2004: 35). Spatial plan making is also once again in vogue with the reintroduction of regional and national spatial strategies while a renewed emphasis upon design and quality has sought to focus on place making across the professions. Both English and Scottish administrations have sought to reintroduce an emphasis on space and place through a variety of policy and administrative mechanisms. Most notable is the new emphasis upon regional level planning (or national level planning in Scotland) which was a strong theme of planning in both nations up until the late 1970s.

- **Policy integration.** Since 1997 a key theme of government has been to develop ways to tackle complex social problems through cross-cutting and integrative policy (Sullivan, 2001). However, planners and planning have a long history of seeking both vertical and horizontal policy integration (Bolton and Leach, 2002). Integration has been given a fillip through the emphasis placed on sustainable development and sustainability as both a process and outcome of planning. While there are clear question marks over the reconciliation of social, environmental and economic concerns inherent within the notion of sustainable development the process of resolution provides a mechanism for formalising the process. In part, policy integration has also been re-emphasised through the prominence given to emerging plans and strategies at the regional level in England (particularly Regional Spatial Strategies) and Scotland (the National Spatial Strategy and City Visions).

These and other themes have framed the debates and scope for change while also providing planners and policy makers with doctrines that guide their approach to spatial planning. But more significantly, as both the English and Scottish Labour administrations have shied away from offending both business and environmental interests it has led to more ambiguous policy advice or mixed signals concerning the objectives of planning. Changes introduced to appease one set of interests are balanced by other new initiatives. This has left planners in the position of having to reconcile or interpret ambiguities in national policy direction.

The trajectories of spatial planning post devolution

There has been a broad approach in the evaluation of policy change post devolution that has privileged 'top down' perspectives and overlooked other influences upon state regulation in the evaluation of policy divergence post devolution. Nevertheless, the broad themes identified above have shaped the trajectories of planning in Scotland and England post devolution. Fundamental revisions of spatial planning are periodically undertaken though were initiated in England[1] following the Labour victory in the UK general election in 1997 and devolution in Scotland in 1998. Prior to devolution a review of spatial planning in Scotland was launched by the (then) planning minister Callum McDonald (Scottish Office, 1999) that drew heavily upon the issues, debates and trajectories of the future of planning under way in England. The broad scope of change was established prior to devolution including the abolition of structure plans, the introduction of a national planning framework and the refocusing of national planning guidance.

The analyses of these reviews and their identification of what aspects of planning needed to be addressed north and south of the border were remarkably similar. The Review of Strategic Planning (Scottish Executive, 2001c; 2002b; 2002c) acknowledged the difficulty in undertaking development planning in an era where, 'a greater number and variety of agencies and organisations whose policies and spending priorities have the potential to impact significantly on planning' was coupled with an 'uncertainty about the relationship of development plans to other plans and strategies, in particular community planning' (Scottish Executive, 2001c: para. 8). In England, the Green Paper on planning highlighted the 'complexity' of the system and the difficulty in coordinating plans and strategies: 'the multi-layered structure of plans with up to four tiers in some areas – at national, regional, county and local levels. Plans are often out of date and can be inconsistent with one another and with national planning guidance' (DTLR, 2001: para 2.3).

The solution of the Executive, largely mirroring that south of the border (ODPM, 2002), was to both rationalise development plans through the abolition of strategic planning outside the four main cities and integrate the new plan structures with Community Plans. Community Planning in Scotland and Community Strategies in England are seen as mechanisms by which spatial planning – a phrase widely used in England and Scotland since devolution to describe the new integrative and cross-cutting nature of planning – can be coordinated and delivered at the local level (Scottish Executive, 2000a: 17; ODPM, 2003).

The multi-layered hierarchy of development planning both north and south of the border has been a characteristic of planning since introduced in the 1960s though planners had long been calling for a more strategic overview of planning at the regional or national level. Structure plans were widely felt by many planners to be neither strategic nor local enough: 73 per cent of Local Planning Authorities (LPAs) in Scotland supported the proposal to abolish the 'two tier' development plan system outwith the four main city-regions (though the introduction of a national spatial development framework effectively introduced a further tier of plan in the four city-regions) (Scottish Executive, 2003). The Western Isles Council, for example, felt that:

A single tier of development plan should reduce production time and costs, be easier to keep up-to-date and would reduce the need for prospective developers and other interested parties to look at two documents in order to ascertain the planning policies for their area of interest.

(Scottish Executive, 2003: 89)

The Chief Planner for Scotland also thought there were too many plans and not enough planning: 'One senior Director of Planning commented that too many plans were little more than compendiums of pious hopes' (Scottish Executive, 2003: 2).

The broad aims and objectives of planning prior to devolution north and south of the border were similar. The prospect of devolution raised expectations of a divergence of planning systems from a variety of quarters and in a variety of directions. Since 1998 both the English and Scottish administrations have reformed the aims of planning. Rather than diverging both demonstrate a continued similarity (see Boxes 10.1 and 10.2).

While the analysis of changes required to 'modernise' planning north and south of the border are similar the proposals for change and the actual changes, with one significant exception, also echo each other. Three broad themes can be identified (Table 10.2). Initially, there has been a significant *rescaling* of planning. There are four dimensions to this rescaling. First, there has been a shift 'upwards' in the spatial scale of planning with the (re)introduction of national/regional level plans across the UK. As discussed above this shift has involved the loss of structure plans in England and parts of Scotland though the newly devolved administrations including the English regions have gained important strategic planning and economic development functions. While regions and nation regions have been privileged in the scale of planning there has also been a resurgence in sub-regional planning. The Scottish Executive has begun to issue strategic spatial plans such as the West Edinburgh Planning Framework while in England Regional Planning Bodies have introduced a range of functional sub-regional planning areas.

Box 10.1 Post devolution aims of the Scottish planning system

The planning system guides the future development and use of land in cities, towns and rural areas in the long term public interest. The aim is to ensure that development and changes in land use occur in suitable locations and are sustainable. The planning system must also provide protection from inappropriate development. Its primary objectives are:

- To set the land use framework for promoting sustainable economic development;
- To encourage and support regeneration; and
- To maintain and enhance the quality of the natural heritage and built environment.

Source: Scottish Planning Policy 1: The Planning System. Scottish Executive, 2002: para 1.

Box 10.2 Post devolution aims of the English planning system

Planning should facilitate and promote sustainable patterns of urban and rural development by:

- Making suitable land available for development in line with economic, social and environmental objectives to improve the quality of life.
- Contributing to sustainable economic growth.
- Protecting and where possible enhancing the natural and historic environment and the quality and character of the countryside, and existing successful communities.
- Ensuring high quality development through good design.
- Ensuring that development supports existing communities and contributes to the creation of safe, sustainable and liveable communities with good access to jobs and key services.

Source: Consultation Paper on Planning Policy Statement 1: Creating Sustainable Communities. ODPM, 2004: para 1.5.

The second dimension to rescaling involves the changing role of different actors or the scope of planning. National planning guidelines in both England and Scotland are to be distilled and refocused. This is linked to the third dimension of rescaling: the centre (English, Scottish and UK) is taking on a subtly different role in shaping the trajectories of policies and plans. This new role involves identifying and 'green lighting' more strategic (mostly economic) developments and frameworks which (according to some) contradict the spirit of devolution in bypassing the newly created arrangements.

Finally, while planners have long been charged with policy interpretation and delivery under the UK system the re-scaling and emphasis upon a more horizontal and vertical policy integration in both nations places an even greater responsibility on the part of planners to ensure that plans and policies are coordinated across the public and private sectors.

The second broad theme is *policy delivery*. A range of changes both to local government generally and planning specifically have provided a continuation and expansion of the system of performance indicators introduced during the John Major Governments. In England such performance related policy delivery has been linked to resources through the Planning Delivery Grant (PDG). PDG is intended as an incentive to improve planning performance and to improve the resourcing of planning authorities. Such under-resourcing was delaying development and economic growth. In Scotland there are subtly different performance indicators though there are similar concerns about the resourcing of local authorities' planning function which may lead to a similar initiative.

Finally, there has been a focus on *involvement*. Both Westminster and Holyrood have been keen to promote the notion of inclusiveness and involvement. However, the push to inclusion seems to represent less of a commitment to democratic principles and more of an acknowledgement that effective and efficient policy delivery is more likely to be achieved through including a wide range of stakeholders.

Table 10.2 Common themes of change in Scottish and English spatial planning

Main theme	Sub-theme	Proposals or changes to date in Scotland	Proposals or changes to date in England	Doctrine(s)
Re-scaling	Plans	National Spatial Strategy prepared. Structure plans abolished though four main city-regions have strategic development plans (proposal)	Sustainable Communities Plans (e.g. Milton Keynes) prepared for a number of areas. Regional Spatial Strategies to be prepared. Structure Plans abolished. Research under way on national spatial strategy.	Policy integration Sustainability Economic competitiveness Space and place
	Scope of planning	From National Planning Policy Guidance to Scottish Planning Policy – reduced content and a 'sharper' focus on policy.	From Planning Policy Guidance (PPGs) to Planning Policy Statements (PPSs) – reduced content and focus and the identification of major infrastructure projects.	Economic competitiveness
	Central role	Greater central 'steering' role through National Spatial Strategy, Scottish Planning Policies and other planning frameworks such as the West Edinburgh Planning Framework.	Greater central 'steering' role through Sustainable Communities Plans, Planning Policy Statements and (if taken forward) a National Planning Framework.	Economic competitiveness
	Integration	Community Planning and Community Plans as mechanisms to better integrate policy.	Regional Spatial Strategies and Community Planning	Policy integration Social welfare policy
Delivery		Action plans to be published bi-annually to indicate plan progress and implementation (proposal).	The Local Development Scheme of the Local Development Framework will set out the programme for the preparation of the local development documents.	Economic competitiveness Policy integration
Involvement		Third Party Rights of Appeal (consultation)[2]		Community involvement

Clearly, a range of common themes underpin approaches to planning in Scotland and England post devolution. While the policy detail differs in many respects the driving forces and scope for change are common. As Table 10.2 highlights, the most common theme is that of economic competitiveness and this underpins many of the changes to planning in England and Scotland post devolution. The notion of the economic, social and environmental that underpins the idea of sustainability is not an even one in its influence upon policy. However, there are some differences in the policy processes followed that distinguish the approaches to planning in England and Scotland. The most significant potential difference was the proposal to introduce a third party right of appeal. Such a power would break the applicant–local planning authority nexus and allow anybody to appeal against a decision to approve a proposal. The call for third party appeals has been strongly advocated by environmental groups and equally strongly resisted by business groups and developers. However, it is interesting to note that at the time of writing (June 2005) it is looking likely that the Scottish Executive would reject the idea on the grounds that it would slow down the planning system and inhibit development. The theme of economic competitiveness is again being given primacy.

Conclusions

While the focus in this chapter has been on the post-devolution trajectories of Scottish and English planning there are clearly a range of influences and drivers of divergence and convergence that 'fill in' the space carved out by state restructuring – party politics, common or shared culture and ideas about the nature of society, its problems and possible solutions, the limits to liberal democratic politics provided by economic globalisation, etc. Nevertheless, it is important to recognise the significance of individuals and ideas and path dependencies in tempering the expectations and likelihood of change.

One significant difference between the Scottish and English trajectories of planning concerns the role of the Treasury and Department of Trade and Industry in England in dictating change. Both departments have been instrumental in commissioning research, establishing inquiries and directing change in English spatial planning. While the planning function in Scotland has been subsumed within a wider social justice agenda in England the driving forces have been more concerned with economics and competitiveness. The dominance of direction in the trajectories of English planning are themselves reflective of the wider drivers of globalisation.

According to Mitchell, the simplest form of policy to change is regulatory and planning covers a range of such functions (Tiesdell and Allmendinger, 2005). Policy output change in spatial planning can relate to a range of functions and expectations. Business interests might expect a reduction in regulatory controls while other groups might be expecting a change to the redistributive functions. As Mitchell (2004: 11) has stated, 'decisions and practices determined in the early life of an institution (or indeed a policy) can have profound long-term implications'. While supporters of devolution were quick to point to the role of a Scots Parliament in stopping the imposition of policy from Westminster following the experiences of the poll tax there were fewer clear arguments about what the Parliament could achieve and planning was no exception to this. There was never a 'big vision' for land use planning in Scotland post devolution. The changes to date demonstrated broad parallels with those in England and, as significantly, have tended to follow changes introduced in England with few modifications.

However, while the focus post devolution has been on the trajectories and distinctiveness of planning regimes in the nation-regions less attention has been paid to differences between regions within nations. Divergence in policy output or process between places is a common feature of UK planning. Brindley, Rydin and Stoker (1996) identified a range of planning 'styles' that existed across the UK within different localities. The scope for discretion at the local level in spatial planning means that variations in such styles between regions within England could be as significant as those between nation-regions.

Notes

1 A review of planning was also initiated in Scotland prior to devolution by the then Scottish Office though was superseded.
2 A Private Members' Bill, the Third Party Planning Rights of Appeal (Scotland) Bill, was proposed by SNP MSP Sandra White in 2003. However, press reports in Spring 2005 based upon a leaked Executive memorandum claimed that Ministers would resist proposals.

References

Allmendinger, P. (2001a) 'The head and the heart: National identity and urban planning in a devolved Scotland', *International Planning Studies*, 6(1): 33–54.

Allmendinger, P. (2001b) 'The future of planning under a Scottish parliament', *Town Planning Review*, 72(2): 121–48.

Allmendinger, P. (2002) 'Prospects for a distinctly Scottish planning in a post-sovereign age', *European Planning Studies*, 10(3): 359–81.

Allmendinger, P. and Tiesdell, S. (2005) 'Planning tools and markets – towards an extended conceptualisation', in D. Adams, C. Watkins, and M. White (eds) *Planning, Public Policy and Property Markets*, Oxford: Blackwell

Baillie, J. (2004) 'Building inclusion communities', *Scottish Planner*, April.

Barnett, A. (1997) *This Time: Our Constitutional Revolution*, London: Vintage.

Bolton, N. and Leach, S. (2002) 'Strategic planning in local government: a study of Organisational Impact and Effectiveness, *Local Government Studies*, 28(4).

Boyack, S. (1997) 'Fruits of modernity', *Town and Country Planning*, November, 308.

Brindley, T., Rydin, Y. and Stoker, G. (1996) *Remaking Planning: The Politics of Urban Change*, Second Edition, London: Routledge.

Brown, G. (1998) The Chancellor's Statement on the Economic and Fiscal Strategy Report, HM Treasury News Release, June 11th.

Byers, S. (2001) Speech to the Institute for Public Policy Research, 26 July, copy available from the author.

Connolly, M. and Loughlin, S. (1990) *Public Policy in Northern Ireland: Adoption or Adaptation?*, Belfast: Policy Research Institute.

Curtice, J. (2003) 'Public attitudes and identity, in nations and regions: the dynamics of devolution', Quarterly Monitoring Programme, Scotland, Quarterly Report August 2003, The Constitution Centre, UCL.

Curtice, J. (2004) 'Public attitudes, in nations and regions: the dynamics of devolution', Quarterly Monitoring Programme, Scotland, Quarterly Report May 2004, The Constitution Centre, UCL.

Department of Transport, Local Government and the Regions (2001) Planning Green Paper, *Planning: Delivering a Fundamental Change*, London: DTLR.

Fawcett, H. (2004) 'The making of social justice policy in Scotland', in A. Trench (ed.) *Has Devolution Made a Difference? The State of the Nations 2004*, Exeter: Imprint Academic.

Griffiths, D. (1995) *Thatcherism and Territorial Politics*, Cardiff: University of Wales Press.

H.M.Treasury (1998) Pre-Budget Report, London: HMSO.

Hague, C. (1990) 'Scotland: back to the future for planning', in J. Montgomery, and A. Thornley, (eds) *Radical Planning Initiatives. New Directions for Urban Planning in the 1990s*, Aldershot: Gower.

Haughton, G. and Counsell, D. (2004) *Regions, Spatial Strategies and Sustainable Developments*, London: Routledge.

Hayton, K. (1996) 'Planning policy in Scotland', in M. Tewdwr-Jones (ed.) *British Planning Policy in Transition: Planning in the 1990s*, London: UCL.

Hayton, K. (1997), *Town and Country Planning*, July/August: 33.

Healey, P. (1997) *Collaborative Planning: Shaping Places in Fragmented Societies*, London: Macmillan.

Hill, D. (2000) *Urban Policy and Politics in Britain*, Basingstoke: Macmillan.

Johnson, T. (1993) 'Expertise and the State', in M. Gare and T. Johnson (eds) *Foucault's New Domains*, London: Routledge.

Keating, M. (1998) 'The New Regionalism in Western Europe,' in *Territorial Restructuring and Political Change*, Cheltenham: Edward Elgar.

Keating, M. and Midwinter, A. (1981) *The Government of Scotland*, Edinburgh: Mainstream.

Kellas, J. (1989) *The Scottish Political System*, Cambridge: Cambridge University Press.

Loughlin, J. (2000) *Subnational Democracy in the European Union: Challenges and Opportunities*, Oxford: Oxford University Press.

Macdonald, R. and Thomas, H. (1997) (eds) *Nationality and Planning in Scotland and Wales*, Cardiff: University of Wales Press.

McGarvey, N. (2002) 'Central-local relations in Scotland post devolution', *Local Government Studies*, 38(6): 154–77.

McKinsey Global Institute (1998) *Driving Productivity and Growth in the UK Economy*, London: McKinsey Global Institute.

Midwinter, A., Keating, M. and Mitchell, J. (1991) *Politics and Public Policy in Scotland*, Basingstoke: MacMillan.

Mitchell, J. (2001) 'The challenges to the parties: Institutions, ideas and strategies,' in G. Hassan and C. Warhurst (eds) *The New Scottish Politics: The First Year of the Scottish Parliament and Beyond*, Edinburgh: HMSO.

Mitchell, J. (2004) 'Scotland: expectations, policy types and devolution', in A. Trench (ed.) *Has Devolution Made a Difference? The State of the Nations 2004*, Exeter: Imprint Academic.

Mooney, G. and Johnstone, C. (2000) 'Scotland divided: Poverty, inequality and the Scottish Parliament', *Critical Social Policy*, 20(2): 155–82.

Nolan Committee (Committee on Standards in Public Life) (1997) *Third Report: Standards of Conduct in Local Government in England, Scotland and Wales, Vol. 1*, London: HMSO.

North, D. (1990) *Institutions, Institutional Change and Economic Performance*, New York: Cambridge University Press.

ODPM (2002) *Plans Rationalisation Study Report*, London: ODPM.

ODPM (2003) *Planning Policy Statement 12: Local Development Frameworks*, London: ODPM.

ODPM (2004) *Consultation Paper on Planning Policy Statement 1: Creating Sustainable Communities*, London: ODPM.

Paterson, L. (1994) *The Autonomy of Modern Scotland*, Edinburgh: Edinburgh University Press.

Raco, M (2003) 'Governmentality, subject-building, and the discourses and practices of devolution in the UK', *Transactions of the Institute of British Geographers*, 28: 75–95.

Rowan-Robinson, J. (1997) 'The organisation and effectiveness of the Scottish planning system', in R. Macdonald and H. Thomas (eds) *Nationality and Planning in Scotland and Wales*, Cardiff: University of Wales Press.

Royal Town Planning Institute (RTPI) (2001) *A New Vision for Planning*, London: RTPI.

Scottish Executive (2000a) *Making a Difference: Effective Implementation of Cross-Cutting Policy*, Edinburgh: Scottish Executive Policy Unit.

Scottish Executive (2001c) *The Future of Strategic Planning*, Edinburgh: Scottish Executive.

Scottish Executive (2002a) *Review of Strategic Planning: Analysis of Responses to the Consultation*, Edinburgh: Scottish Executive.

Scottish Executive (2002b) *Review of Strategic Planning: Digest of Responses to Consultation*, Edinburgh: Scottish Executive.

Scottish Executive (2002c) *Review of Strategic Planning: Conclusions and Next Steps*, Edinburgh: Scottish Executive.

Scottish Executive (2003) Speech by Jim Mackinnon to the Scottish Planning and Environmental Law Conference, 21 March, Edinburgh: Scottish Executive.

Scottish Executive (2004) *Rights of Appeal in Planning*, Edinburgh: Scottish Executive.

Scottish Office (1999) *Land Use Planning Under a Scottish Parliament*, Edinburgh: Scottish Office.

Storper, M. (1997) *The Regional World: Territorial Development in a Global Economy*, New York: Guidford Press.

Sullivan, H. (2001) 'Modernisation, democratisation and community governance', *Local Government Studies*, 27,(3): 1–24.

Swyngedouw, E. (1997) 'Neither global nor local: 'Globalisation' and the politics of scale', in K. Cox (ed.) *Spaces of Globalization*, New York: Guildford Press.

Taylor, M and Sim, D. (2000) 'Social inclusion and housing in the Scottish Parliament: prospects?', *Critical Social Policy*, 20(2): 183–210.

Tewdwr-Jones, M. and Lloyd, M.G. (1997) 'Unfinished business', *Town and Country Planning*, November, 302–4.

Thornley, A. (1993) *Urban Planning Under Thatcherism: The Challenge of the Market*, Second Edition, London: Routledge.

Wannop, U. (1980) 'Scottish planning practice: four distinctive characteristics', *The Planner*, May, 64–5.

11 London

A Millennium-long battle, a millennial truce?

Peter Hall

Introduction

Devolution to London, together with Scotland and Wales, was a first-stage element of the Blair government's devolution project, completed with elections for the London Mayor and the Greater London Assembly in 2000. This was novel in several important respects: as well as devolving power to an executive Mayor (the Assembly being given only weak powers of scrutiny), it required the Mayor to develop a series of strategic plans – for economic development, for transport, and for the environment – which would then be integrated into a spatial development strategy. This last was essentially the first attempt by the UK government to implement the recommendations of the European Spatial Development Perspective (ESDP), finalised in Potsdam in 1999, which recommended that all EU member countries seek to move their planning systems toward the development of regional spatial development strategies; it was followed by the Planning and Compulsory Purchase Act 2004, which applies the new system generally across England. Consequently, the Mayor's London Plan, which was published in draft in 2002 and (after Examination in Public in 2003) in final form in 2004, is the first English example of such a regional spatial strategy.

The strategy essentially seeks to steer development from the more prosperous and economically overheated western side of London towards the eastern side, which suffers from multiple deprivation but also offers greater opportunities for physical development and regeneration. It thus accords with central government's Thames Gateway strategy, continued by the Blair government from the previous Major administration and further developed in the Sustainable Communities strategy of 2003, which reinforces the Thames Gateway strategy but adds two other development corridors north of London, one of which (London–Stansted–Cambridge) includes the Lea Valley, another priority of the London Plan. But, as Tony Travers states in opening his book on the subject:

> Governing London is a complex business. The city's vast population, its geography and history conspire to make the British capital an unusually difficult place to govern. The election of a city-wide Greater London Authority in May 2000 brought into being the fourth system of metropolitan government within 35 years. By contrast, New York City has had a single system of government since 1898. The regularity with which London's government is reorganized suggests there is something unusual about the pressures that affect successive systems.
>
> (Travers, 2004: 1)

A capsule history: London 1888–1986

In considering why this should be so, it would be obsessive to go back to Boadicea. But it is helpful to go back over a century, because the history of London government contains themes that recur again and again, for the good reason that they reflect deep and abiding political realities. Ken Young and Patricia Garside's classic study of local government reform in London (Young and Garside, 1982), covering a century and a half from the Municipal Corporations Act of 1835 to the 1980s, makes startlingly clear both the permanence and the intractability of these basic political considerations. The calculations of party advantage were essentially the same in the 1880s as in the 1950s or 1980s or 1990s, the four key decades of major change in London government. Only the basic social and political geography has changed. And the politics are big politics: as Young and Garside remind us, the primary fact about the government of London is that it represents a serious potential threat to the government of the United Kingdom. That has been true ever since the medieval City and the medieval King stood eyeball to eyeball, symbolised by William the Conqueror's construction of the Tower of London soon after the Norman Conquest.

In the 150-year period covered by their book, two great upheavals occurred in London local government. The first, in 1888–1901, created first the London County Council and then the 28 Metropolitan Boroughs; the second, in 1963, replaced them by the Greater London Council and the 32 London Boroughs. Young and Garside show that in both, exactly the same political forces were at work: the Conservatives feared a strong Labour-dominated London government and sought to weaken it through creating and enhancing borough power (in 1901) and by creating a Greater London Council in 1963. The latter was secured through the report of the Herbert Commission on London government in 1960, which found an overwhelming deficiency in the lack of an overall authority responsible for strategic planning, including transport planning (G.B. Royal Commission Local Government Greater London, 1960).

The two basic questions, always intertwined, concerned the size of the top-level authority, and the relationship between it and the boroughs. Young and Garside show that the creation of the GLC was a gigantic historic anomaly, arising from Harold Macmillan's typically audacious judgement that it would fall to the Conservatives (Young and Garside, 1982: 308). But, in response to powerful interests in Surrey and Hertfordshire, the area was cut back to produce a delicate political balance that in practice produced a change of control at every election. And, in the process, several outer London boroughs surprisingly became solid areas of Labour support.

Had Young and Garside waited a few more years to extend their story, they could have described Ken Livingstone's banner-bedecked County Hall of 1983–6, shouting defiance at a Conservative government on the other bank of the river. This was the same deeply traditional battle between a quasi-independent London and a central state that had been fought ever since the City burghers first stared down the Norman kings, and it evidently so alarmed Mrs Thatcher and her government in the 1980s that they took their fateful decision to abolish the body their predecessors had created only twenty years earlier. The Greater London Council, which assumed power on April 1 1965, was officially pronounced dead 21 years later.

The official reason for its demise, as of the six Metropolitan County Councils for the provincial English conurbations simultaneously abolished, was given in a government White Paper of 1963: these bodies had been set up during 'the heyday of a certain fashion for strategic planning, the confidence in which now appears exaggerated' (G.B. Secretary of State, 1983: 2); they had been engaged in 'a natural search for a "strategic" role which may have little basis in real needs' (1983: 3).

This statement seems in itself to date from a distant age. Yet at the time, most dispassionate observers agreed that, in some important respects, the Greater London Council was a 'flawed design' (Flynn, Leach and Vielba, 1985: 64–5): it did the things it was supposed to do badly or not at all, and it tried to do too many things it should never have tried to do (Hall, 1989: 170). It was created as a slim, strategic authority to coordinate land use planning and transport. To do that effectively, it would need authority over the 32 London boroughs, through binding directives on their own plans and development control decisions. It would need direct powers to build highways, to manage traffic, and to direct the development and management of London's public transport system. In practice it did these things either inadequately, or not at all. It took over a decade, from 1965 to 1976, to draw up and win approval for a Greater London Development Plan, and even then it lacked real teeth. On transport there were difficult and complex demarcation disputes with the boroughs and with central government, which retained highway powers in outer London. But at least, after it gained control over London Transport in 1969, the GLC did achieve some success in investment – the first stage of the Jubilee Line, opened in 1978, and the improved North London Line of 1985 – and in revenue subsidy, with the development of the celebrated Fares Fair policy of 1981–3. Ironically, it was this policy, and the resulting clash with government, that was the key reason for the government's decision to abolish the GLC. But during the GLC's brief history it almost never pursued a balanced transport policy: it swung from an emphasis on large-scale highway building in the 1960s, to abandonment of that strategy and emphasis on public transport subsidy in the 1980s. But even then, it had little influence over British Rail's commuter services.

This relates to another critical fact: the GLC's circumscribed geographical remit. Even at its birth in the early 1960s, Greater London was no longer great enough (Hall, 1963): London's population was declining, growth was occurring in a wide belt outside, and was rolling progressively farther out. By the 1980s, the question was whether it made sense to attempt strategic planning over such a limited area – even though, from 1983, London's population again began to grow. Already, by 1970 the Strategic Plan for the South East demonstrated that the key planning decisions concerned the entire South East region,[1] stretching as far as 80 miles from Central London. Yet the 1970 Plan had to be produced on an *ad hoc* basis, by a joint central–local government team specially created for the purpose. Against that broad frame, it could be argued that Greater London was an irrelevant unit.

In 1986, at abolition, the GLC's powers went three ways. Most, including virtually all planning powers, went to the City and the 32 boroughs, which were charged with producing so-called unitary plans and were also given responsibility for 825 miles of the former metropolitan roads and for traffic management; though in both planning and transport, the respective Secretaries of State were given wide powers of direction. Direct responsibility for London Regional Transport (as it was then, somewhat inaccurately, known), and for some 70 miles of metropolitan road, went to the Department of Transport. Finally, some powers, such as the London Fire and Ambulance Services, were assumed by new *ad hoc* bodies. The one anomalous exception to the rule was the establishment of a London Planning Advisory Committee (LPAC), composed of borough representatives with a small but able staff, and responsible for a vestigial system of strategic planning for the capital.

It worked, after a fashion. Basic services still operated, much as they always had. Major new developments even occurred: major regeneration of London Docklands, a new Docklands Light Railway, a new road in east London. It was not easy to see that the disappearance of the GLC had made a substantive difference. Yet the major problems, which had propelled the GLC into existence, had not gone away. In the early 1990s, the Conservative government responded to pressures, especially from London's business community, by setting up a Government Office and a Cabinet sub-committee for London and encouraging business-led initiatives. But it resolutely refused to establish a 'new GLC'.

By the mid-1990s, debate about London's future reached fever pitch. There was a frantic, ceaseless networking among a whole series of *ad hoc* authorities, some (such as London First, the business organisation) created by government, others (such as Vision for London) created by grassroots initiatives. As Tony Travers puts it:

> The period from late 1992 till 1997 was characterized by the development of dozens of partnerships within London. In the absence of city-wide government, the capital's political and business class indulged in an orgy of power breakfasts, canape-laden receptions, seminars, conferences and report-launches. Capital-wide creations such as London First and the London Pride Partnership were vastly outnumbered by large numbers of local partnerships, many of which were stimulated by government funding programmes. London boroughs became more cohesive following the creation of the Association of London Government.
>
> (Travers 2004: 34)

It seemed at the time as if there was an event every night of the week.

London after 1986: the boroughs take command

Meanwhile, the boroughs soldiered on. In the absence of a metropolitan authority they individually became more used to the idea of deciding their own fate, and collectively came to play a more strategic role. Since they had only the most limited strategic control or guidance, they increasingly showed divergent styles of governance and even quality of performance. The one major exception was in London Docklands, where in the 1980s Thatcher had imposed a non-democratically elected Urban Development Corporation (the London Docklands Development Corporation – LDDC) over the east London boroughs, brutally stripping them of their planning powers. Here Thatcher's minister, Michael Heseltine, was openly contemptuous:

we took their powers away from them because they were making such a mess
of it. They are the people who have got it all wrong. They had advisory committees, planning committees, inter-relating committees and even discussion
committees – but nothing happened ... UDCs do things. More to the point
they can be seen to do things and they are free from the inevitable delays of
the democratic process.

(Quoted in Thornley, 1991: 181)

In fact, after the mid-1980s, the relationship between the LDDC and the boroughs moved in stages from open hostility through peaceful coexistence to a form
of cooperation as the boroughs realised that the corporation would soon be
wound up and they would re-inherit its powers. The LDDC, however powerful,
required the democratic imprimatur of the local authority; while the councils
progressively came to appreciate the advantages of focused, professional intervention backed by ring-fenced funding.

In *Working Capital* (Buck *et al.*, 2002) we tried to assess the performance of six
London boroughs at a time when they were being impelled into new developments: the pursuit of economic competitiveness and social cohesion, and the
drive for new ways of working and organisation accompanied by radical changes
in the accountability of local government to its electorate and central government. Councils, we discovered, found themselves performing three key functions:

- The *traditional role of providing basic public services and goods*: maintaining
  highways and public spaces, collecting rubbish and disposing of waste, regulating the environment, running libraries, cultural and sporting facilities.
- *Promoting economic competitiveness*, by providing infrastructure, planning
  land use, attracting inward investment, providing business services, and supporting enterprise through education and human capital development.
- *Engendering social cohesion*, by combating social exclusion and disorder: much
  of this comes through established services such as education and social housing, but with new aspects such as fighting crime and disorder, promoting
  equal opportunities, raising labour market skills and assisting community
  development.

Borough councils, the study found, had experienced problems in extending their
remit from the first to the second and third areas. Competitiveness and cohesion
involve strong positive and negative spatial externalities, which lap across local
authority boundaries. In London, they require policies and actions minimally at
the metropolitan level, frequently at the level of the functional urban region or
even the nation. So, left with the boroughs, activities with large positive
spillovers might be ignored or underprovided and inefficiently organised, while
those with negative spillovers – such as competitive 'city marketing' – may get
overemphasised. During the 'interregnum' of 1986–2000, London business, particularly larger more global business represented by London First, filled this gap.
But they strongly supported the creation of the Greater London Authority
(GLA), primarily to advance competitiveness. Similarly with cohesion: residential segregation (itself one way of dealing with local cohesion issues) frequently
requires redistributive action across local boundaries.

There is less conflict over the spatial scale of the first, basic, function, which is generally accepted to be intrinsically local. Providing basic services involves fewer and smaller spillovers, and addresses needs and preferences that can differ from one area to another (see Janice Morphet's contribution on the new localism in this volume). But some services need to be provided London-wide, sometimes directly (transport), sometimes indirectly (affordable housing). Moreover, service delivery is central to the modernisation agenda, with an extraordinary bombardment of legislation, regulation, inspection, monitoring and micro-management from the centre. Post-2000, a clean distinction between a strategic GLA and boroughs focused on local service delivery was never on the cards.

It was impossible for the *Working Capital* team to reach a definitive verdict on borough performance on each of these dimensions. So we sought merely to assess the degree to which key local authorities were engaging with the wider agenda of competitiveness and cohesion. To what extent, we asked, were London boroughs adapting to the modernisation agenda and its new remit? Under this remit, a council is supposed to act as an 'enabling authority', that 'leads the provision of services – "steering not rowing" – by working with and through a wide range of public, private and voluntary organisations' via a 'network of partnerships, providing a mixed economy of services' (Hill, 2000, 180; Savas, 1987; Osborne and Gaebler, 1992). The study asked whether six London boroughs – Wandsworth, Southwark, Greenwich, Redbridge, Newham and Hounslow – were moving from 'rowing' to 'steering' or, two less optimistic possibilities, 'waving' – for example, multiplying strategies and visions that lacked resources and action – or, worse, 'drowning' – sinking beneath the weight of their problems. They sought to find evidence for the emergence of coalition-based 'urban regimes' (John and Cole, 1998; Dowding *et al.*, 1999).

Politically, there were important contrasts. In autumn 2001 Greenwich, Newham and Hounslow all had solid Labour majorities; in Redbridge and Southwark, Labour was the largest party, but could be outvoted by the other parties; Wandsworth was dominated by the Conservatives. Attitudes to the modernisation agenda correspondingly varied: non-Labour authorities were less enthusiastic, Wandsworth was highly critical. But enthusiasm also varied across and within the Labour authorities; depending on whether they had 'New Labour' leaderships, able to carry the rest of their party with them.

More than this: these boroughs varied according to the problems they faced. Using the government's deprivation indices (Department of Transport, Local Government and the Regions, Indices of Deprivation, 2000), and combining income and employment deprivation, of 354 local authorities in England and Wales, Newham was the 5th most deprived, Southwark ranked 14th and Greenwich 44th. Here, demands on local services are greatest – as is the salience of the competitiveness and cohesion agenda. Wandsworth (148th) and Hounslow (153rd) were much better-placed, though not without problems.

These factors – powers, politics and position – affected both the borough's agenda and its response to issues of economic and social development. Hounslow and Redbridge had some concern about deprivation and social exclusion, but confined to limited areas and not yet priorities. In Newham, Southwark and Greenwich, competitiveness and cohesion had far greater importance. They were important for Wandsworth too – but here there had already been a marked change in fortunes, as demonstrated by its declining deprivation score. Improving service delivery was important everywhere, but with variable results.

Labour's modernisation agenda aims not just to improve traditional local government services but also to create an effective delivery vehicle for local aspects of the core concerns of competitiveness and cohesion, especially through partnerships and network governance. But this agenda is less relevant where the local economy is dynamic, unemployment is low and deprivation confined to a small minority – as in Redbridge and Hounslow. Here little seems to be needed beyond good basic service provision. So the key boroughs to test local government's capacity to respond to the competitiveness–cohesion agenda are Wandsworth, Newham, Southwark and Greenwich. They show four conclusions.

First, the contrast between an activist approach, changing the social mix of the area and relying on trickle-down from growth to open up opportunities and reduce exclusion, and one that tries to balance economic development and social inclusion through policies that address the tensions between them. The contrast here is between Wandsworth and Southwark, on the one hand, and Newham and Greenwich, on the other. In terms of impact so far, the first approach had probably achieved more. However, these boroughs were more advanced in implementing their strategies than the other two. And both were helped by location and market forces that are missing in the latter.

Second, the contrast between Wandsworth and Southwark suggests that a 'social engineering' approach to competitiveness and cohesion, rather than serving both, may improve the first at the expense of the second. In Wandsworth, social change had already been induced, through consistent policies pursued over two decades. In Southwark, planned social change was at an early stage and more contested. Moreover, while Wandsworth's gentrification policies had clear electoral advantages for the Conservative administration, this was not the case in Labour-controlled Southwark. In Wandsworth, importantly, improvements in service delivery, including services serving low-income residents (at least those still remaining in the borough) had played a role. But Southwark had been unable to combine its development strategy with delivering satisfactory services to residents, rendering it open to the accusation that it had turned its back on local communities, fostering conflict and division.

Third, there is a contrast concerning the leadership, control and culture of the local authorities. Establishing and implementing the new local governance agenda requires leadership, strong and stable control and a shared, supportive culture across the authority. Wandsworth had all these; in Newham they seemed to be emerging, while Greenwich was less advanced. Southwark showed strong corporate leadership at officer level and support for policy innovation and new thinking, but (at least in 2001) it lacked secure and sustained long-term leadership linking politicians and officers.

It proved difficult to reach well-founded conclusions about the overall impact of local programmes to address the competitiveness–cohesion agenda. How could changes initiated by the governance system be separated out from the effects of market-driven change in those places where policy was most evident – in Southwark and Wandsworth? It was easier to see the limitations imposed by a weak market position in Newham and Greenwich. Here, one concern is that locally controlled policies may seek to 'solve' economic and social problems by moving them, altering the class or ethnic composition of an area.

A fourth conclusion concerned the shift from government to governance. The team found little sign of it. Of the three functions of local government earlier outlined, service provision is the one being most effectively (if unevenly) discharged. And, especially in health, education, housing and social services, it proves to be a crucial basis on which more specific regeneration-related initiatives depend, as well as being the basis on which local government establishes its credibility with residents, communities and businesses. Thus, in pursuing new objectives, there is a danger that some authorities may neglect traditional functions. And it proves relatively hard to judge the effectiveness of new programmes to promote competitiveness and cohesion.

This relates to another problem: within each borough, a proliferation of government-ordained partnerships and specially focused programmes actually made it harder to achieve 'joined up' governance. Respondents from business, the councils, voluntary organisations and community groups despaired of the fragmentation that partnerships created, the overload placed on partners and councils; the lack of long-term financing; and the organisational costs of the system. The research team concluded that while partnership had a role, its virtues had been overstated and its costs downplayed or unrecognised.

Finally, then: is London's local government shifting from 'rowing' to 'steering', from government to governance? Or is it possibly 'waving' – multiplying 'strategies' and 'visions' without resources and action – or even 'drowning' – sinking under the weight of local problems and its own ineffectiveness? First, any shift from rowing to steering is limited within each locality and also varies across them. Second, while no borough was sinking, for some – notably Greenwich, Southwark and Newham – avoiding this fate presented an acute challenge. Third, any proactive local government cannot avoid a degree of waving, which should not simply be dismissed as evidence of failure. Indeed at the London-wide scale, some form of waving – the politics of presentation, publicity and promotion – may be essential to achieving policy goals.

Millennial changes: the Mayor, the GLA and the RDAs

The new metropolitan government, which New Labour established and which came into office in 2000, is no repeat version of the GLC. While the GLC employed several thousand staff, the GLA directly employs only 400. The Mayor oversees London-wide authorities for policing, fire and emergency services, transport (Transport for London) and economic development (this, the London Development Agency, is the equivalent of the Regional Development Agencies outside London). He sets the budget for the GLA and the functional agencies, directs the activities of the economic development and transport agencies and makes key appointments to the other two. The Assembly is extremely weak by comparison: its elected members scrutinise the Mayor's work, consider the budget (but need a two-thirds majority to change it) and can investigate issues of importance to London.

Five years after, it is still too early to reach a considered view on the new system. But it is useful to recall the conclusions of *Working Capital* (Buck *et al*. 2002) and add some further observations on the record since that study was published.

First, it is clear that very few of the city-wide partnerships that arose in the gap between 1986 and 2000 disappeared after 2000. Most survived, by continuing their previous role, by reinventing themselves, or by making themselves useful to the Mayor. Second, the weak powers of the Mayor and the few staff directly working for him mean that he is dependent on other agencies to implement his strategies. A key issue facing the Mayor is, as a respondent remarked, 'how do you influence the agenda of others to deliver your policies?' To do so the Mayor must engage with partnerships, business organisations and sub-regional entities – but above all with the boroughs. In the early years some Mayor–borough relations were antagonistic, some more harmonious. The Mayor's level of spending, his frequently expressed desire to see much more affordable housing, his support for tall buildings and his successful introduction of the congestion charge on cars coming into central London, together with his general pro-growth agenda have emerged as areas of conflict, actually or potentially. Much less ambiguous is the position of the London Assembly, widely seen as weak and ineffective: as one informant stated, 'you tend to hear about them in negatives rather than positives – when you hear about them at all'. Another referred to the Assembly as 'a reactive scrutiny committee that occasionally makes mischief'. Its major success during the early years was in forcing the Mayor to cancel a firework display. Just possibly, the changed political complexion after the May 2004 election, when Labour lost overall control, which deprived the Mayor of a built-in approval for his budget, may change matters. But, given the growing involvement of the boroughs with the cohesion–competitiveness agenda, it is they rather than the Assembly that tend to provide the 'checks and balances' to the executive power of the Mayor.

In the first years the main political issue was the Mayor's titanic battle with the government – specifically with the Chancellor of the Exchequer, Gordon Brown – over the management and financing of the Underground, a battle that the latter eventually won. Ken Livingstone hired an American with long experience of the New York and Boston transit systems, Robert Kiley, as Transport Commissioner to head Transport for London (TfL); they wanted to modernise and renew the Underground themselves via the American solution of a bond issue. The Treasury proposed to split the tube, with private consortia receiving long franchises to upgrade the track and train infrastructure while TfL simply ran the system. Eventually Livingstone gave way as part of a deal that readmitted him to the Labour Party after standing as an independent against the official candidate in 2000. In addition the Mayor was developing (as he was required by law to do) a series of strategies, the most important being those for transport, economic development (via the LDA) and spatial development. By late 2001 the first two were completed; the last appeared in draft in 2002, went through Examination in Public in 2003 and was published in final form early in 2004 (see also David Goode and Richard Munton's discussion of the strategy in this volume).

The economic development strategy takes a strongly pro-business approach, unlike that taken by the Mayor and the Chief Executive of the LDA when they led the GLC in the 1980s, thus remaining remarkably close to the government's competitiveness and social exclusion agenda. Its main concerns include: improving the skills of Londoners, especially school leavers; using regeneration programmes to combat social exclusion; focusing large-scale physical regeneration on certain corridors – Thames Gateway, Lea Valley, West London Approaches and Wandle Valley – through supporting the partnerships already in place there; promoting and supporting business development and growth, with a particular emphasis on minority ethnic business; fostering higher and further education and knowledge transfer; and working with business to help improve London's schools.

In 2001 the London Assembly published a critical response to the draft strategy, stating that it was based on 'existing priorities and patterns of activity, offering a fragmented "pick and mix" approach to economic development'. It lacked an overall vision and a balanced approach to economic development across London. The role of the suburbs and local employment centres needed greater consideration and the principles of sustainable development, health and equal opportunities were not properly integrated throughout the strategy. The draft EDS – and, even more so, the subsequent Spatial Development Strategy or London Plan – focused on developing London as a 'world city' and thus on the inner city and CBD. This perhaps was inevitable; it seems to support an observation of the Mayor of Denver at a pre-GLA London seminar that 'elected mayors inevitably come to see their city through the warped crystal of its Central Business District – and airport'.

A MORI (2002) poll found that 77 per cent of respondents could name the Mayor, and 35 per cent were satisfied with his performance, with only 14 per cent dissatisfied. Peter Kellner commented in the *Evening Standard*, 'Most politicians would sell their grandmother for such a rating' (22 January 2002). Knowledge about the Assembly was more limited: less than one third could express any view. Most respondents had an idea (not always accurate) of the Mayor's responsibilities; in 2002, only 41 per cent could even attempt to say what the Assembly did (MORI, 2002).

The London Plan 2002–4

The Mayor's *Draft London Plan* was published in June 2002 (Mayor of London, 2002). It was only the third official plan ever to have been produced for Greater London, following the 1944 *Greater London Plan* and the 1969 *Greater London Development Plan*. In tone and content it differs greatly from both of these earlier models, reflecting major differences in authorship, in the situation that it addresses, and in the nature of the authority producing it. The 1944 Plan bore the stamp of a visionary professional planner, Patrick Abercrombie, while the GLDP (as its successor was appropriately known) was the product of the Greater London Council bureaucracy and political compromise; in contrast to both, the new *London Plan* appears in the name of the first directly elected Mayor and clearly represents his personal aspirations for the capital. In terms of situation, the Abercrombie Plan foresaw strong post-war growth pressures on available space in London, to be dealt with via planned decentralisation into the Outer Metropolitan Area (OMA); by the time of the GLDP, however, planned dispersal had been greatly overtaken by voluntary out-movements into a booming OMA, and people were starting to get worried about the effects of unbalanced population and job loss within London. Again in sharp contrast, the Livingstone Plan (or 'Ken's Plan' as it was soon called) appeared at a time of substantial growth in London's population and employment, and enthusiastically adopted that growth as its key theme. Further decentralisation or expansion into the surrounding region, which it identifies as the chosen path during much of the nineteenth and twentieth centuries, is rejected as an option, because it would be both environmentally unacceptable (particularly in that region) and contrary to central government policy.

A key theme of the Plan is the inevitability of large-scale growth: 809,000 additional residents (54,000 each year) and 636,000 extra jobs (42,000 a year) are forecast for the year 2016 (Mayor of London, 2004, Table 5A1, 224). The issue, the Draft Plan argues, is simply whether this is specifically planned for and supported by appropriate infrastructure provision, or is simply allowed to overtake the city, which would entail 'a deepening of the many adverse economic, social and environmental problems already facing London' (Mayor of London, 2002: 35). Thus the Plan is driven by a strong determinism about the scale of growth – which it justifies by projections for population based 'on long run and deeply rooted trends' (2002: 17) and for the economy on the 'persistence of strong structural trends over a period of three decades … driven by deeply rooted changes in the international and UK economies and society' (2002: 23).

At the subsequent Examination in Public, this view was questioned. In the case of employment, the record has been highly volatile over the past twenty years – with two booms, one bust and another bust quite possibly on the way – making trend-spotting a very uncertain business. The authors of *Working Capital*, and other critics, suggested that the long-term trend is flatter than the Plan assumes, and that it should be applied to a lower base position than the exceptionally high level of demand observed in 2002. But whatever is technically the best forecast now should be regarded as having a very large margin of error attached, rather than as representing a pre-set growth path.

In the case of population, there is a more specific source of uncertainty: the contribution to the rapid growth in the years 1997–2002 of asylum seekers, a factor not mentioned in the Plan. This flow to London could well shrink in an economic downturn, or in the face of strong diversionary policies – though it could also conceivably grow. A slowing of current high rates of natural increase is also possible, as immigrants became more like established Londoners – though potential householders and job-seekers in 2016 have already been born. And there is also a very important question about the extent to which other Londoners will choose to live within the more densely developed city which the Plan envisages, rather than move out in much larger numbers into less dense developments in the counties around London. If that proves to be their preference, then there may be little that planners can do to prevent it happening, since out-movers do not have to go into newly constructed dwellings requiring planning permission; they can buy older houses, forcing out less-well-paid local competitors in the property market.

Within the Plan's framework of thought, however, this possibility is simply unthinkable. One reason, as well as the Plan's commitment to the 'compact city' ideal, is that the prospect of very large-scale growth underpins the claim for much higher investment in London infrastructure, implicit in the draft Plan but becoming more explicit in the final version. In other words: the Plan says that London's growth is inexorable, that it is a good thing anyway, and that the resources have to be found to underpin it.

The Plan's strategy for accommodating growth can be summed up in two phrases: 'go east' and 'raise densities'. The East and Central sub-regions together are supposed to take 60 per cent of all additional residents (484,000 of 809,000), 57 per cent of all new homes (196,000 of 345,000) and 77 per cent of the new jobs (488,000 of 636,000), of which the East alone would take 39 per cent (there is some statistical sleight of hand here: the City of London and the City fringe, with two of the 'major opportunity areas', are counted in the East). The main development thrust is eastwards: 100,000 new jobs in the Isle of Dogs, 30,000 at

Stratford (with a European business quarter around the new Eurostar station), another 19,500 in the Lower Lea Valley and the Royal Docks: 150,000 in all, nearly one quarter of the London total, in what is effectively a huge eastward extension of London's central business district. Likewise with the new homes, most of which are planned to be even farther east, 10,000 at Barking Reach and 7,500 on the Greenwich Peninsula (Mayor of London, 2004, Table 5C1, 247).

This represents a dramatic reversal of the long historical westward drift of development in London – though it builds on an early reversal in the Docklands project. There are two mutually reinforcing justifications: land is available here, and this is the zone of major multiple deprivation. To achieve it, the plan proposes fundamental improvements in physical accessibility in the east, particularly into central London, through major transport schemes: the east–west express Crossrail line (and another, north-east to south-west, in the more distant future); two new road crossings of the Thames below Tower Bridge, plus an extension of the Docklands Light Railway to Woolwich; and two new bus transit schemes for east London, one north and one south of the river.

But there are two problems. First, the transport links are not closely tied to the major opportunity areas: Crossrail will not serve Barking, a key site for large-scale new housing, and to remedy this a further extension of the DLR, not included in the Draft Plan, had later to be added. Second, there is no assurance that the links will be available in time to support the new jobs and homes: Crossrail cannot be completed before 2013, towards the end of the Plan period, and still had no guarantee of financing in summer 2005 even after the announcement of London's winning Olympic Games bid for 2012 with its centring in this very area. Other schemes – the East London Line extension (making possible an outer circle or Orbirail), extensions of the Docklands Light Railway, the east London busways – have a better chance of being realised in time. But the question is whether in themselves they could provide the necessary capacity to underpin major new development.

Crossrail is symptomatic of a general issue: the Mayor and central government share the view that private capital could and should be attracted to make a major contribution to such major infrastructure schemes, and on this basis the government are introducing a hybrid Crossrail Bill into Parliament in the 2004–5 session. But there is presently no guarantee that the money will be forthcoming: the expert review conducted under the chairmanship of Adrian Montague makes it clear that the funding gap is as high as £700–800 million, and no obvious solution is in sight. At the conclusion of the Examination in Public, the panel accepted the Mayor's argument that most of the proposed growth could be supported on the basis of more modest investment in so-called intermediate modes such as busways or DLR extensions. But this would seem to call in question the case for investment in Crossrail.

All this is related in turn to the argument for increasing densities, since these are held to help generate the demand for new fixed-route transport links. But two-thirds of the extra housing capacity will have to be found outside the 28 designated opportunity areas – meaning a very rapid and sharp increase in densities right across London. It is quite unclear how this could be achieved, given the fact that extensive areas of inner London are designated conservation areas and that most housing consists of owner-occupied homes where occupiers would resist large-scale redevelopment. The plan places weight on reducing the numbers of empty dwellings – but it is quite inexplicit on how to achieve this. The plan would like to see densification around key town centres in Outer London with good access to public transport. But this would depend on cooperation from the boroughs, whose voters are likely to be very unenthusiastic, and on better accessibility to them of bus services – for which no provision is made in the transport strategy. Here, as elsewhere, there seems to be inadequate liaison between the Mayor's planners and his own agency, Transport for London.

All this adds up to a serious question about housing targets, which are hugely ambitious: a minimum of 458,000 new homes from 1997 to 2016, representing 23,000 a year. Even this figure is well below estimates of actual need, but well above recent construction levels. The Mayor's Housing Commission calculated the need at 40,000 units per year up to 2016, of which 28,000 a year should be 'affordable'; the actual figure for 2000 was 12,000, 30 per cent of target, of which a minuscule 2,743 were affordable (Joseph Rowntree Foundation, 2002: 20). The plan estimates the affordable housing need at 25,700 dwellings per year – but, considering capacity and economics, reduces this to 10,000 a year, four times the estimated 2000 rate, requiring an estimated £6 billion in subsidy.

In summary, the plan is vulnerable to changes – especially to economic downturn. Its 'Little London' strategy is driven by a campaign to win funds for infrastructure. Its eastern thrust is generally supported, but will need funding on a scale that is not guaranteed. And there is a sense that its development proposals and its transport proposals do not entirely hang together. The big question is whether the plan can be implemented, when crucial levers are in the hands of central government, the boroughs and the private market. That will take twenty years to answer.

The unaddressed regional agenda

Meanwhile the outer part of the wider London Region has been partitioned between two RDAs and two Regional Planning Bodies – the South East England RDA and RPB (SEEDA and SEERA) and the East of England RDA and RPB (EEDA and EERA). The latter covers the counties to the north and north-east of Greater London, SEEDA/SEERA those lying from Buckinghamshire round to Kent, so including Reading/Wokingham and Dartford. The RDAs, like the LDA, exist to further economic development and regeneration; promote competitiveness and employment; improve the skills of the regional labour force; and contribute to sustainable development. In line with the government's overall approach to social exclusion, RDA concerns here centre on raising employment and skills. By 2000 all RDAs had produced economic strategies. At first their relatively small budgets were taken up by inherited commitments (the SRB and other programmes), though this has now begun to change.

The balkanisation of regional government in the South East will make an effective strategy for the functional region hard to achieve. Over the past 25 years governance in this region has been characterised by the politics of growth, with conflicts over where growth should occur (away from the west and to the east) and, in the growth areas, over further housing development. The government's 2003 Sustainable Communities statement aims to develop an overall strategy for three major development corridors that cross the boundaries of these two regions, but it remains to be seen how effectively the delivery can be coordinated.

In a huge, complex, polycentric place like London, governance will always be fragmented. But the Mayor's strategies compound this by assuming that 'London' ends at the GLA boundary, which (we have shown) is far from the case – while those for the South East and Eastern regions turn their back on London. The fact that both central and decentralised government each operates at the Greater London level (through the Government Office for London and the GLA) exacerbates this situation and also leads to a confusion of roles at this level – especially when the GLA's major 'diplomatic' concerns are either with 'ministries' which are not themselves regionalised (i.e. the Treasury) or with ministers personally. There remains a huge potential gap in strategic capacity and political leadership in relation to major issues such as planning, housing, transport and major public investments. That gap is the need for some single unit of governance at the level of the entire wider Functional Urban Region. This would seek to shape the cross-regional distribution of economic activity, to address the resulting infrastructural and environmental needs, and generally to help share out more widely the benefits of a dynamic and competitive regional economy. It is perhaps unrealistic to envisage the creation of a political authority for such a vast urban region, given its scale relative to the UK and the sensitivity of any national government to developments in its own city-region. But it is realistic to think about an effective form of governance for such a region.

Conclusion: moving towards consensus?

For Tony Travers, London government represents an abiding paradox, that of 'Governing an Ungovernable City' (Travers, 2004). And this remains as true of its latest manifestation, as of previous prescriptions:

> The first three years of the Greater London Authority suggest that London's government remains balkanized and weak. London's history up to 1965, the creation and abolition of the GLC, the 'interregnum' years from 1986 to 2000 and the first administration of the Mayor of London point determinedly in one direction: the largest city in Europe simply defies all efforts at giving it an effective and consistent system of government. The best that could be said of this fragmentation and change is that the atomized and constantly reformed system of government had much in common with the habits of individualism recognized by urbanists in London's social development.
>
> (Travers, 2004: 182)

Yet, remarkably, London seems to have been highly successful: it has successfully made the transition from an imperial to a post-industrial global city, and has remained largely liveable in the process. No previous model seems to have been very successful in guiding the process finally – and this may be equally true of the latest prescription (2004: 183) – because as in the past, central government will not let go:

> Overall, the GLA can be counted as only a limited success in terms of the government's original objectives. The reason for this heavy qualification is that the objectives set in 1997 and 1998 were simply too ambitious for a

government that was unwilling to cede much power to the new London government ... London government reform in 2000 was only partly successful – in the government's own terms – because the government itself was unsure just how far it was prepared to trust the people of London to govern themselves.

(Travers, 2004: 199)

Partly, this harsh verdict may reflect the highly unusual circumstances of the Mayor's first four years. Elected as an independent in the face of official Labour opposition, he was forced to govern though a rainbow alliance, on the one hand with fellow left-wing idealists in the Green Party, but on the other – somewhat anomalously – with the financial and economic power of the City, across the river from his City Hall, to promote a 'global city'-oriented regime. Consequently, the scripture might perhaps have been written differently only a year later. For at least three important shifts occurred in 2004, quite fundamentally strengthening the Mayor's role.

First, he effectively conceded defeat in his long battle to prevent the part-privatisation of the Underground. It had been an honourable battle in which, remarkably, he had won support of Londoners across the political spectrum. Hence his concession appeared almost as conferring heroic status on him. Second, it allowed him to make peace with the Labour mainstream again, allowing him to fight a successful election for a second term as an official Labour candidate. Third, his successful introduction of the Central London congestion charge, in which he ignored dire predictions from government politicians who clearly thought he would fail, was another personal triumph which enhanced this heroic status. And finally, all these things came together in an agreement with government that allowed him to borrow substantial funds to finance major transport improvements, incidentally underpinning London's bid to host the 2012 Olympic Games. From all this, Ken Livingstone emerged at the end of 2004 as a much stronger Mayor, no longer an independent maverick but an ally of a Labour government, and incidentally freeing him in part from reliance on his former coalition with the Green Party – a freedom which he exploited in his critical casting vote in favour of construction of a new Thames Gateway Bridge carrying the North Circular Road across the river.

Thus, from a first-term position as challenger to the hegemonic role of central government, recalling the heroic days of 1983–6, the Mayor has moved into a much more established role as part of a broader central–local government alliance. That may well set the scene for a new episode in the long struggle between state authority in Westminster and the dynamic city power a short distance down the river to the east. It is a rich irony, perhaps, that the Mayor's view, across to the sky-scrapers of the City, looks over and across the Norman Tower of London – an abiding token of a thousand-year power struggle that might now, at last, reach some kind of resolution. Or, at least, for a time.

Notes

1 This was the 'old' South East Region, a ring surrounding London. In 1994, through an ill-advised change, the north-eastern sector of this ring – the counties of Essex, Hertfordshire and Bedfordshire – were transferred to join East Anglia in a new East of England region.

References

Buck, N., Gordon, I., Hall, P., Harloe, M. and Kleinman, M. (2002) *Working Capital: Life and Labour in Contemporary London*, London: Routledge.

Dowding, K. et al. (1999) 'Regime politics in London local government', *Urban Affairs Review*, 34: 515–45.

Flynn, N., Leach, S. and Vielba, C. (1985) *Abolition or Reform?: The GLC and the Metropolitan County Councils*, Local Government Briefings, 2, London: Allen and Unwin.

G.B. Department of the Environment , Transport and the Regions (2000b) *Measuring Multiple Deprivation at the Small Area Level: The Indices of Deprivation 2000*, London: Department of the Environment, Transport and the Regions.

G.B. Royal Commission on Local Government in Greater London (1960) *Report* (Cmnd. 1164), London: HMSO.

G.B. Secretary of State for the Environment (1983) *Streamlining the Cities: Government Proposals for Reorganising Local Government in Greater London and the Metropolitan Counties* (Cmnd. 9063), London: HMSO.

Hall, P. (1963) *London 2000*, London: Faber and Faber.

Hall, P. (1989) *London 2001*, London: Unwin Hyman.

Hill, D. (2000) *Urban Policy and Politics in Britain*, Basingstoke: MacMillan.

John, P. and Cole, A. (1998) 'Urban Regimes and Local Governance in Britain and France: Policy Adoption and Coordination in Leeds and Lille', *Urban Affairs Review*, 33(3): 382–404.

Joseph Rowntree Foundation (2002) *Britain's Housing in 2022: More Shortages and Homelessness?* York: JRF.

Mayor of London (2002) *The Draft London Plan: Draft Spatial Development Strategy for London*, London: GLA.

Mayor of London (2004) *The London Plan: Spatial Development Strategy for Greater London*, London: GLA.

MORI (2002) *Annual London Survey 2001: Londoners' Views on Life in the Capital*, London: MORI.

Osborne, D. and Gaebler, T. (1992) *Reinventing Government*, Reading, MA: Addison-Wesley.

Savas, E. (1987) *Privatization: The Key to Better Government*, New York and London" Chatham House Publishers.

Thornley, A. (1991) *Urban Planning under Thatcherism: The Challenge of the Market*, London: Routledge.

Travers, T. (2004) *The Politics of London: Governing an Ungovernable City*, Basingstoke: Palgrave Macmillan.

Young, K. and Garside, P.L. (1982) 'Metropolitan London, Politics and Urban Change, 1837–1981', *Studies in Urban History*, 6, London: Arnold.

Part 3

Institutions of governance and substantive policy roles

The bonding power of a republican community of free and equal members is supposedly insufficient to assure the political stability of a state. I consider this assumption – that democracy needs to be backed up by the bonding energy of a homogenous nation – to be both empirically false and politically dangerous.

(Jürgen Habermas)

12 Pressure for housing in the English regions

Back to the future

Mark Baker and Cecilia Wong

Introduction

The pursuit of distinctive regional planning agendas has gained unprecedented momentum since the election of the Labour government in 1997. A weightier emphasis on the regions has been matched by moves to 'modernise' the planning system through an overhaul of planning policy guidance (DETR, 1999a, 1999b, 2000a, 2000b) and more recently the new spatial planning reforms (HM Government, 2004). Government policy has assumed a dual focus: first, emphasising the importance of achieving geographical sensitivity and, second, promoting integration (or interconnectivity) within and between policy instruments. However, such planning policy guidance has been criticised in the past for its failure to focus clearly on practical detail (House of Commons Select Committee, 1998) and the lack of sensitivity to significant inter- and intra-regional variations in the nature and scale of planning issues.

The resurgence of a regional and spatial policy perspective in the last few years has occurred at a time when concern, both public and professional, has also been sharply focused on housing issues. The Department of the Environment's 1992-based household projections (DoE, 1995) suggested that no fewer than 4.4 million additional households would form in England between 1991 and 2016. These figures were subsequently revised downward to a projected growth of 3.8 million households in the 1996-based projections (DETR, 1999c). However, when re-examining housing demand and need in England between 1996 and 2016 (based on 1998 population projections), Holmans (2001) suggested that the official figures underestimate the scale of future growth within this period by 413,000 households and 46 per cent of these extra households will form in London, accentuating the current access and affordability problems being faced in the capital (Hamnett, 2001). Subsequently, ODPM's latest interim 2002-based projections (ODPM, 2004a) show a projected increase of 189,000 households per annum to 2021 compared with 150,000 implied by the previous 1996-based evidence. Some 55 per cent of this household growth is projected to occur in London, the South East and the Eastern regions.

It is against these wider policy contexts that this chapter aims to assess the extent to which national policy guidance has been sensitive to the spatial complexity inherent in key housing/planning issues and the extent to which the regional housing policy framework appears to add up to a coherent national picture for England as a whole to meet with the challenges. This is achieved by reconstructing the historic contextual and policy backgrounds which led to the housing land requirements as set out in the latest round of Regional Planning Guidance (RPG) in the late 1990s and to contrast and compare the different approaches and outcomes of housing land allocation policies in the nine English regions. This analysis of the complete cycle of housing land allocation exercises in the latest round of RPG production offers a one-off snapshot of the nature and scope of the national and regional planning frameworks for addressing housing issues as well as identifying potential lessons for the preparation of forthcoming Regional Spatial Strategies (ODPM, 2004b).

This chapter will, therefore, first outline a number of the key housing issues that bore significant regional differences in the early 1990s. It will then assess the extent to which the institutional arrangements, planning processes and policies for regional housing have addressed the differential regional housing issues. This will be discussed within the particular context of the Planning Policy Guidance Note 3: Housing (DETR, 2000a). The chapter will then examine inter- and intra-regional variations over the spatial frameworks adopted to meet with housing needs and demand. This is informed by an empirical analysis of the discrepancy between housing requirements set out in the various RPGs and their referenced projection statistics – the 1996 trend-based household projections at the regional and sub-regional level. Finally, implications from the analysis will be drawn out to help shed light on the latest development of Regional Spatial Strategies (RSS). Current issues such as the planned restriction of housing supply, regional targets for brownfield development, the interface between housing demand and supply, the availability of affordable housing and the severe problems of vacancy and low demand will be examined in the light of the emerging institutional frameworks from the Sustainable Communities Plan (ODPM, 2003) and the Barker Review (Barker, 2004) in the post-RPG era.

The context of Regional Planning Guidance preparation throughout the 1990s

Any such analysis of the spatial frameworks of regional housing requirements has to be grounded in the wider national spatial landscape of key housing/planning issues and the policy and institutional frameworks within which they operate. This section thus provides a summary of these key issues: first the spatial context, then the national policy frameworks and guidance.

Spatial context

The national household growth figure has tended to dominate headlines whilst obscuring the fact that the distribution of this growth is socially and geographically complex. The spatial dimension has been assigned particular importance amongst academics and policy makers, whose concerns have tended to focus on the necessary extent of future house building in the regions (e.g. Champion *et al.*, 1998; Holmans and Simpson, 1999). Indeed, four major issues have been identified in recent literature, which would appear to support the notion of regional spatial differences, most notably in terms of a readily defined north–south divide in housing needs and demand. Each of these is now examined in turn.

Regional growth trends

It was noted that the 1996-based household growth projections (based on which current RPG housing requirements were made) pointed to a deceleration in the number of households forming across the whole of England over a 20-year period. However, this change was projected to have an uneven impact. The figures presented in Table 12.1 show that the national downward revision implies a far greater impact on the North East (−36.1 per cent) and the North West (−29.4 per cent). In London and the South East on the other hand, the 1996-based calculations show an increase of 15.8 per cent (across these two regions) compared with the 1992-based figures. The figures suggest that more people will choose to cohabit and hence the number of single person households, as well as the global figure, is reduced. In the northern regions, this revised assumption only partially explains the fall in projected growth. Some of this deceleration is also the result of overall population change as well as patterns of inter-regional migration.

Champion *et al.* (1998) have demonstrated that population movement has a greater impact on regional totals than the net increases which result from births and deaths. It is also clear that the flows between different regions vary considerably in both nature and extent. The study reveals, for instance, that the northern regions (the North East, North West and Yorkshire and the Humber) have tended to lose population through domestic (UK) migration and have gained very little as a result of international movements. In contrast, the South West, East Anglia and the East Midlands all have a history of making net gains from both domestic and international migration. The 'behaviour' of the South East (including London) differs significantly from all these other regions: whilst it appears to feed the rest of England with a steady stream of out-migrants, it invariably receives the lion's share of international newcomers. To some extent, the West Midlands region displays characteristics similar to those of the South East.

Table 12.1 Comparing the projected household growth (1991–2016) by the 1992- and 1996-based projections

Government office region	1996-based projections ('000)			1992-based projections ('000)			Growth differentials: 1992- and 1996- based figures (%)
	1991	*2016*	*% change*	*1991*	*2016*	*% change*	
North East	1048	1154	10.1	1047	1213	15.9	−36.1
North West	2720	3061	12.5	2720	3203	17.8	−29.4
Yorkshire and the Humber	1993	2322	16.5	1993	2380	19.4	−15.0
East Midlands	1596	1973	23.6	1596	2014	26.2	−9.8
West Midlands	2042	2354	15.3	2043	2410	18.0	−15.0
Eastern	2035	2602	27.9	2035	2617	28.6	−2.6
London	2841	3520	23.9	2842	3471	22.1	7.9
South East	3035	3905	28.7	3036	3843	26.6	7.8
South West	1903	2421	27.2	1903	2448	28.6	−5.0
England	19213	23313	21.3	19215	23598	22.8	−6.5

Source: DETR, 1999c; DoE, 1995

Most recently, the ODPM's latest interim 2002-based projections (ODPM, 2004a) show a projected increase of 189,000 households per annum to 2021, compared with 150,000 implied by the previous 1996-based evidence, but again significant differences at the regional scale are evident. In particular, some 55 per cent of this household growth is projected to occur in London, the South East and the Eastern regions.

Dwelling vacancy and low demand

Another concern that has gained much prominence in recent years is that of vacancy and low demand within the northern regions. Almost a quarter of a million dwellings (243,557) situated across the North East, North West and Yorkshire and the Humber were recorded as vacant in the 1997 Housing Investment Programme returns. Population decline within these regions (a product of out-migration) has resulted in a rising number of empty and difficult-to-let dwellings. Generally 'slack' demand also leads to higher tenancy turnover rates (as people move or switch between units or tenures because of the abundance of available accommodation). Holmans and Simpson (1999) demonstrated that the drift south from northern towns and cities is becoming increasingly pronounced, and this coincides with a sharp rise in departure and vacancy rates affecting local authority housing in the north and, to a lesser extent, the Midlands. Their analysis also seeks to explain this drift in terms of the favourable balance of employment opportunities currently being enjoyed by the south of England.

However, the reduction in housing demand in the northern regions has been neither general nor uniform; rather, it is the least popular locations that have suffered most whilst more popular and relatively prosperous areas continue to thrive. Crook (1998) argued that the general easing of the housing market in the north has enabled marginally better off households to leave the poorest and most deprived neighbourhoods. This has resulted in increased market demand for additional and better quality housing in low-density developments and more popular neighbourhoods. At the same time, void rates in higher density, poorer, urban neighbourhoods have risen sharply. The recent creation of nine housing market renewal (HMR) pathfinders, backed by £500 million of funding, not only represents an acknowledgement by government that these are serious problems but also reinforces the geographical divisions since all nine lie within the three northern regions and the West Midlands.

Stock condition

Similarly, and perhaps unsurprisingly, overall regional geography also remains extremely important in relation to the distribution of unfit dwellings. Leather and Morrison (1997) have detected a clear 'Wash–Bristol Channel' cleavage in terms of housing conditions. In the north – and particularly in northeast Lancashire and industrial cities such as Manchester and Liverpool – there appear to be far greater concentrations of unfit housing. Much of this housing is older (pre-1919) and arranged in terraces, beating the post-war slum clearance programmes, but now in desperate need of repair. The 1996 English House Condition Survey showed that 51 per cent of all unfit dwellings in England were built before 1919 (DETR, 1998). Quite clearly, there is a strong (and not

surprising) link between stock age and condition, and hence the older stock in
the north tends to be in a worse condition than the newer stock in the south. A
further problem is that because housing in the north generally has a lower value
than the same (or similar) housing in the south, the cost of repairs as a proportion of total value is greater (CPRE, 1998). For property owners – or the private
sector in general – this begs the question as to whether it is actually worth
spending money on repairs and refurbishment. The financial benefits of maintaining property in the south might appear clear cut, but not so in the north.

Land recycling

The fourth way of differentiating the English regions in planning/housing terms
relates to the capacity of different regions to achieve the government's 60 per
cent global target for the provision of new housing on previously developed land.
Here, however, there is perhaps a less obvious north–south split with those
regions incorporating large metropolitan areas (London, West Midlands, North
West) contrasting with others such as the South West, South East and Eastern
regions. The data presented in Table 12.2 reveals that in the early 1990s, there
were significant regional variations in terms of new house building on 'brownfield' sites. The lowest figure achieved was 37 per cent whilst the 'best
performing' region reached 87 per cent: had the government's current national
target of 60 per cent been in place during this period, then only the North West
and London would have made the grade. Research by the Civic Trust (1999)
demonstrates that variations in economic, social and environmental conditions
across the different regions invalidate the whole notion of a global target, and
point to the need for specific regional strategies aimed at addressing the brownfield housing issue. The difficulties are almost ubiquitous: the report of the Urban
Task Force (2000) suggested that, at best, a figure of 37 per cent will be achiev-
able in the South East. Although there is a greater abundance of brownfield sites
in the North and the Midlands, these are far more likely to be former industrial

Table 12.2 New housing on previously developed land, 1991–1993

Government office region	% new dwellings on previously developed land
North East	48
North West	60
Yorkshire and the Humber	47
East Midlands	35
West Midlands	52
Eastern	48
London	87
South East	52
South West	37
England	52

Source: HM Government, 1998, 14

and heavily contaminated sites. Variations in market conditions mean that it is easier to develop brownfield sites successfully in the Midlands. The general lack of demand in these inner areas in the North, coupled with the constraints of land contamination, mean that the prices achieved for market housing in these locations are not high enough to absorb the added development costs associated with land remediation (Civic Trust, 1999). It is evident that the availability of brownfield land (in terms of numerical capacity) is not equivalent to actual deliverable capacity (Wong and Madden, 2000).

If so-called RPG 'brownfield land' targets for future housing development are considered, a more clearly defined north–south split again emerges. Against the overall national target of 60 per cent of new housing development to be on previously developed land, the individual targets on a region-by-region basis can be seen to vary from as low as 50 per cent in the South West and East Anglia to targets of 65 per cent in the North West and North East and 68 per cent in the West Midlands. The remaining regions, East Midlands, Yorkshire and the Humber, and the South East (excluding London) all stick with the national target of 60 per cent. Interestingly, a quick calculation of the planned numbers of dwellings to be built on previously developed land for England as a whole, based on the cumulative annual housebuilding requirements and brownfield targets for each region (excluding London) as set out in RPG, results in a planned brownfield supply which almost exactly matches the national target of 60 per cent. Whether this is accidental or the result of considered Government intervention to ensure overall compatability is unknown.

National policy frameworks and guidance

As discussed in our earlier work (Baker and Wong, 1997), the Government has long adopted a top-down approach towards housing land supply. The allocation of additional housing land is largely the outcome of a top-down approach that begins at the national/regional interface with the preparation of Regional Planning Guidance (RPG), which is currently being replaced by Regional Spatial Strategies (RSS). The factors which have been used to derive the housing requirements set out in these statements of regional spatial planning policy generally fall into two categories: technical and policy-based. As far as the technical side of the process is concerned, the nationally produced household and population projections are frequently the most important considerations and tend to be used as a starting point, or baseline, before other technical assumptions and any policy-driven or political considerations are incorporated. These 'technically driven' assumptions are usually supplemented by 'policy-based' considerations, although the nature and extent of these appear to vary rather more on a region-by-region basis. In broadest terms, economic factors or environmental constraints were particularly common, and the relationship between metropolitan areas and the adjacent shires in terms or urban regeneration and housing capacities is also evident. Since population and household projections are found to be very sensitive to different underlying assumptions which are themselves heavily influenced by the recent scale and direction of change, we argued that a more policy-led approach was needed at the local/regional level to assess the effects of different policy decisions both on the projections and on their outcomes.

Following the publication of Planning for the Communities of the Future White Paper (HM Government, 1998) and Planning Policy Guidance Note 3: Housing (DETR, 2000a), the Government introduced a revised notion of 'plan, monitor and manage', as a means of handling the housing land allocation process. Further elaboration was made in the White Paper of the need to have housing provision indicators such as house and land prices, housing standards and local housing needs in RPGs (HM Government, 1998, para 31). This signified a departure from the government's long-standing practice, allowing a greater de-standardisation and de-institutionalisation of national household projection figures from local and regional housing provision policy (Wong, 2000). The White Paper also stated clearly that such projection figures '... were for guidance rather than, in effect, prescriptive ...' (HM Government, 1998, para 27). As a result, it became the task of local authorities and regional planning bodies to put forward their individual cases to justify their housing allocation policies. This sea change in government approach signified a positive, democratic move, apparently allowing more scope for local and regional stakeholders to debate local issues relevant to their housing needs.

Whilst welcoming this less mechanistic approach towards local planning issues, much of the more general ideas set out in the White Paper were criticised for their failure to focus clearly on practical detail (House of Commons Select Committee, 1998). For instance, how should local planning authorities ensure sufficient supply of affordable housing (Perry and Simpson; 1999; Gallent, 2000), deliver the 60 per cent national target of new build on brownfield land (Civic Trust, 1999; House Builders Federation, 1998), or implement the 'sequential approach' in the preparation of development plans (Lainton, 1999; Pike, 1999)? These questions have been framed in both general terms, and in relation to geographical, and particularly regional, differences.

Regional spatial frameworks for housing requirements: RPGs

The discussion above shows that, despite the effort made by central government to develop some kind of action programme to alleviate the housing pressure, by and large such general policy guidance is not too sensitive to the spatial complexity inherent in key housing and planning issues. Following the publication of PPG3 (DETR, 2000a) and PPG11 (DETR, 2000b), a more bottom-up policy-based approach of housing land allocation was supposed to take place. It is, thus, interesting to consider the extent to which regional planning bodies exercised their newly found flexibility to tackle housing issues within their regional spatial frameworks and whether there are significant variations across different parts of the country in their approaches to meeting housing requirements. These issues are examined empirically by comparing the annual housing requirement figures between 1996 and 2016[1] in RPG and the 1996-based DETR household projection[2] figures covering the same period of time. The analysis provided covers both inter- and intra-regional variations in order to detect any emerging patterns of development.

Inter-regional comparisons

The 1996-based household projections provided the main technical information considered by regional planning bodies when devising the spatial distribution of

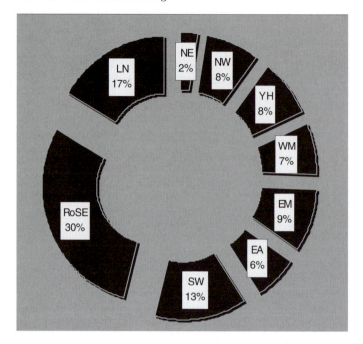

Figure 12.1 Regional share of projected household growth in England, 1996–2016

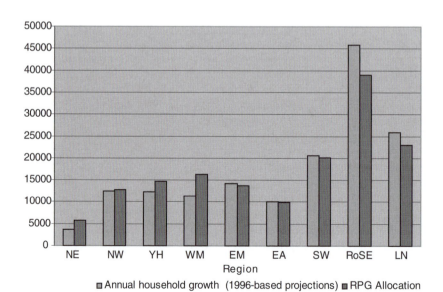

Figure 12.2 Annual projected household growth and RPG housing requirements, 1996–2016

housing requirement in their last round of RPG preparation. Figure 12.1 shows the regional share of the total projected household growth in England between 1996 and 2016. It is clear from this that the lion's share of household growth was expected to take place in the South East, London and the South West; together they account for 60 per cent of projected household growth in England. This rises to 75 per cent when the East Midlands and East Anglia are included. On the contrary, growth in the West Midlands and the three northern regions is seen as somewhat sluggish.

Whilst the 1996-based household growth projections are balanced out with the global housing requirements across the nine English regions, a very different picture is seen when the respective figures for individual regions are compared. The absolute difference between projected annual household growth and the RPG housing requirements in each region is plotted on Figure 12.2. The most striking pattern that emerges from this diagram is the under-supply (compared to projected levels) of housing in London and the Rest of South East. At the opposite end of the spectrum, significant over-supply is found in the West Midlands, Yorkshire and the Humber, and the North East. When such differentials are expressed in percentage terms (see Figure 12.3), the north–south division over housing land policy is clearly evident. The relative differentials between policy allocation and projected growth are most remarkable in the North East (an over-supply of nearly 55 per cent), followed by the West Midlands (39 per cent) and Yorkshire and the Humber (20 per cent). At the other end of the spectrum, there are problems of under-provision in the South East (by 15 per cent) and London (by 11 per cent).

Intra-regional comparisons

Apart from the broad regional variations discussed above, the detailed sub-regional spatial frameworks within each region, upon which the housing requirements are based, also vary significantly. In spite of the encouragement in PPG3 for regional planning bodies to take a more flexible approach to the allocation of housing

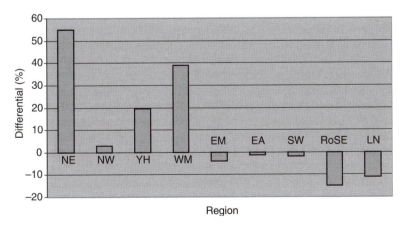

Figure 12.3 Differentials between annual projected household growth and RPG housing requirements, 1996–2016

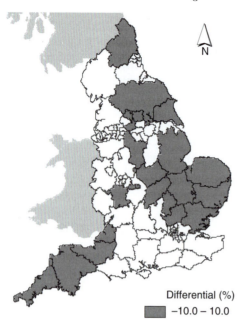

Figure 12.4a
Differentials between annual projected household growth and RPG housing requirements, 1996–2016
– no major differential

Differential (%)

−10.0 – 10.0

50 0   50 100 Kilometres

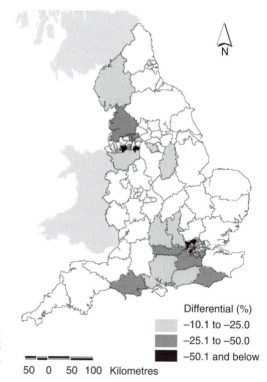

Figure 12.4b
Differentials between annual projected household growth and RPG housing requirements, 1996–2016 – under-allocation

Differential (%)

−10.1 to −25.0
−25.1 to −50.0
−50.1 and below

50 0   50 100 Kilometres

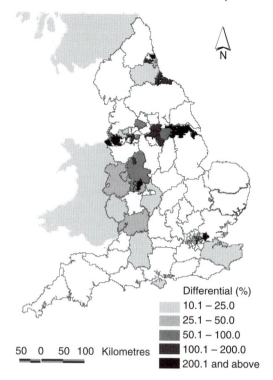

Figure 12.4c
Differentials between annual projected household growth and RPG housing requirements, 1996–2016 – over-allocation

Differential (%)

10.1 – 25.0
25.1 – 50.0
50.1 – 100.0
100.1 – 200.0
200.1 and above

50 0 50 100 Kilometres

requirements at the sub-regional level, our analysis finds that most regions still stuck with a traditional approach of allocations based around shire counties and metropolitan districts (see the boundaries shown in Figure 12.4) during RPG preparation. Such a conservative approach is especially evident in the South East, London, South West, East Anglia and the East Midlands. More interesting sub-regional spatial frameworks are, however, used in Yorkshire and the Humber (based on metropolitan districts and three other sub-regions), and the West Midlands (based on metropolitan districts, shire counties and major townships). It is also interesting to note that, in spite of the creation of unitary authorities after the 1996 local government reorganisation and the availability of independent household projections for them, many RPG documents still remain focused on the 'old' shire counties for housing requirement purposes. Examples of this include the housing requirement figures for Lancashire (including the unitary authorities of Blackpool and Blackburn) and Nottinghamshire (still including Nottingham City itself).

When examining the differentials between projected household growth and allocated requirements in these areas, some interesting patterns emerge. These patterns are mapped in Figures 12.4a to c. The discussion starts with the northern regions, beginning with the North East. Here, all but Newcastle upon Tyne (−24 per cent) have a higher housing allocation than that suggested by the projected household growth, and this is especially so in North Tyneside (+291 per cent) and Tees Valley/Cleveland (+ 127 per cent). The result overall is a significant over-supply for the region as a whole compared to the 1996-based household projections for the region. Interestingly, earlier draft versions of the emerging RPG for the North West also planned for greater levels of housebuilding than the household projections implied. However, here the Government intervened post-public examination to reduce the overall housing figures compared to those set out in the draft plan. The result is that the adopted figures for the North West are quite closely aligned to the levels of projected growth for the region as a whole. However, in terms of intra-regional distribution, there are some major discrepancies that indicate a major policy-based spatial manipulation exercise. As shown in Figure 12.4c, the M62 corridor is identified as a major growth area, covering Merseyside, Halton, Manchester and Salford. This concentration of housing development in the urban core is therefore matched by significant under-supply, compared with what the projections would suggest, in the more rural shire areas. Given the levels of commitments (e.g. sites with planning permission or already under construction) this has subsequently left large chunks of the region with effectively no need for further housing allocations or permissions to meet regional targets. The resulting freeze, or 'moratorium' on new housing development in a significant (and growing) number of districts within the North West has been a source of much recent controversy and debate. Meanwhile, the housebuilding industry faces a slightly more welcoming reception in Yorkshire and the Humber, where the RPG's overall housing requirements represent a 20 per cent increase on the projected growth. At the more local scale, such over-allocations were especially noticeable in North/Northeast Lincolnshire, Rotherham, Wakefield and Barnsley.

Turning to middle England, the West Midlands is seen to have an additional housing requirement of 39 per cent when compared with the projected household growth. Higher growth areas are particularly focused in Sandwell and Walsall. The East Midlands, in contrast, has planned its housing requirement at levels slightly below the projected mark and hence this is generally reflected in the distribution of housing requirements across its sub-regions, though there is a relatively low requirement (compared to projections) apparent in Leicestershire/Leicester and Rutland (–16 per cent).

When examining the equivalent figures in the southern regions, it is clear that both East Anglia and the South West have planned their requirements just below the projected figures and their spatial frameworks of distribution largely follow the projected household figures at the sub-regional level. In the South West, however, there is a notable juxtaposition between Dorset (under-supply by 28 per cent) and Gloucestershire (over-supply by 28 per cent). The spatial framework behind housing provision in the South East is interesting as the allocation is more or less at a lower level than the projections imply across the board, with the exception of Kent (+25 per cent) and Isle of Wight (+19 per cent). Berkshire (–39 per cent) and East Sussex (–38 per cent) are the two sub-regions that have the largest under-allocation. The situation in London is very similar to the South East (as shown in Figure 12.4b) in that most boroughs have housing requirements below the projected growth. Major growth areas, however, are identified in the central and east sub-regions, especially in Southwark, Havering, and Barking and Dagenham.

The differential between housing requirements and projected growth can be broadly classified into three groups that are mapped in Figures 12.4a to 12.4c. Figure 12.4a shows areas that have housing allocation requirements similar to their projected growth. These areas tend to be clustered around the eastern side of England and the South West and they tend to be in more rural areas. Areas that have a lower housing allocation than the projected growth are mapped in Figure 12.4b. As discussed earlier, these tend to be areas in the South East and London and the shire counties of the North West. Finally, areas with an above projected growth housing requirement are shown in Figure 12.4c. It is very interesting to note that these areas tend to be found in the northern regions (many of these areas possess significant amount of brownfield land), the West Midlands and some London boroughs and Kent. The map also clearly illustrates the very strong steering of the spatial framework of housing provision in Merseyside and along the M62 corridor; through the coalfield areas in South Yorkshire to North Lincolnshire; in North Tyneside and the Tees Valley/Cleveland; in the Black Country; and in the central and east of London.

Further institutional development: post RPG era

By the early years of the new millennium, with house prices soaring and consequent impacts in terms of demand for new housing, affordability, particularly for first-time buyers and so-called 'key workers', the Deputy Prime Minister had given serious attention to setting out a long-term programme of action to accelerate the provision of housing. This was articulated via the launch of the Sustainable Communities: Building for the Future (ODPM, 2003) programme in February 2003. As stated in the foreword by the Deputy Prime Minister, 'A step change is essential to tackle the challenges of a rapidly changing population, the needs of the economy, serious housing shortages in London and the South East and the impact of housing abandonment in places in the North and Midlands. This Action Programme sets out the policies, resources and partnerships that will achieve this step change' (2003: 2). This action plan makes substantial investment in housing improvements and the provision of affordable housing as well as a particular focus on four identified major growth areas, all located within London, the South East, the East and the (southern part of) the East Midlands. It also provides a broad spatial framework for tackling housing issues by which, it can be argued, the government seems to have back-tracked towards a more top-down approach of planning for housing. As a result, in areas identified as growth corridors (the Thames Gateway; Milton Keynes/South Midlands; Ashford; and London–Stansted–Cambridge), regional planning bodies have had to revisit their housing requirement figures upwards from those previously established in the last round of RPG.

This Sustainable Communities Plan has, however, been seriously criticised as a plan for the southern regions as a significant amount of resources is devoted to the identified growth areas in the South. In comparison, the attention paid to the Northern regions and rural areas is somewhat less noticeable. This was indirectly acknowledged by the Deputy Prime Minister when introducing the *Creating Sustainable Communities: Making it Happen: the Northern Way* (ODPM, 2004b) report, which stated that '… the first progress report on Sustainable Communities Plan published in July 2003 showed how we are creating new sustainable communities in the wider South East and South Midlands. This new report focuses on how we are "making it happen" in the North, where there are different challenges which demand different solutions' (2004, 3). A glance at the initial Northern Way documentation, however, suggested that there was no real coherent thinking on the

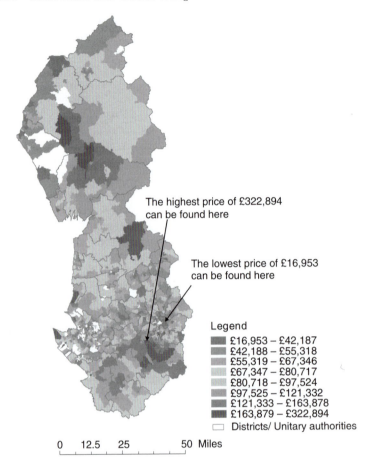

The highest price of £322,894 can be found here

The lowest price of £16,953 can be found here

Legend
- █ £16,953 – £42,187
- █ £42,188 – £55,318
- ▓ £55,319 – £67,346
- ░ £67,347 – £80,717
- ░ £80,718 – £97,524
- ▓ £97,525 – £121,332
- █ £121,333 – £163,878
- █ £163,879 – £322,894
- ☐ Districts/ Unitary authorities

0   12.5   25   50 Miles

Figure 12.5 Average house price for semi-detached properties in the North West by post-code, 2001

planning and delivery of such a northern growth corridor, other than providing a compilation of the government's regeneration schemes and initiatives already relevant to the area, including the new market renewal pathfinders. It could, therefore, be argued that the Northern Way initiative is more of a 'sop' to the northern regions in the light of the southern growth strategy than a significant new approach to strategic planning in northern England (see also Deas' chapter in this volume).

Meanwhile, the gravity of housing pressures, particularly in southern regions, prompted the Treasury and the ODPM to commission Kate Barker to conduct a review (Barker, 2004) of issues underlying the lack of supply and responsiveness of housing in the UK. This is an authoritative review of the issues that surround the operation of the housing market and the reform required on the planning system and the development industries. However, the rationale that underpins this work is based on macro-economic assumptions of housing demand and supply. The Report concluded that the supply of new homes constantly lags behind demand, and that the overall numbers of houses being built in Britain needed to rise substantially if house price inflation was to be reduced and the number of affordable houses available for people wishing to buy or rent was to be increased. The bulk of the recommendations are, therefore, aimed at finding ways to stabilise the volatile housing market and to resolve the housing affordability issue. These again tend to make a lot of sense in regions such as London and the South East where there has been uniform housing pressure. Hence, a more responsive approach to releasing land could help to alleviate the problem of housing affordability. However, in the northern regions where there are pockets of high prices and low demand within a few miles of each other, such a blanket approach can potentially create a downward spiral within the less popular locations. Figure 12.5 shows the contrasting house price distribution in the North West region. In such a region, a spatial framework, rather than macro-economic approach, is needed to provide a sensible balancing act for the provision of new housing land to release housing pressure without compromising the regeneration potential of the vulnerable areas. This once again demonstrates the lack of geographical sensitivity of general guidance.

Given that the lion's share of household growth lies in the South East and London, the collective housing strategies, as set out in the respective RPGs for these regions and analysed above, do not seem to address the serious pressures of household growth and the issue of affordability, instead focusing on environmental protection and, less charitably, a strong element of 'Nimbyism'. It is, thus, not surprising to find that the Sustainable Communities Plan (ODPM, 2003) has emerged from a central government perspective to fill a strategic gap in releasing housing in these regions. The Barker Review also serves a similar function in providing a macro-strategic overview of the situation. Without a proper national spatial framework, what the regional policy frameworks do, at best, is to stagger up the housing requirements to meet the global figures for projected growth in England without providing a realistic mechanism of meeting housing needs and demand in spatial terms. The logic currently underpinning the housing distribution framework seems to be to set the housing allocations first, with the hope that the population will move accordingly. Hence, there is the irony that there are the concomitant problems of low demand and a theoretical over-supply of properties in the northern regions (in reality, there may not be over-supply if significant demolition of poor stock is taking place); and the problems of serious affordability and under-supply of properties in the south. If there is a deliberate central government vision and strategy that the overcrowded southern regions should lose population to their northern counterparts, then other strong macro-economic and planning policies have to be in place to support this. As it currently stands, the Sustainable Community Plan does not set out to achieve such an inter-regional balance, but rather appears to work more on a piecemeal basis, area by area.

The realisation of spatial vision and the target housing provision allocations relies on the integration between housing, planning, transport and urban regeneration. Housing provision has to relate to changing demographic trends and the housing demand and needs of the population. The Government's planning system reforms, as expressed in particular through PPS11 (ODPM, 2004c) and the moves to new forms of regional spatial strategies (RSS), lay much emphasis on improved spatial vision and integration in the quest for sustainable communities.

For this to occur at the regional level, there is a need for the regional development agencies (RDAs), regional planning bodies and the Housing Corporation
to work closely together. The move, firstly, towards the creation of a Regional
Housing Strategy and, subsequently, the proposals post-Barker for merging the
regional housing forum with the regional planning body may well be a move in
the right direction in terms of such enhanced integration. Additionally, the
HMR pathfinders initiative and the strategic vision for the Northern Way do at
least start to address, more specifically, the particular problems and opportunities
of the northern regions. Whether all this will be enough to address the important
housing challenges raised at the start of this chapter, especially in the continuing
absence of an explicit national spatial planning framework for the country as a
whole, remains to be seen.

Conclusions

This chapter serves two main purposes: first, to reconstruct the historic case in
providing a snapshot of the latest round of RPG housing land allocation
approaches adopted in English regions; and second, to examine the national and
regional policy and institutional frameworks in addressing the different socio-
economic spatial contexts of planning for housing.

The historic analytical exercise does not aim to promote a technical-based or
the 'predict and provide' approach of planning for housing. Indeed, it can be
argued that the analysis shows that some regions are too wedded to a conservative
approach of sticking closely to structure plan boundaries and closely tracking the
household projection figures in their spatial allocations. Having said that, there is
a question that needs to be asked – to what extent should the spatial framework of
allocations substantially depart from the projections? Since the projections are
trend based, any such discrepancies represent a major challenge to market forces in
respect of regional and sub-regional housing markets. At the national level, the
RPG requirements analysed above suggest a general trend towards under-supply
within the overheated southern regions and over-provision in the less prosperous
north. However, considerable deviations between projections and requirements
are also evident between parts of the same region, with an explicit attempt to steer
new housing into the urban core and the wider conurbation. This raises concerns
over whether such a radical spatial strategy steering housing provision away from
the projected distribution will succeed. In any event, the impacts, both direct and
indirect, brought by such strategies need to be closely monitored.

The publication of the revised PPG3 (DETR, 2000a) was heralded at the time
as a new approach to planning for housing; the need for sustainable residential
development, integrated with other policy considerations, was to be the defining
principle of the new guidance. However, in retrospect this was essentially a
national sectoral guidance framework that aimed to deal with policy issues with a
blanket geographical coverage. Such an aspatial approach is clearly reflected from
paragraph 2 of the guidance that advises that local planning authorities should
achieve eight identified objectives without any differentiation being made in
terms of the appropriateness of all these eight objectives in geographic terms. The
only spatial elements taken into account in PPG3 are the distinction between
'town and country' and 'urban and rural' areas, and there is no explicit admission
of a serious north–south divide in the nature of housing demand and supply or of

any significant intra-regional differentials based on socio-economic or environ-mental contexts. The only caveat provided in PPG3 to deal with spatial variations and differential circumstances was that regional planning bodies and local planning authorities are advised to identify and assess regional and sub-regional trends and factors in preparing regional and local planning policies.

This, however, offers no consolation to local and regional actors who are strug-gling to strike a balance between their local housing circumstances and the adherence to the universal framework set out in the guidance. Whatever politi-cians might say about the reality of north–south contrasts, or the scale at which physical and social variations are manifest, it is clear that the challenges facing the regions in housing and planning terms are significantly different. The sheer scale of projected growth in the south of England has resulted in tremendous pres-sure to allocate sufficient land for housing whilst at the same time preserving environmental quality. Severe restrictions on new housing development are now a feature of local and sub-regional planning policy in large parts of the country, including much of the South East, though some are starting to doubt the wisdom of blankly refusing to open up new greenfield sites to development as the schism between housing supply and demand widens daily. The resulting rise in house prices, and increasing problems of affordability, have subsequently led to calls by Barker (2004) for a substantial increase in housing supply, beyond even that envisaged in the growth areas that lie at the heart of the Sustainable Communities Plan.

In the north of the country, the story is somewhat different. The focus here is on urban regeneration, poor quality housing and economic revival. But it is not all gloom and doom. Within all three northern regions are also areas of affluence and intense housing pressure. Indeed, the general upturn in land and house prices over the last three or four years, and the resulting additional housing demands, have led to the stated RPG housing requirements for the North West being met across large swathes of the region. The paradox here is that these mixed condi-tions arguably create an even more uncertain and complex policy arena than in the south. The challenge is to encourage much needed redevelopment and eco-nomic growth to the regions as a whole, and especially the worst-hit areas such as those targeted under the housing market renewal (HMR) pathfinders initiative, whilst simultaneously dealing with growing affordability problems in the more affluent suburbs and rural areas.

Behind this consideration of regional housing provision lie more fundamental issues of regional governance and devolution. The move away from the old 'pre-dict and provide' approaches to regional housing provision, as heralded by PPG3, implied a significant shift from centrally determined housing figures (based on OPCS projection models) to regional requirements linked much more to region-ally and sub-regionally derived visions and planning strategies and the analysis above suggests that several regions embraced these opportunities in their latest RPGs, particularly so in the northern regions. At the time, such a shift was fully in line with the Government's proposals for greater levels of regional devolution as set out firstly in the White Paper on regional devolution, *Your Region, Your Choice: Revitalising the English Regions* (DTLR, 2002) and subsequently in the Regional Assemblies Bill. By the early years of the new millennium, however, this picture had radically changed. The overwhelming rejection in the autumn of 2004 of the idea of a directly elected regional assembly in the North East has stalled the regional devolution agenda in its tracks. At the same time, concerns over the impact of restricted housing supply and rapidly rising house prices on national economic performance have led to Treasury intervention, via Barker, in matters more normally left within the remit of the ODPM.

Both the response to Barker and the ODPM's own HMR initiative and Community Plan, including its Northern Way spin-off, reflect a greater direct re-engagement of central government in matters of national and regional housing supply. They also tend to emphasise the importance of the inter-regional and cross-boundary dimensions to housing provision; something that has generally been conspicuously lacking from previous rounds of RPG preparation. Thus, although there is still no explicit national spatial framework in place, the current round of RSS preparation can be expected to be more constrained by national planning policy than the last round of RPG. Local and regional planning policies in the southern regions will be expected to embrace the Government's growth areas whether they like it or not; whilst the success of the HMRs and the Northern Way agenda imply a continued targeting of future housing growth to the core city regions and ever-more tight restrictions on new housebuilding else-where. Yet, by way of contrast, the provisions of the recent Planning and Compulsory Purchase Act 2004 in relation to RSS preparation and the abolition of county structure plans appear to strengthen the role of the regional planning bodies in relation to the sub-regional dimension of RSS (at least compared with its immediate RPG predecessor). Though there may be little prospect of further moves towards the establishment of directly elected regional assemblies in the third term of the 2005 re-elected Labour Government, the tensions and com-plexities of the current institutional frameworks and structures of fragmented regional governance thus look set to remain for the foreseeable future.

Acknowledgements

We would like to acknowledge Philip Jeffery for compiling the data and Alasdair Rae and Andreas Bäing for drawing the maps.

Notes

1 Data for the West Midlands is the annual figures between 1996 and 2011. This is due to the fact that only broad sub-regional units have figures beyond 2011.
2 The 1996-based household projections were provided by the Population and Housing Research Group at Anglia Polytechnic University that developed the projections for the DETR. These projections were first released in October 1998.

References

Baker, M. and Wong, C. (1997) 'Planning for housing land in the English regions: a cri-tique of household projections and Regional Planning Guidance mechanisms', *Environment and Planning C: Government and Policy*, 15: 73–87.

Barker, K. (2004) *Review of Housing Supply: Delivering Stability: Securing our Future Housing Needs*, London: HM Treasury.

Champion, T., Fotheringham, S., Rees, P., Boyle, P. and Sitwell, J. (1998) *The Determinants of Migration Flows in England: A Review of Existing Data and Evidence*, Newcastle upon Tyne: Department of Geography, University of Newcastle upon Tyne.

Civic Trust (1999) *Brownfield Housing – 12 Years On: Findings and Recommendations*, London: Civic Trust.

CPRE (1998) *Anybody Home? Empty Homes and Their Environmental Consequences*, London: Council for the Protection of Rural England.

Crook, T. (1998) An Apparent Paradox, *Housing Agenda Newsletter*, March.

DETR (1998) *English House Condition Survey 1996*, London: The Stationery Office.

DETR (1999a) *Revision of Planning Policy Guidance Note 12 Development Plans*, London: Department of the Environment, Transport and the Regions.

DETR (1999b) *Revision of Planning Policy Guidance Note 13 Transport: Public Consultation Draft*, London: Department of the Environment, Transport and the Regions.

DETR (1999c) *Projections of Households in England to 2021: 1996-Based Estimates of the Numbers of Households for Regions*, London: Department of the Environment, Transport and the Regions.

DETR (2000a) *Revision of Planning Policy Guidance Note 3 Housing*, London: Department of the Environment, Transport and the Regions.

DETR (2000b) *Revision of Planning Policy Guidance Note 11 Regional Planning*, London: Department of the Environment, Transport and the Regions.

DoE (Department of the Environment) (1995) *Projections of Households in England to 2016*, London: HMSO.

DTLR (2002) *Your Region, Your Choice: Revitalising the English Regions*, London: Department of Transport, Local Government and the Regions.

Gallent, N. (2000) 'Planning and affordable housing: from old values to New Labour', *Town Planning Review*, 71(2): 123–48.

Hamnett, C. (2001) 'London's housing', *Area*, 33(1): 80–4.

HM Government (1998) Planning for the Communities of the Future, White Paper, London: HMSO.

HM Government (2004) *Planning and Compulsory Purchase Act 2004*, Norwich: HMSO.

Holmans, A. (2001) *Housing Demand and Need in England 1996 to 2016*, London: Town and Country Planning Association.

Holmans, A. and Simpson, M. (1999) *Low Demand: Separating Fact from Fiction*, Coventry: Chartered Institute of Housing.

House Builders Federation (1998) *Urban Life: Breaking Down the Barriers to Brownfield Development*, London: HBF.

House of Commons Select Committee of the Environment, Transport and Regional Affairs (1998) *Committee Report on Housing*, London: House of Commons, HM Government.

Lainton, A. (1999) 'A workable approach to housing?' *Town and Country Planning*, 68(5): 150.

Leather, P. and Morrison, T. (1997) *The State of UK Housing: A Fact File on Dwelling Conditions*, London: The Policy Press.

ODPM (2003) *Sustainable Communities: Building for the Future*, London: Office of the Deputy Prime Minister.

ODPM (2004a) *Interim Household Projections in England to 2021*, London: Office of the Deputy Prime Minister.

ODPM (2004b) *Creating Sustainable Communities Making it Happen: The Northern Way*, London: Office of the Deputy Prime Minister.

ODPM (2004c) *PPS11: Regional Spatial Strategies*, London: Office of the Deputy Prime Minister.

Perry, J. and Simpson, M. (1999) 'Are planners part of the problem?', *Town and Country Planning*, 68(5): 154–5.

Pike, D. (1999) 'Predict the sequence and provide the housing – is this what the Government really wants?', *Town and Country Planning*, 68(5): 143.

Urban Task Force (2000) *Financing an Urban Renaissance*, London: E&FN Spon (supplementary leaflet).

Wong, C. (2000) 'Indicators in use: challenges to urban and environmental planning in Britain', *Town Planning Review*, 71(2): 213–39.

Wong, C. and Madden, M. (2000) *The North West Regional Housing Need and Demand Research*, London: DETR.

13 Modernising transport planning in the English regions

Geoff Vigar

Introduction

In seeking to manage contemporary mobility trends, governments often look to innovate in terms of policy mechanisms and/or associated institutions. In the UK, a new emphasis has emerged in the last decade on the regional scale for much governance activity both in relation to transportation and other policy areas. This chapter examines progress toward using the regional scale as an effective level through which to deliver transport policy goals, focusing particularly on a new policy mechanism, the Regional Transport Strategy. The chapter notes the difficulties arising from intervening at the 'regional' scale in the UK principally: the institutions for strategy-making as currently configured; a continuing centralisation of governance activity; and, relatedly the difficulties of constructing and operating policy processes given this institutional context.

The strategic policy context

Accommodating the mobility expectations of citizens in traditionally developing economies has become an issue of prime political concern. For politicians, few policy areas bring into sharp focus perceived relationships between economic growth, environmental protection and social equity more than dealing with transport issues. The political difficulties that arise from trying to manage the demand for personal and freight transportation have led governments to search for innovative *institutional* solutions to such difficulties. This has led many governments to transfer and share responsibilities both between public and private sectors and between different geographical scales. Such efforts are part of, or often tied to, wider attempts to shift the relationships between markets, regulation and the state, or at public service reform. As part of these attempts, rescaling processes have shifted attention to the capacity of *regions* to respond to policy challenges:

- as a way of escaping problems of centralisation (see Vigar and Healey, 1999);
- to avoid the 'lowest common denominator' approach that typifies policy if left to individual municipalities;
- to allow regions to generate their own responses to problems and generate their own priorities in dealing with them; and,
- to achieve a better fit between policies and the specific demands arising in particular regions.

The increasing importance of the regional scale is evidenced in a number of new mechanisms to develop transport policy at this level, in line with efforts in other policy areas. Such processes are widely thought of as a rescaling of governance activity as responsibilities are transferred between spatial scales. In the UK, this consists of transfers of responsibilities up from the local level and down from the central level. However, as the discussion below will demonstrate, while a redistribution of power to regional level has occurred, such activity also serves to highlight the continued centralised nature of transport policy-making. In the absence of elected regional Assemblies it also serves to highlight increased elitism in policy-making, most notably in concentrating power in the hands of bureaucrats, appointed consultants and an economic development policy community in particular.

This chapter mostly focuses on the Regional Transport Strategy (RTS) mechanism in England. There are other new mechanisms, however, and discussion of Multi Modal Studies, other policy mechanisms, and changes – such as to funding regimes for example – will also follow.

Regional transport strategies

The revised arrangements for regional planning in England introduced in 2000 highlighted the significance of transport issues by elevating them from just one component among many in past regional planning guidance (RPG) to an explicit free-standing strategy within new Regional Spatial Strategies (RSS) (DETR, 2000). Since April 2003 RTS is developed by designated Regional Planning Bodies (RPBs) who in all regions outside London are the Regional Chambers (also known in some territories as Regional Assemblies – RA). RPG preparation was typically heavily dependent on the efforts of a small body of officers on secondment from local authorities and thus to a great extent on the goodwill of local authority officers and members prepared to divert resources to it. More recently as RAs have developed greater capacity of their own to develop policy, this dependence is likely to be lessened but is still significant (DfT, 2004). However, the perceived highly technical nature of RTS has led many RAs to devolve development to transport planning consultancies (see Vigar, forthcoming, for a discussion of the implications).

The main aims of RTS are outlined in Box 13.1. These are partly aimed at getting regions to develop their own priorities and ownership of them and thus avoid the generation of 'wish lists' of infrastructure schemes that have typified such efforts in the past. The RTS is also supposed to provide a framework for the development of Local Transport Plans. While these are the sorts of aims one might expect of a regional transport strategy some issues are notable by their absence. There is no overall steer on strategic objectives, such as managing overall travel demand to meet environmental objectives. The emphasis on local demand measures seems to miss this point and reinforces a perception that demand management is a micro-level concept to be applied to specific network elements at particular times and may not provide an overall strategic discourse to guide policy (see Vigar, 2002, for more on this).

That said, such issues, it could be argued, can now be left to regional expression. However, while the new arrangements are intended to allow for such articulation, central government maintains a great deal of control over both

Box 13.1 Main aims of the Regional Transport Strategy

The RTS should provide:

- regional objectives and priorities for transport investment and management across all modes to support the spatial strategy and delivery of sustainable national transport policies;
- a strategic steer on the future development of airports and ports in the region consistent with national policy and the development of inland waterways;
- guidance on priorities for managing and improving the trunk road network, and local roads of regional or sub-regional importance;
- advice on the promotion of sustainable freight distribution where there is an appropriate regional or sub-regional dimension;
- a strategic framework for public transport that identifies measures to improve accessibility to jobs and key services at the regional and sub-regional level, expands travel choice, improves access for those without a car, and guides the location of new development;
- advice on parking policies appropriate to different parts of the region; and
- guidance on the strategic context for local demand management measures within the region.

Source: ODPM, 2004: Annex B

process and product, first, providing guidelines above which variable structures RSS, and second, through a gatekeeper, policing, role devolved to its regional offices.

The very existence of an explicit transport strategy at regional level could be construed as an attempt to coordinate policy at the regional scale following efforts at national level through the brief merger of transport and planning agendas in the same ministry and the development of some collaborative policy and research in the 1990s. The treatment of RTS as part of RSS rather than separately as another strategy possibly reflects a concern not to create a proliferation of separate free-standing documentation and a concern to consider spatial planning and transportation issues together as far as is practicable. By producing RTS with RSS greater coordination between them should be facilitated than if they were free-standing documents. This seems a sensible way of achieving synergy without diluting the debate surrounding either strategy that may occur through full integration of the two documents.

Progress in regional strategy development

The considerable changes to the English planning system as it maps on to the processes of regional plan preparation have generated a complex policy picture. The current state of play in England as at the end of 2004 is detailed in Table 13.1.

As can be seen from Table 13.1, a range of different approaches to the presentation of RTS has been adopted. In the East Midlands RTS is merely a section in a chapter. This kind of approach seems to go against the spirit of

Table 13.1 Progress with RTS development in England (as at end 2004)

Region	Current RPG/ RTS	RSS position/Planned RTS
North East	RPG adopted Nov 2002 containing limited RTS	Draft RSS produced Nov 2004, RTS policies embedded as a theme
London	Contained in Plan for London, Feb 2004 (RTS not explicit)	Unknown
East of England	RPG 'banked' to be submitted as RSS	Draft RSS submitted to ODPM Nov 2004
East Midlands	RPG final Jan 2002 with RTS as chapter	Draft RSS with RTS as chapter section, April 2003. Final RSS expected Spring 2005
South East	RPG adopted March 2001	Draft expected end 2004. RTS published separately July 2004
South West	RPG adopted Sept 2001 with RTS as chapter	Consultation document Summer 2004.
West Midlands	RPG adopted June 2004, published as RSS, with RTS as chapter	Adopted June 2004
Yorkshire and the Humber	Draft revised June 2003, transport as chapter	Consultation August 2004, draft expected April 2005.
North West	RPG adopted May 2003	Consultation Autumn 2004, draft expected Sept 2005.

upgrading transport issues within an explicit RTS; although it also suggests a close 'integration' between spatial planning and transportation avowed in PPS11. Certainly there is no clear picture as to the articulation of RTS within an RSS and we might expect continued difference across England in future also.

Other regional developments in transport policy

The increased attention to transport issues at regional level is bolstered through the devolution of more authority over trunk roads planning at the regional level, greater levels of finance available for transport initiatives generally, and through the development of 'Multi-Modal Studies' (MMS) for which regions have responsibility (see Headicar, 2002, for more detail). MMS were to consider the future demands for transportation in a series of strategic corridors throughout England. They feed into spatial planning policy as well as regional and national transport policy by highlighting where development might occur that could lead to more sustainable transport outcomes. In practice these have been fraught with difficulties, not least in their implementation where the roads elements have been progressed but other measures have proved far more difficult due in part to the pursuit of separate objectives by other stakeholders, not least in the rail industry (CPRE, 2004). The timing of the production of MMS, whereby they were too late to inform the RPGs of the late 1990s, has also proved problematic

(Haughton and Counsell, 2004). The adoption of their findings into the next round of RTS has also proved difficult with stakeholders wanting to unpick the assumptions contained in them before approving their findings for translation into regional policy (Vigar and Porter, 2005).

Substantive policy change

It is also worth noting at this stage the over-arching shifts in UK transport policy. Overall, policy turned in the 1990s from an emphasis on road-building and facilitating infrastructure supply to one of managing demand for travel and a greater emphasis on modes other than the private car (Shaw and Walton, 2001; Doherty and Shaw, 2003; Vigar, 2002). This is not, however, some large-scale paradigmatic change (Vigar, 2002). Rather it is a shift in emphasis that is continually being negotiated at the margins. This in itself highlights the absence of a single powerful discourse in UK transportation policy. The 10 year plan developed in 1998 (DETR, 1998) did promote a strategic discourse of demand management but was effectively abandoned to short-term political pressure. The follow-up White Papers (DETR, 2000; DfT, 2004) took a more pragmatic response with a renewed emphasis on infrastructure supply and in the latter case a process of softening up public opinion in the idea of road pricing was begun. Overall, the White Papers adopted a 'balanced' approach, redolent of the Conservative approach to transport policy in the 1990s. The White Papers did, however, highlight that the regional scale would develop in significance by providing guideline budgets for English regions in the 2005 Budget. This was to be supported by a new Transport Innovation Fund, which could fund packages of measures in regions, rather as three demonstration towns had been chosen to do at a smaller scale in 2003. Again, mechanisms for deciding on allocations from this Fund were to be published in Budget 2005.

So, there is increased attention to the regional scale, albeit within processes structured by government priorities and ideas about process. A case study is mobilised below to assess how far this attention is affecting regional policies and practices in actuality.

Regional transport strategy-making in North East England[1]

A case study of the development of an RTS provides a window on the issues involved in RTS development and the implementation of new regional institutions and policy processes more generally. This allows conclusions to be drawn as to the operation of new policy mechanisms, especially where much change has occurred in institutional landscapes.[2]

The North East regional context

The North East of England is a region with a strong identity, in part fostered by a sense of isolation from London and central government. In comparison with other UK regions, it is small in population terms (2.5 millon in 2001) but covers a large area (850,000 hectares). It consists of three major conurbations: Tyneside, centred on Newcastle; Wearside, centred on Sunderland; and Teeside, centred on the city of Middlesbrough. To the north and south of Tyneside are two former coalfield areas where regeneration presents a major challenge. The remainder of the region is mostly rural with varying degrees of remoteness and a number of small towns. The region contains two National Parks and two Areas of Outstanding Natural Beauty.

The North East Region depends to a large extent upon a main north–south transport corridor featuring both road and rail infrastructure. There are two smaller-scale transport routes to the west, one consisting of road access only, and one with both road and rail access. A further north–south axis of road, with some limited parallel rail service, connects the three main conurbations and the two coalfield areas. The region also contains a number of seaports and two airports with passenger and freight capacity. Long-distance walking and cycle networks complete the regional transport infrastructure picture.

Public transport services are delivered by private companies with the exception of the Tyne and Wear Metro. Great North Eastern Railways and Virgin currently operate the intercity rail links on a long-term franchise basis. Bus services are dominated by national operators – Arriva, GoAhead, and Stagecoach. As in much of the rest of the UK this fragmentation of services creates its own difficulties, not least in the formulation and implementation of strategic policy.

The RTS process

The North East RTS was prepared in a changing institutional landscape. The Association of North East Councils (ANEC) and the regional development agency, One North East, oversaw its early preparation. Latterly responsibility was handed to the North East Assembly, which, while subsuming much responsibility from its predecessor, was restructured and re-staffed in anticipation of an unrealised elected regional government. The Assembly was therefore very much an emerging organisation while the RTS was being developed.

An initial attempt at a North East RTS was incorporated into draft Regional Planning Guidance submitted to central government in February 2000. A Public Examination in July 2000 considered the transport strategy incomplete for a number of reasons, including:

- It lacked information about parking standards and public transport accessibility criteria that at the time were about to be included in central government guidance.
- The results of a variety of multi-modal studies (in particular that for Tyne and Wear) had not been available for incorporation into the strategy.
- A new North East 'vision for transport' was thought to be needed.
- There was a need to reflect central government's 10-year transport plan.

(Department for Transport, 2002)

At the Examination, a commitment was given to review the RTS to include it in revised Regional Planning Guidance so that it could guide a round of Local Transport Plans for the period 2006–2010. The recommendations of the RPG Panel Report (GONE, 2000) emphasised that the RTS objectives should prioritise accessibility, support the priorities of the Regional Economic Strategy (ONE, 2000), and include Regional objectives for the environment, biodiversity and conservation. Such criticisms were common of other RPGs at the time as the institutional and policy landscape was changing around RPBs through the RPG preparation processes of the late 1990s/early 2000s.

The preparation process

A 'Transport Co-ordinating Group' (See GONE, 2000, para. 8.61) involving the local authorities, Government Office North East, and the regional development agency was established to oversee the development of the RTS. Consultants were then appointed to produce a draft RTS for incorporation into the next Regional Planning Guidance (see Table 13.2).

Phase One consisted primarily of a visioning event where a group of stake-holders representing the main transport, commercial, economic, environmental and local government interests explored key issues and discussed alternative transport and communications infrastructures. During Phase Two the consultation process was extended through the establishment of a Wider Reference Group (WRG). The WRG included local politicians, Members of Parliament, Members of the European Parliament, representatives from seaports, airports, transport operators, local authorities, government and quasi-governmental agencies, trades unions, health authorities, research agencies/universities, pressure and user groups, and business – including organisations such as the Confederation of British Industry and local chambers of commerce.

Participation consisted of breakout groups from the WRG where various policy options were discussed. While such groups were necessary to facilitate detailed discussion, in practice they had two perverse effects. First, strategic priorities became lost in the detail as participants used the fora to state their preconceived positions over individual schemes. Second, because many participants chose to take part in breakout groups that were closely related to their own interests, this may have led to a polarisation of the breakout groups along existing 'urban versus rural' and 'environment versus economy' fault-lines, resulting in limited opportunity for learning and exchange of ideas. In addition, some groups were facilitated differently from others. In some the facilitator contributed knowledge and views; in others they merely noted the views of the participants. This contributed to a situation whereby statements were later presented as 'fact' with contestable issues and views being left unchallenged.

Table 13.2 The North East RTS Process

Phase	Task	Timescale (planned)
1	Inception Report and Visioning event	January–February 2001
2	Opportunities and Constraints	January–April 2001
3	Develop the options	April–August 2001
4	Draft Regional Transport Strategy	August–October 2001 (Actual: June 2002)
5	Revised Regional Transport Strategy	Basic version incorporated into RPG Nov 2002; fuller version merged into RSS Nov 2004

Participants were also asked to complete an exit questionnaire requesting a score on a scale of 1–5 for a variety of regulatory options and approaches.[3] These proved problematic in terms of the issues covered and the limiting nature of the method given the outlined results of consultation and a set of priorities focused largely upon upgrading infrastructure such as river crossings and strategic roads.

Phase Three of the RTS process contained a further WRG event. Four alternative strategic approaches were offered as a basis for further study and comment. For each strategic approach, a set of generic policy measures and specific infrastructure developments was presented in tabular form for, respectively: city centres; conurbations; former coalfields; market towns and rural areas; and strategic links. There was little opportunity to discuss the relative merits of these various measures. Nevertheless, a detailed questionnaire requested first a quantified opinion as to what the balance of the four strategic approaches in each of the above locations should be. A further, and much lengthier, element of the questionnaire requested an opinion (on a scale 1–5) on each of the generic policy measures and specific infrastructure developments. For each question, an opinion was sought as to whether the facility was needed in the next five years, five to 15 years, or beyond 2016, each on a scale of 1–5.

Each of the three strategic approaches, excluding the future base, was favoured in roughly equal measure for all areas except the coalfields where an economic development-led approach was favoured to a significant extent. Two issues can perhaps be raised in this context. First, two approaches, labelled 'sustainability' and 'accessibility' were similar, and split the vote for what in both cases amounted to a 'public transport-led' approach with varying degrees of traffic restraint and new infrastructure. Second, it polarised the debate between economy–accessibility–environment, failing to capitalise on the links between them. It ignored, for example, the notion that there is more to 'sustainable development' than 'balancing' or trading off economic, environmental and social factors, and that there are in many instances 'triple wins' to be had.

Phase Four included the publication of a Draft RTS. It was presented to the WRG with an opportunity for questions in plenum in July 2002. The main strategic elements of the Draft RTS were proposed as follows:

- Reducing the overall need to travel
- Managing travel demand
- Making the best use of existing infrastructure
- Using more sustainable means of travel
- Improving regional gateways (highways, rail, ports, airports)
- Priority regional transport schemes.

The discussion at the WRG meeting was dominated by conflict between the views of an 'economy first' camp, who saw large-scale investment in new infrastructure as a key component in the renaissance of the North East economy, and a 'sustainable development' camp who wanted the strategy to focus on managing the demand for transport, in part to achieve the same economic renaissance. The strategy was vigorously and well defended against charges from both sides by the consultants charged with preparation.

Much of what appeared in the draft RTS was reflected in the draft RSS that incorporated it. This in turn went out for consultation in November 2004 with a series of public consultation events. The draft RSS failed to resolve some of the tensions inherent in the RTS process focusing on 'effective access' as a theme to unite its belief in the links between economic development objectives and a lingering belief in the necessity of fast road connections to underpin this. To this end, a list of road schemes that had been thrown out by central government found their way back into the document albeit with a realistic sense of when they might actually come forward – in many cases beyond 2020 (the RSS itself covered the period to 2021 but would clearly be revised several years before that date).

Evaluating the North East RTS

This section is structured around four common issues, which have resonance beyond the case study documented above. Many contemporary discussions suggest that: strategy consistency; dealing with the relationship between transport and economic development; representing issues of social equity; and, attention to the micro-dynamics of policy processes, are four of the major issues in securing sustainable transport policy (see for example: Doherty and Shaw, 2003; Low and Gleeson, 2003; Vigar, 2002).

Consistency of the strategic approach

The RTS consultation process was useful in publicising the very existence of the strategy. It was able to send clear signals to funding bodies about agreement in the region over, for example, priorities for certain infrastructure upgrades. It was also able to instigate a process of brokering agreement among local authorities on issues such as region-wide parking standards. There was a partial commitment to managing demand on infrastructure networks (proposals for specific spaces, journeys and times rather than adopted as a strategic network(s)-wide principle). However, the document, despite leading with demand management as its principal focus, proposed a great deal of new road infrastructure that appeared to undermine this emphasis.

Such issues were also present when the RTS was translated into the draft RSS. The strategic objectives of enhancing network capacity, improving journey time reliability and tackling congestion 'hotspots' were likely to conflict with the aim of reducing and managing the need to travel as the focus of such measures was very much on the road network with no attempt to prioritise particular journeys. Such issues are always likely to be in tension in such documentation but here the measures that flowed from the objectives were principally focused on enhancing overall network capacity. As such they would inevitably encourage traffic growth across the networks and thus impinge on the primary national transport and spatial planning objective of supporting sustainable development and the strategy's aim of managing and reducing the demand for travel. In parts the draft RTS strongly exhibited the widely discredited 'predict and provide' approach to transport planning, justified by a belief that improving infrastructure capacity would tackle congestion, create economic wealth and improve journey time reliability. While there are likely to be small-scale localised points where such efforts may lead to the desired outcomes, at more strategic levels such an approach is unlikely to work due principally, in the absence of mechanisms for prioritising 'essential business traffic' and public transport journeys, to the phenomenon of induced traffic (SACTRA, 1999). The strategy thus suggested investing in public transport, roads, *and* demand management without any acknowledgement of the synergistic relationships between the three. While such a policy enabled the document to maintain a fragile consensus there remain considerable doubts as to whether such an approach can actually work, not least as investment in the road network would induce more car ownership and usage, undermining attempts to grow public transport markets.

Transport infrastructure, demand management and economic growth

Extensive investment in heavy infrastructure, particularly roads, was justified by reference to the North East's economic difficulties. Few would disagree with the strategy's overarching emphasis on regeneration and economic competitiveness. Nevertheless, there is little evidence put forward as to which improved links would lead to what economic benefits and for whom, let alone what costs might accrue in social and environmental terms. There was little discussion in the WRG or the draft RTS itself of the efficacy of using transport infrastructure investment to deliver economic growth or to demonstrate that the North East economy was held back by poor infrastructure supply. The infrastructure-led model of regional development has been dominant in the region for 80 years and the region currently has the lowest GDP of any English region.[4] Indeed, a One North East commissioned report concluded that, 'the investment-development relationship ... is likely to be weak' (Ove Arup, 2002: 24). Despite this, a senior ONE representative continued to lobby for more infrastructure 'improvements' at subsequent meetings to publicise the Draft RTS.

More generally, while a link between infrastructure investment and economic development seems probable, such intuitive solutions, in transport planning as in many fields, often do not turn out to be the case (such as the link between road-space supply and congestion relief for example – see Black, 2001, for a discussion). This seems to be particularly problematic in the transport field where everyone considers themselves an expert despite the fact that it is as complex and evidence-driven as any other field (see Vigar, 2002, Chapter 9). The development and dissemination of a solid evidence base, as suggested in Phase One of the process should, therefore, be given greater prominence in policy-making exercises such as these.[5]

In this regard, a need was identified in the draft RTS for more studies into links between economic development and transport. This could be useful, particularly at sub-regional level but the policies currently ignore the evidence that overwhelmingly concludes that such relationships are negligible in areas with mature transport networks (see for example SACTRA, 1999; Banister and Berechman, 2000; Black, 2001; Ove Arup, 2002). Indeed, the relationship between traffic growth and GDP is decoupling as economic growth in the 'knowledge economy' becomes much less focused on the ease of movement of people and goods by roads infrastructure (CfIT, 2002a; DfT, 2002). Given the apparent reluctance of the RTS to accept this conclusion, some of the schemes presented in the draft appear premature. (In assessing priorities all the road schemes in the document were allocated a score of '5' out of 5 for their contribution to economic development.)

The need to refocus on people and everyday life

One of the primary justifications for having a regional transport strategy at all is to get agreement on policy principles to avoid the lowest common denominator problem. More infrastructure appears likely to fuel a culture of automobility and merely defer the time when the region will have to seriously intervene to prevent quality of life in its urban and rural areas from being further undermined. New roads infrastructure would also have the effect in many instances of undermining attempts to maintain local services and encourage movement by public transport, bicycles and on foot, further fuelling an automobile culture and undermining the health of its citizens, notably the region's children (see Whitelegg, 1997, for more on this argument). Thus, regardless of the merits of new road construction in specific places, as a principle such an approach may fail to achieve the strategy's wider objectives unless space can be given to sustainable modes or priority given to 'essential business traffic' on the upgraded infrastructure.

The RTS process also rather quickly got bogged down in long-standing debates about the merits of specific infrastructure upgrades. There is of course good reason to debate such issues but the principles that should have structured these and other discussions were absent. This was reflected in the RTS' treatment of social inclusion. While it rightly pointed to infrastructure improvements as a way of promoting inclusion, it largely ignored the fact that many quality of life issues are dependent on place qualities and, with direct reference to transport, on the effects of road traffic on neighbourhood quality and public health. Expanding road networks could exacerbate such conditions by undermining both the quality of life in heavily trafficked neighbourhoods and the viability of local services and public transport networks (see SEU, 2002). New infrastructure tends to benefit those in society who are already highly mobile and does little for lower income groups. In this regard the proposals in the RTS were deeply inequitable: there was an over-emphasis on providing new bits of infrastructure (rather than new services) and on expanding (rather than managing) networks. The RTS should aim to minimise the impacts of transport externalities on quality of life and do this in an equitable fashion. In this regard the links to other parts of the RSS and spatial planning policy were poor. Making neighbourhoods more permeable – using the tools of urban regeneration, village design statements and spatial planning and accessibility criteria where appropriate – could play a major role in making sure that sub-regions are well connected internally, thus maximising the social and economic opportunities within them. There is little acknowledgment of the role of the RTS to guide and perhaps to set targets for such activity.

The importance of process

Two conclusions can be drawn on issues of policy process design. The first relates to the strategic level. The absence of institutions, and indeed key individuals, to guide and facilitate a process that was legitimate in the eyes of participants was a major issue. The RTS resembled a free-for-all of views with little sense that there was expertise that could be drawn on to shape and structure the debate and policy outputs. Second, workshops were facilitated poorly in some instances. Many were conceived almost as a one-way information transfer exercise with passive facilitators, often very junior staff faced with very senior participants, doing little to challenge views and provide information that might lead to any *learning*. This failure to conceive of the RTS preparation process as a way to share and develop knowledge among participants, but rather as a way of soliciting views, caused problems later when the draft RTS emerged. Thus focus group discussions and questionnaires might in fact have been counter-productive in the development of policy.

Conclusions and future prospects

A number of issues arise from the above analysis for the future of regional strategy-making, both in relation to transport and more generally.

The explicit recognition of the need for a transport strategy within the emerging panoply of regional strategy documents is a positive recognition from central government of the significance of this issue. However, its relegation in implementation within RSS documentation may undermine this emphasis. Further, the exposure of the RTS to rounds of consultation through RSS in the North East case seemed to undermine much of what went before, especially as this process was not referred to in the RSS, a statement that would have at least conferred legitimacy to what was presented.

The regional level has proved a successful level to broker certain issues such as parking standards, contrary to some expectations about the likelihood of local authorities to sign up to this. However, other issues have proved more difficult to reach agreement over and this also relates to questions surrounding the evidence base. Given the Multi-Modal Studies and other technical work in all regions this evidence is now considerable. And yet in implementation many policy priorities became lost as some proposals came forward but others did not. Even where research was commissioned, such as above into the relationship between economic development and infrastructure, it was often ignored in policy development. Thus, attention again needs to focus on the processes of policy development to ensure that such evidence is not relegated in the push and pull of high profile, media-friendly but ultimately unhelpful discourses. This was particularly problematic in the North East where a narrow regeneration/economic competitiveness discourse squeezes out other agendas of environmental sustainability and social equity. This needs to be accounted for in process design and policy outputs, and while Local Strategic Partnerships have been given responsibility to provide input on 'hard to reach groups', their long-term ability to respond, not least to quite technical issues within policy sectors, is not clear at present.

Thus, issues as to who to involve and how in processes of regional strategy-making will continue to be problematic. In the North East case the range of stakeholders involved was broad but, partly as a consequence, processes were too simplistic to capture the complexity of the issues, given that they were oriented toward solution finding and not information gathering. Greater depth of deliberation is needed if such an approach were pursued in future. At present, however, the skills base and the institutional capacity related both to process design and technical knowledge within policy sectors such as transport, particularly as regards the Regional Assemblies, is patchy regionally.

To conclude, something of a vicious circle exists whereby, in the absence of significant institutional reform such as through the implementation of elected regional assemblies, continued centralisation will inhibit the development of the capacity to overcome many of the problems above (see also CfIT, 2002b; DfT, 2004; Headicar, 2002). And yet while such issues remain central government is understandably wary of giving up its gatekeeper role to allow regions to spend as they please even though this would then give an incentive for stakeholders to take participation in strategy development seriously. At present, therefore, difficult decisions over transport priorities at regional level continue to emerge from the centre and a key policy component of spatial strategy, and indeed the new regional governance, is largely sidelined in terms of influencing policy outcomes.

Notes

1 This section draws extensively on work conducted with Geoff Porter (see Vigar and Porter 2005 for more detail on the case).
2 The author undertook participant observation of the process as part of a Wider Reference Group involved in the RTS. This was accompanied by documentation review and a semi-structured interview with a key stakeholder.
3 This included: Review of parking standards (regulation of parking, bus provision, park and ride); Pricing measures (road use charges, etc.); Public transport (zones, routes, interchanges, rail services); Sustainable measures (cycle lanes, pedestrian zones, development in the Green Belt (sic)).
4 Road traffic growth in the North East was the highest of any English region in the period 1990–99 and yet its growth in GDP was the lowest of the regions (CfIT 2002a).
5 A commitment to 'evidence-based' policy-making is a feature of the UK Labour government's approach and despite being often rhetorical there is evidence of it filtering into some governance practices.

References

Banister, D. and Berechman, Y. (2000) *Transport Investment and Economic Development*, London: Spon Press.

Black, W.R. (2001) 'An unpopular essay on the subject of transportation', *Journal of Transport Geography*, 9: 1–11.

Commission for Integrated Transport (2002a) *CfIT's initial assessment report on the 10 Year Transport Plan*, London: CfIT. Online at: http://www.cfit.gov.uk/reports/10year/index.htm

Commission for Integrated Transport (2002b) *Organisation, Planning and Delivery of Transport at the Regional Level*, London: CfIT. Online at: http://www.cfit.gov.uk/reports/regionaltransport/index.htm

Council for the Protection of Rural England (CPRE) (2004) *Back Together Again*, London: CPRE.

Department for Transport (DfT) (2002) *Delivering Better Transport: Progress Report*, 17th December (2002), London: DfT.

Department for Transport (DfT) (2004) *The Future of Transport*, London: DfT.

Department of Environment, Transport and the Regions (DETR) (1998) *A New Deal for Transport: Better for Everyone*, Cm. 3950, London: HMSO.

Doherty, I. and Shaw, J. (eds) (2003) *A New Deal for Transport? The UK's Struggle with the Sustainable Transport Agenda*, Oxford: Blackwells.

GONE (2000) *Regional Planning Guidance for the North East – Panel Report*, September 2000, Norwich: The Stationery Office.

Haughton, G. and Counsell, D. (2004) *Regions, Spatial Strategies and Sustainable Development*, London: Routledge.

Headicar, P. (2002) 'Regional Transport Strategies: fond hope or serious planning?' in T. Marshall, J. Glasson and P. Headicar (eds) *Contemporary Issues in Regional Planning*, Aldershot: Ashgate.

Low, N. and Gleeson, B. (eds) (2003) *Making Urban Transport Sustainable*, Basingstoke: Palgrave Macmillan.

ONE North East (2000) *Unlocking our Potential: Regional Economic Strategy for the North East*, Newcastle: ONE North East.

ODPM (2004) *PPS 11*, London: HMSO.

Ove Arup with Centre for Urban and Regional Development Studies (2002) *Transport and the Economy*, Newcastle: One North East.

Shaw, J. and Walton, W. (2001) 'Labour's trunk roads policy for England: an emerging pragmatic multimodalism', *Environment and Planning A*, 33: 1031–56.

Standing Advisory Committee on Trunk Road Assessment (SACTRA) (1999) *Transport and the Economy*, London: HMSO.

Vigar, G. (2002) *The Politics of Mobility: Transport, the Environment and Public Policy*, London: Spon Press.

Vigar, G. and Healey, P. (1999) 'Territorial integration and 'plan-led' planning', *Planning Practice and Research*, 14(2): 153–71.

Vigar, G. and Porter, G. (2005) 'Regional Governance and Strategic Transport Policy: a case study of north east England', *European Spatial Research and Policy*, 12(1):89–108.

Whitelegg, J. (1997) *Critical Mass*, London: Pluto Press.

14 The Welsh Assembly and economic governance in Wales

Philip Cooke and Nick Clifton

Introduction

This chapter explores an important field of action in the devolved administrations, namely economic development, for which Wales and the other devolved territories are each responsible in policy terms. From 1997 New Labour pursued a strong sterling policy outside the eurozone. That, supported by globalisation effects as foreign and domestic firms outsourced and 'offshored' investments to cheap labour zones in Eastern Europe, North Africa, India and China, seriously eroded manufacturing employment. To these woes were added pressures for new industries caused by the end of the Internet stock market bubble from March 2000, and the ensuing global economic recession. So serious was the haemorrhaging by 2005, estimated at 1,000,000 manufacturing jobs in the UK since New Labour came to power in 1997, that concerns from industry and trade unions were voiced as that milestone was passed. This meant that, since devolution each administration has been struggling to modernise its economic governance mechanisms inherited from the preceding administratively but not politically devolved regimes. In this chapter, it is suggested that Wales, in particular, has perforce relied heavily on public sector job-generation as its main economic contribution to economic welfare in Wales. Comparisons with Scotland and Northern Ireland as well as regions of England demonstrate the case for 1998–2003. However, since 2003 it has become apparent that most job growth across the UK is accounted for by public sector employment growth and that what showed up first in Wales has become more widespread.

New Labour industry policy has, since the first Competitiveness White Paper, been to deliver a knowledge-driven economy (DTI, 1998). Subsequently the devolved administrations issued versions echoing this aspiration in the *Wales for Innovation* plan, Northern Ireland's *Think, Create, Innovate* consultation paper, and the Scottish Executive's *A Smart, Successful Scotland* strategy (Welsh Assembly Government, 2002a; DETI, 2003; Scottish Executive, 2001b). These are broadly in line with the framework laid out in DTI's first White Paper in 1998 under Secretary of State Peter Mandelson. From 1999 until 2000 a joint Ministerial committee reported to the Prime Minister on the Knowledge Economy, and policies and actions in support of improved science infrastructure and better-funded scientific research, industrial innovation, regional innovation and innovative cluster-building can all be traced to that initial impulse.

The chapter proceeds by analysing some of the key moments in economic policy evolution in Wales, comparing them where relevant with events in Scotland

and Northern Ireland. It is important from the outset to note that Northern Ireland's experience is only partly comparable with what happened in the other two devolved territories. This is not least because Northern Ireland's devolution has been suspended four times since 1999 and at the time of writing (July 2005) remains in suspension since October 2002. The most recent political developments in Northern Ireland make it difficult to see any swift resumption of devolution because the polity has become more polarised between republicanism and loyalism, the middle-ground has become attenuated, and questions of criminality of various kinds have entered the political discourse in ways that make resolution of profound political disagreements highly problematic. Nevertheless, some economic governance issues have been at times mirrored in each territory, notably the one at the centre of this chapter.

Economic governance policy modes

This concerns shifts in the mode of economic governance from a traditional grant-led 'handouts to the periphery' model of regional policy practised by the UK government throughout the post-war era, towards a more enlightened, more market-facing approach. In the latter, other instruments than regional development grants or Regional Selective Assistance (as it became known in its last manifestation as regional policy) are deployed (Welsh Assembly Government, 2002b). They include public or private venture capital purchase of equity in firms, especially smaller technology-based firms of the kind beloved by promoters of the 'knowledge driven economy', specific economic development loans from public funds that firms repay to an agreed timetable, and specialised, small-scale grants for smaller firms not in advanced technology sectors. These are intended to introduce for smaller firms and science-based start-up businesses financial disciplines regarding management of the company that make the 'begging-bowl' model, often seen as rewarding financial incompetence in the past, redundant. Such financial instruments might be provided as a package mixing more than one type of investment.

To varying degrees, such economic governance modes were put in place in each devolved territory, with setbacks, on occasions, difficulties and compromises. These problems are traceable to relative weaknesses in formulating a broader policy model. Modern policy management stresses the importance and impact on action processes of vision-led strategy formulation. An example of this is often taken to be Japan, pursuing an internationally expansive consumption goods strategy from the 1950s, outcompeting the West on quality, reliability *and* price criteria, a policy subsequently emulated by the Asian 'Tigers'. However, this rather bold approach may often not be pursued because of constraints of a financial or political kind. A *constrained* approach characterises contemporary German economic policy, hemmed in by economic history to the pursuit of counter-inflationary, social market, and strong welfare state policies when present imperatives for creating growth demand their relaxation, mainly at a politically unsustainable social cost which massively *constrains* room for manoeuvre. Furthermore, to attempt to moderate economic instabilities a *precautionary* policy may be preferred.

During the Asian growth period and even more recently it can be argued this was practised by the UK. Until the 1980s, the UK was widely characterised conservatively hanging on to Victorian industry and industrial practices, reluctant to invest in research and innovation, and failing adequately to modernise institutionally, either in general constitutional terms or in more specific economic governance. Until the 1990s Japan and its emulators thrived; even Finland was spoken of as Europe's Japan for its systemic commitment to support the electronics industry and Nokia as a 'national champion'. Meanwhile the UK economy languished, losing ground to many it had historically superseded. Taking policy precautions not to stray from historic economic pathways meant the UK eschewed 'envisioning' its future – rather it belatedly discovered 'Europe' having lost 'Empire'.

Scotland is developing economic policies that seek to transcend old, and even not so old, path dependencies. Northern Ireland is constrained to taking policy steps to build from an anachronistic industrial base that, as it erodes, exacerbates inter-ethnic tensions, in a context where the cushion of increased public employment is impossible due to its 'bloated' state after decades of armed political conflict, and its embryonic 'new economy' is assailed by the global technology downturn. Wales, faced with a legacy of 'smokestack' deindustrialisation, and rapid erosion of Asian transplants that brought semi-skilled alternative employment, unlike Scotland but less so Northern Ireland takes precautions against strategies that require exercising foresight to embrace possibly 'faddish' abstractions like the Knowledge Economy. Given Scottish Executive's early embrace in 2001 of a *Science Strategy* for Scotland to identify basic science expertise and funding gaps, it was striking that asking the Welsh Development Agency (WDA) and Invest in Northern Ireland (INI) executives about the possible value of a Science Council to advise Assembly members elicited the identical response that 'we don't need yet another level of bureaucracy'. Theses could no doubt be written on the etymology and interpretation of this sentiment. Interestingly, *innovation*, which is familiar from a decade or more of EU and more recently UK injunctions, often directed at manufacturing industry, invokes far less opprobrium, and as we have seen is widely promoted in economic governance, especially in Northern Ireland and Wales. The link with (traditional) manufacturing is almost certainly the political reason for this, although it supplies a decreasing share of regional wealth.

Manufacturing meltdown in Wales

After 1998 when it actually grew slightly in its share of UK manufacturing, the Welsh economy experienced a significant turnaround. As shown in Table 14.1, there was a loss of 44,000 private manufacturing jobs between November 1998 and the same month in 2002. This was quantitatively more than compensated for by a simultaneous rise of 67,000 public administration jobs, overwhelmingly in health and education. However, these replaced higher value adding, higher productivity, export earning jobs for lower productivity jobs increasingly reliant on financial transfers from Whitehall. Under devolution, due to an absence of *envisioning* policy making to tackle changed global economic realities, Wales looks to be becoming more dependent, not less, on London for the underwriting of its economic future. As a *precaution* against rising net job loss the Welsh Assembly Government (WAG) has used its own block grant resources, growing as UK expenditure on health and education burgeons, rapidly to increase employment in those sectors plus direct public administration.

This question was raised in a number of interviews with senior executives in WAG and the Welsh Development Agency (WDA) conducted in summer 2003 for the *ESRC Devolution & Constitutional Change* programme. Three kinds of response were forthcoming, ranging from what might be referred to as 'the new cynicism', to 'new realism' and finally, 'new public management'. The first of these, conditioned by numerous hard knocks as the clothing industry shed 2,600 jobs, steel a further 6,000 and the South Korean electronics giant LG Electronics some 900 in 2002–3, takes the form of insisting upon the success of policy in keeping such employment in being for, in the case of LG, seven years while blaming private management for poor anticipation of market movements. The second is a dawning recognition of a 'new realism' about the limitations of traditional economic governance, particularly that involving economic development agencies. This arises from awareness that the 'generalist' skills their functionaries possess are often no longer adequate, that where such organisations were set up to be integrated 'powerhouses' as was officially the case with WDA post-devolution, nowadays they may require partial 'decommissioning'. Furthermore this implies that advising and assisting by signposting firms to be more demanding customers of private business support services and less reliant on a grant-dependency culture is no bad thing. The 'new public management' position is based on trickle-down economic philosophy that rests not on foresight or scenarios but rather on efficient execution of actions that are within the sphere of responsibility of the devolved governance mechanism. Thus if increased funding flows in to Wales' block grant from Whitehall, as it has for health and education, efficiency is measured by the speed with which it is allocated to spending organisations. This can set up certain effects such as increased employment in public services. It brings trained nurses and teachers into the public sector from less-skilled jobs, freeing up labour market spaces for long-term economically inactive persons whose inactivity contributes to low GDP per capita indicators. Evidence that precisely this is happening would, if true, vindicate this approach in the short run because of its 'win, win, win' nature regarding increasing employment, improving public services, improving GDP per capita while hitting EU Structural Funds targets, and lessening long-term unemployment. Whether this analysis leads logically to promotion of an economy that is innovative and competitive in terms of growing exports and global market shares for the private sector remains to be seen.

The first of the above suspicions about the suitability of variants of the National Enterprise Board model of economic intervention circa 1976, which is what the WDA was when set up at that date, gained strength on 14 July 2004 when the First Minister announced the cessation of the WDA as an arm's length economic development agency. It is, by April 2006, to be absorbed into the WAG Ministry of Economic Development along with the Wales Training Agency (ELWa) and the Wales Tourist Board in a 'bonfire of the quangos', a term widely employed by the First Minister at the time.

Regarding the employment change referred to earlier, Table 14.1 suggests the question needing investigation is what happened and why to reverse the upward trajectory of Welsh manufacturing. Comparative statistics of manufacturing job change 1991–2001 reveal Wales until 1998 as the UK's only increasing source of manufacturing employment. Briefly, Table 14.1 shows three relevant things. First, note the growth in Welsh manufacturing 1991–8. Second, note the higher than average percentage job loss in manufacturing 1998–2001 (which nevertheless translated into a relatively modest 9,287 jobs). Accordingly, third, we see Wales' slippage from third to fourth in regional manufacturing employment share in Britain.

In the UK, large firms accounted for two-thirds of the losses and Wales is unlikely to be much different. Table 14.2 then shows us what happened until November 2002, the last date for which Labour Force Survey statistics were available at the time of writing. This table reveals three important features for the 1998–2002 period. First, although not the largest magnitude in absolute numbers, the Welsh percentage decline in manufacturing was, at 4.6 per cent, the steepest. Second,

Table 14.1 Manufacturing employment change in Great Britain, 1991–2001 (March)

Region	2001 (%)	1998 (%)	1991 (%)	% Change 91–01	% Change 91–98	% Change 98–01
E. Midlands	20.5	21.5	26.7	–6.2	–1.2	–1.0
Eastern	14.5	17.1	19.7	–5.2	–2.6	–2.6
London	6.4	7.8	10.0	–3.6	–2.2	–1.4
North East	16.5	21.1	21.6	–5.1	–0.5	–4.6
North West	16.4	20.2	22.1	–5.7	–1.9	–3.8
South East	11.2	13.9	15.5	–4.3	–1.6	–2.7
South West	14.2	13.9	15.5	–3.2	–0.7	–2.5
W. Midlands	19.9	25.7	28.0	–8.1	–2.3	–5.8
Yorks. and H.	17.8	21.3	23.0	–5.2	–1.7	–3.5
Scotland	12.8	15.4	17.6	–4.8	–2.2	–2.6
Wales	**17.1**	**21.7**	**21.6**	**–4.5**	**+0.1**	**–4.6**
GB	14.1	17.4	19.3	–5.2	–1.9	–3.3

Source: Office of National Statistics

Table 14.2 Regional manufacturing employment change, 1994–2002 (November)

Region	2002 (000s)	%	2001 (000s)	%	2000 (000s)	%	1998 (000s)	%	1994 (000s)	%
E. Midlands	434	21.0	453	21.9	455	22.6	481	24.1	494	26.4
Eastern	430	15.5	460	16.6	444	16.2	465	17.6	475	19.0
London	287	8.0	284	8.0	282	8.1	319	9.4	310	10.1
North East	194	17.6	210	19.1	220	19.9	233	21.7	205	19.6
North West	557	17.4	569	18.2	594	19.1	622	20.4	665	22.6
Scotland	336	13.9	337	14.1	368	15.3	375	16.1	380	16.7
South East	569	13.6	582	14.0	567	13.8	656	16.3	600	16.2
South West	366	14.7	364	14.8	385	15.8	378	16.2	377	17.1
Wales	**206**	**15.8**	**220**	**17.4**	**223**	**17.7**	**250**	**20.4**	**237**	**19.9**
W. Midlands	563	22.5	572	22.8	567	23.2	639	25.8	629	26.7
Yorks. and H.	444	18.7	440	18.7	479	20.3	477	20.8	471	21.3
GB	4,386	15.7	4,491	16.2	4,584	16.7	4,893	18.2	4,843	19.1

Source: Office of National Statistics
NB. Northern Ireland statistics are separately held. Manufacturing Employment Change for Northern Ireland over a roughly equivalent period is:
Northern Ireland 93,530;103,220; NA; 105,750; 102,700
Northern Ireland Dates: 2003, 2000, 1999, 1995

the two-to-one ratio of large firm to SME job loss suggests that large firms accounted for approximately 30,000 of the 44,000 jobs lost from 1998 to 2002. Third, Wales slipped from fourth to sixth in regional manufacturing employment share in approximately one year.

Moving on, Table 14.3 shows how fast employment in 'public administration' has grown in Wales of late. Wales now has the highest share of employment accounted for by public administration in the land. Reflecting back on the 1998–2002 period that saw a major downturn in the manufacturing labour market, the 67,000 rise in public administration employment more than made up for the 44,000 manufacturing jobs lost in that period. It is well known that during the mid-2000s it was the policy of New Labour in London to operate a Neo-Keynesian demand management strategy by pumping significant amounts of public expenditure into health and education as counter-cyclical measures.

This worked well at the macro-economic level with the UK suffering no recession or stagnation such as that experienced by many economies in the EU eurozone. However it created absorption problems in Wales where the Barnett formula squeezes expenditure out of the system if it cannot be contained within the block grant cap of some £13 billion. Hence in 2004 cuts of £300 million were necessitated, mainly in the economic development budget, to make room for, especially, substantial increases in much-needed healthcare expenditure.

Scotland and Northern Ireland: brief comparisons of economic governance approach

Briefly, Scotland's Barnett formula settlement is more favourable than that of Wales. Accordingly the problem associated with the 'Barnett squeeze' did not apply. Returning to Table 14.2 we also see Scotland's new deindustrialisation hit

Table 14.3 Regional public administration employment change, 1994–2002 (November)

Region	2002 (000s)	%	2001 (000s)	%	2000 (000s)	%	1998 (000s)	%	1994 (000s)	%
E. Midlands	502	24.3	488	23.6	467	23.2	438	21.9	400	21.4
Eastern	640	23.1	633	22.9	637	23.2	570	21.6	475	19.0
London	850	23.8	851	24.0	771	22.1	769	22.7	731	23.7
North East	326	29.6	327	29.7	301	27.3	283	26.4	263	25.2
North West	874	27.4	856	27.4	853	27.4	756	24.8	720	24.5
Scotland	690	28.6	671	28.1	649	27.0	635	27.2	583	25.6
South East	1,004	24.0	974	23.4	987	24.0	960	23.8	864	23.3
South West	653	26.1	626	25.4	642	26.4	588	25.2	564	25.6
Wales	**415**	**31.8**	**368**	**29.1**	**368**	**29.2**	**348**	**28.4**	**311**	**26.1**
W. Midlands	606	24.2	621	24.8	568	23.2	559	22.5	516	21.9
Yorks. and H.	634	26.7	592	25.2	604	25.6	552	24.1	513	23.1
GB	7,193	25.7	7,008	25.3	6,846	24.9	6,459	24.1	5,964	23.5

Source: Office of National Statistics
NB: Northern Ireland 234,300 (2003); 296,170 (2000); 291,530 (1999)

earlier so that although it lost a lot of manufacturing jobs between 1998 and 2002, its share was lower and fewer were lost than in Wales. Northern Ireland shed 12,220 between 1999 and 2003. Scotland's economy has typically performed at or near the UK norm in recent years. Accordingly, Scotland has only temporarily attracted the kind of Structural Funds support, notably Objective 1, now experienced by Wales and from 1988 to 1999 by Northern Ireland. Of course, the Highlands and Islands region was a recipient of Objective 1 funding but that also ceased, or rather, tapered off after 1999. Moreover, the support was more for peripherality and rural policy to counter demographic decline than for the restructuring of heavy industrial economies in decline. As in Northern Ireland, its implementation was associated with an increase in GDP that brought that measure above the 75 per cent of EU GDP per capita that triggers Objective 1 funding that, from 2006, will mainly focus on the EU Accession economies.

Nevertheless the weakening of manufacturing, including a spate of closures by foreign electronics firms in Silicon Glen, contributed to broader concerns that led the Scottish Parliament to commission *Scotland's Science Strategy* (Scottish Executive, 2001a). This reviewed basic scientific research, costed it, assessed it in relation to world-class benchmarks, and prioritised three fields for which extra resources and attention would be forthcoming. The fields are Biosciences, Medical Science and E-Science. Activities to develop closer networking among public and private research laboratories, to stimulate technology transfer *from* the Scottish health system and to promote a science-based economy were begun.

Regarding the last, the Scottish Executive then produced an economic strategy document charging Scottish Enterprise and economic actors generally, to espouse their vision of *A Smart, Successful Scotland*. This emphasised the need to position Scotland to exploit to the full the Knowledge Economy and proposed actions to: enhance knowledge inputs and outputs among global businesses in or relevant to Scotland; hasten the rate of spin-outs from scientific research; make Scotland's 'talent' base more 'sticky' and augment it by stimulating a more cosmopolitan image.

One of the key ways in which this is expected to be implemented is through Scottish Enterprise's *Global Connections* strategy. This sets out Scotland's strategic direction for taking advantage of the opportunities in the knowledge economy and ensuring it is a globally integrated economy. The two overarching objectives of the strategy are:

- Helping Scotland realise value by attracting knowledge from overseas
- Helping Scottish knowledge generate value abroad for Scotland

This is more simply expressed in the shorthand description of the strategy as focusing on Knowledge-in and Knowledge-out. It is illustrative of the perception of the need for a new approach to set the *Global Connections* strategy in its wider Scottish context. It provides one strand of an overall approach set out in Scotland's economic development vision, *A Smart Successful Scotland*, alongside complementary strands on Growing Business and Learning and Skills. While the main aim of *Global Connections* is to raise Scotland's level of international activity given increasing globalisation, its objectives are key means of meeting Scotland's broader economic challenges.

The aims, means, instruments and obstacles to their implementation, of the new model of economic development and its financing in Northern Ireland were mostly clarified consequent on the first period of devolution and before the first suspension. Modernisation of institutions and approaches came under the Northern Ireland Executive's 'Programme for Government' commitments. Invest Northern Ireland was the principal executive vehicle for economic development in this process, an integrated development agency distinct from but recognisable to observers of such long-established enterprise support organisations as Scottish Enterprise and the Welsh Development Agency. INI is now augmenting this new approach with integrated service delivery, moving from the traditional bias towards provision of incentives to the creation of demand for services. This is testified to in the 2002–5 Corporate Plan, which envisaged that 'Invest NI will play a role in the formation of a local risk capital market' and encourage firms to raise risk portfolios (INI, 2002). Thus for there to be more research and development (R&D) in the economy, firms and knowledge centres must demand more R&D. Similarly for more systemic linkages to be in the economy these must be 'in demand' from innovation actors and firms. Creating a healthy demand for public and private services to aid the restructuring of the Northern Ireland economy is a politically established goal of a reorganised service delivery mechanism.

This, incidentally, was something also under review at the Welsh Development Agency in its internal re-think about the provision of business services. There, a move away from almost archaic *sector* policies in which the economy is strategised according to industrial categories formed by government statisticians in the late nineteenth century (the Standard Industrial Classification) towards market *segment* thinking was under way. Associated with this was the outsourcing of business services advice to the private sector, an admission of the *generic* contract management expertise that now typifies the skills sets of public economic governance functionaries under New Public Management. The latter has caused a major erosion of professional competence and confidence throughout the UK public services under the attacks of the Thatcher and Blair critiques of public sector management inadequacies. The beneficiaries of this New Public Management process are, of course, private consultancies.

Thus a key institutional aim of economic governance in Northern Ireland was to create a WDA-like 'economic powerhouse' in the shape of InvestNI. With the early creation of InvestNI, drawing together the activities of the previous development agencies (and the Tourist Board), a more concerted effort is being made to put in place the elements of a modern economy, including instruments to strengthen the regional innovation system, such as university incubators, spin-out firms, venture capital, exacting technology customers, supply chains, cluster-building programmes, science park facilities and science entrepreneurship support. Each of these is meant to contribute to a more regionalised network of complementary facilities and functions, to ensure that public and private elements of the system work together in a less *dirigiste* and more locally networked way, and that knowledge and resources flow efficiently to meet the needs of firms.

Does the recent experience of seeking to remake a traditional public enterprise support system into a more market-facing, less grant-allocating vehicle offer pointers about Northern Ireland compared to elsewhere? Three key developments currently characterise the concerns of the more adventurous enterprise support systems globally. The first is the *knowledge economy* and how to engage with it, supply important infrastructural and financial preconditions for nurturing it, and try stoically to accommodate its vicissitudes (Power, 1997). Northern Ireland was fortunate to have the Industrial Research & Technology Unit (IRTU) as a research and technology investment arm, now assimilated into InvestNI. The second challenge is *globalisation* and its implications for the evolution of strategic thinking. The third is *entrepreneurship*, a characteristic in great demand given the drying up of new FDI projects and weakening manufacturing employment opportunities contingent upon redundancy announcements such as those of Bombardier-Short's.

It is relatively easy to see the *vision* embodied in the new approach to economic development and its financing in Scotland. It is, first, constructed on recognition of weaknesses in entrepreneurship and, particularly, indigenous knowledge exploitation. Second, it builds on strengths, such as a good scientific performance. Finally, it adopts measures aimed at integrating knowledge flows to exploit strengths and tackle weaknesses. It envisages a strong knowledge generation and commercialisation future for the Scottish economy. It is thus, given some of Scotland's avowed economic weaknesses, an ambitious programme. Moreover in a context of perceived market weakness, it is a markedly *interventionist* programme, controversial enough in ideological terms to have entailed the resignation of Scottish Enterprise chief Robert Crawford on account of an unfavourable press from, inter alia, Andrew Neil's *Scotsman*. In Northern Ireland much of the early policy promise of the new approach to economic governance was overshadowed by suspension of the Assembly from October 2002. Ironically, under Whitehall rule many difficult policy issues concerning hospital closures, local taxation, water rating and use of Private Finance Initiative (PFI) schemes to adaptively reuse former security and military facilities have been addressed and change accomplished.

Post-devolution Wales: assessing progress

In Wales, a process of introducing administrative economies of scale was initially pursued as a number of disparate initiatives were launched, variously listed under the enlarged and centralised bureaucracies of the WDA and (until April 2003) ELWa, the education and training agency. These emerged as disparate measures, mostly dependent upon the designation of much of north and west Wales, plus the former coalfield in the south, as qualifying for EU Structural Funds Objective 1 status (Osmond and Jones, 2003). Not experienced in managing transfers of the £1.2 billion scale plus match-funding that this designation released, the Assembly Government cast around for methods of spending and managing expenditure. Here the *precautionary* principle overwhelmed any pretence at a more *visionary* alternative.

A Task Force to consider a 'national economic development strategy' and the Objective 1 Single Programming Document was established to design the financial structure for programme expenditure. The process of drawing up the priorities on which the money was to be spent had been complex and not entirely successful. Participation by representatives of local government, business and the voluntary sector had led to deadlock with the voluntary sector complaining of being outmanoeuvred by the other parties. Accordingly the new First Minister dismantled the administrative machinery set up by his predecessor and handed the task to the civil service. Time was short as the final submission deadline to Brussels was looming, so they simply allocated the funding in the same proportions as it had been divided in the old Objective 2 programmes. Some of the resulting imbalances were raised in the UK Parliament's Select Committee on Welsh Affairs investigation into The Structural Funds in Wales, and from the evidence given by the First Minister the above account of administrative expedience emerged.

It is instructive to understand from the First Minister's own words how the *precautionary* reeling-in of this process from an 'inclusive' institutional partnership back into the civil service proceeded:

Mr. Caton:[1]   207. You just mentioned that the Assembly added the financial tables.
How did you arrive at the figures for the financial tables in Objective 1?

Mr. Morgan: … there was not a huge amount of prescription from the Task Force to the Assembly … on that, so basically at the end of the day it was the conventional Minister/civil servant relationship, kind of, 'What do you think?' 'Well, I've got to take responsibility for this, so I will come back to you if I get hung, drawn and quartered down in the Assembly over it' … Second, we did use consultants to do a bit of sort of feeling out of the European Commission themselves as to what their views on the priorities were. We also took account of the precedents from the Objective 2 and 5b programmes in Wales and we also took some account of alleged success stories in the Irish Republic which we took with a pinch of salt … On the split between the five priorities … let me try out … Huw here.[2]

Mr. Rawlings: 209. … I think a point I would want to make is that it is perhaps not surprising that the Task Force did not find it easy to make a clear recommendation on these monies. The Task Force is drawn from a wide range of interests and …

Mr. Morgan: Vested or otherwise.[3]

For managing the allocation of monies an extremely complex system of interlocking committees was set up, responsible for each programme area, involving Assembly and other government, business, voluntary and academic representatives and experts who were recruited to fill these committees, whose main task was to judge whether grant applications for funding should be approved. At the end of the first year of this process an unofficial estimate was made by a former European Union senior official who had returned to advise the Assembly on this financial absorption and allocation nightmare, that 1,700 people had been recruited to manage the approval system and support it administratively. Such were the complaints from, particularly, the business community at the glacial progress of implementation of the Objective 1 programme that reforms were instituted, consisting of the insertion of a new layer of committees given a 'troubleshooting' function to break the administrative logjams that kept recurring.

It is possible to construct a logic to key policy tools invented to absorb substantial *tranches* of Objective 1 funding. They fit the EU's standard 'innovation push' view of regional policy. Thus there are tools to promote knowledge transfer, entrepreneurship, venture capital, and incubators – all standard EU fare. This is presumably explicable by reference to the First Minister's reference to 'consultants ... feeling out ... European Commission ... priorities'. However, from the timing and way these initiatives were set up it is likely that the precaution to find some things acceptable to the Commission on which to spend Objective 1 money overrode any idea of constructing a modern, joined-up innovation support architecture. Rather, tools are operated and assessed in their own terms. Not surprisingly they have not performed well. On the face of it WAG acting as a developmental state, intervening to guide economy and society towards a specific model of economic modernisation does not seem an acceptable description of reality. However, that it is interventionist and inclined to be precautionary about running everything itself, including direct job-generation in public administration, seems indisputable. The question of whether this can seriously be described as 'developmental' is returned to towards the end of this chapter.

Formal assessments of performance regarding initiatives such as the Entrepreneurship Action Plan (EAP), Knowledge Exploitation Fund (KEF) and Finance Wales (FW) are seldom published. However, official statistics reported in the *Western Mail* on 16 January 2004 showed that for the financial year 2001/02, in return for an average £80 million per year expenditure in its first three years, the EAP was set a target of providing support to 4,600 new business ventures, but in fact only aided 1,800 – a deficit of 2,800. For 2002/03 EAP was set a goal of supporting 6,300 start-up businesses and 4,000 start-ups were assisted by the WDA from April 2002. Such assistance can include fairly trivial telephone inquiries about eligibility. Part of this expenditure is on *entrepreneurship* modules in colleges.

A report on KEF's own website shows that despite budgets of well over £20 million per year being spent only 5 per cent more entrepreneurship modules were being taught in universities and other higher education institutes, although 25 per cent more were taught in further education colleges. But 75 per cent of the latter had no or few mechanisms for technology transfer, while the statistic for universities was 25 per cent. It can be concluded that there is a significant disconnect in this particular part of the entrepreneurship-driven renewal of the regional innovation system in Wales. So much so, that in the disastrous events that saw the demise of ELWa in March 2003, its separation into the Higher Education Funding Council for Wales (HEFCW) and the National Council for Education and Training in Wales (NCET), KEF was transferred to the WDA. Remarkably, it transpired that as well as issuing training-related contracts illegitimately, ELWa had no legal status, being effectively merely a brand name for HEFCW and NCET. The precautionary principle had collided with the 'public enterprise' aspiration in a bureaucratic nightmare. The Chair resigned in June 2003 amid further evidence of irregularities in the appointment of an acting Chief Executive, himself author of a hundred redundancies in response to the HEFCW divestiture. Thereafter, as we have seen, in the 'bonfire of the quangos' the main economic governance agencies were swept away and their functions absorbed into the Ministry of Economic Development. Poor performance in the world of knowledge economy and New Public Management were the excuses. But also tensions between WAG and Labour's Westminster MPs who are hypercritical of WAG performance generally and were interpreted as disdainful of the

Assembly's lack of powers should not be overlooked in trying to make sense of such decisions. After all, less than a year later a 'quango bonfire' was mooted though not implemented in the Chancellor of the Exchequer's March budget.

Finance Wales, a vehicle designed to supply venture capital to innovative SMEs and start-up businesses because of a perceived market failure in private provision also registers such disconnects in the far lower than targeted number of businesses coming forward in quest of equity investment. Accordingly, public venture capitalists are redeployed on to firefighting co-funding grant packages. Further administrative expediency and risk aversion has resulted in equity now being tied to accessing Regional Selective Assistance, thus incentivising entrepreneurs to becoming 'grant junkies' rather than weaning them off grant-dependence as modern investment theory advocates.

The Technium incubator-building scheme seems equally over-ambitious, with twenty planned to host many more spinoffs that can be legitimately expected to arise from academic entrepreneurship in Wales alone. Programme costs are some £260 million, again funded largely by Objective 1 resources. One Californian flagship technology firm in Swansea's Technium has folded and a privately led media Technium in a west Wales rural setting has failed to progress despite large sums of Objective 1 and WDA funding having been allocated. As well as over-ambition there are a set of design flaws in the policy that include, first an inclination to replicate old incubation approaches that failed to prioritise management assistance, including allocating part-time space to such services as venture capital, legal advice and management accountancy. Second, true to WDA traditions they are properties leasing space, now for SMEs previously for FDI businesses, thus they are not in themselves innovative. Finally, they assume 400 or more incubator spaces can be filled. A study of this question calculated that, from academia in Wales, where there are less than one thousand tenured scientists and engineers, some 20 to 30 spinouts could be anticipated during the lifetimes of those academics if international rates of academic entrepreneurship prevailed. Clearly, a major 'recruitment' effort is required for aspirations to have any chance of being fulfilled, and this at present is not evident as policy or practice.

Wales' economic governance has been precautionary and confined to, initially, reorganisation of the administrative apparatus, with disastrous consequences in the case of ELWa. It was further rendered precautionary by virtue of the windfall receipt of Structural Funds in significant measure that, nevertheless, came with strings attached which meant funds required heavy guidance to met EU allocation criteria. A cumbersome bureaucracy, possibly created from an over-interpretation of EU requirements slowed the flow of investment to a trickle in the 2000–2003 period. Finally, precautions have often been taken in a discretionary manner not to give the appearance of over-spending. The most recent victim of last-minute plug-pulling by WAG has been the National Botanic Garden of Wales, housing another of the ill-fated Technium incubators, and on the verge of bankruptcy awaiting an unlikely private-sector suitor.

We have seen that devolution in Northern Ireland can be constitutionally, historically and politically distanced from that found in Scotland and Wales. For this and other reasons to be discussed we consider use of the adjective *constrained* economic governance appropriate. We saw how political cleavages and alignments internal to Northern Ireland's governance under devolution constrained policy implementation and how under suspension different UK agendas are brought to the fore. In Figure 14.1 we characterise Northern Ireland's economic (and other) governance as triply constrained. This we represent comparatively as the apex of a triangle of visible economic governance modes under UK devolution, and refer to for reasons discussed as *constrained* economic governance in the case of Northern Ireland, *precautionary* in the case of Wales and *visionary* in Scotland.

In Figure 14.1 an attempt is made drawing upon fuller analysis in Cooke & Clifton (2005) to balance for each devolved territory its economic governance style with its policy regime. Thus it is a kind of stylised input–output policy analysis of convergences and divergences among the three. As we saw in the first two accounts, it is not unreasonable to refer as *visionary* to Scotland's relatively coherent grasping of the exigencies of the onset of a Knowledge Economy and demise, in Scotland, of first an elderly Industrial Age economy and then an FDI Manufacturing Transplant modernisation in the last quarter of the last century (Cooke, 2005). We paraphrase Hall and Soskice (2001), in characterising 'varieties of devolution' rather than 'varieties of capitalism'. In their terms, all three devolved territories are *Liberal Market Economies* (LMEs) rather than *Co-ordinated Market Economies*. However, within a broad LME institutional setting at the UK level, the policy regime and its institutional stance of the devolved territories differ quite markedly. Both Scotland and Northern Ireland set up or privatised financial governance functions for economic development, notably venture capital, whereas precautionary Wales set it up within the WDA. While InvestNI is avowedly market-led in its policy outlook it nevertheless has to face reality and intervene considerably. Scottish Enterprise intervenes by fine-tuning of market mechanisms that function relatively less well than they might, but the WDA is being 'nationalised.'

Thus, as variants on an LME theme we propose that the three output (policy) regimes are capable of being understood as in Figure 14.1. The evolution of devolution, whereby three distinctive national identities, sharing peripherality and a history of both an early rise as well as demise of modern industrialisation, is shown to have developed distinctive forms of policy input and economic governance

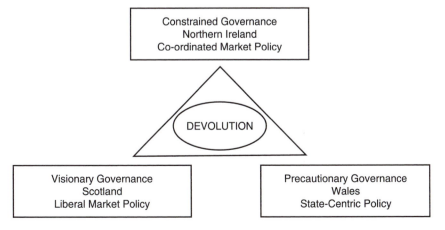

Figure 14.1 Forms of economic governance in the UK's devolved territories

output. This is interesting because devolved economic governance strategies diverged despite adopting similar economic governance leadership through integrated, multi-purpose development agencies. The reasons for the distinctive outcomes of similar governance mechanisms facing similar economic problems in comparably peripheral settings will be discussed in the concluding section of this chapter.

Conclusions

While referring to Scotland's experience of economic development policy experimentation as *visionary* in comparison to that of Wales as *precautionary*, and Northern Ireland's *constrained* form of manoeuvrability in economic governance, it is clear that more attention has been devoted to the often negative recent experiences of the Northern Ireland Assembly's intermittent suspensions and the Welsh Assembly Government's efforts to restructure the nature of financial support for enterprise and innovation. Scotland has had problems connected to political leadership more generally rather than specific problems of economic governance and policy formulation (although the hounding out of *Scottish Enterprise* Chief Executive Robert Crawford by *Scotsman* editorial neoconservatism is an exceptional case). This is because it is intellectually interesting and important to engage in sensemaking of policy failure as well as policy success. More broadly, the kind of limited devolution Wales and Northern Ireland but not Scotland received is likely to be more generic if such powers are applied in, for example, English regions. The title proposes 'varieties of devolution' not merely as a descriptor but as a possible indication of causes of the disappointing performance of new economic development financing tools in Wales as compared to Scotland. It also captures features, like tough auditing in collision with New Public Management constraining governance that lackeys the varying tides of Whitehall and Stormont in Northern Ireland. We can draw three conclusions from the foregoing analysis that gives credence to the idea that *fewer* devolved powers may result in weaker policy perspectives than *greater* devolved powers.

Why should this be? First, as was shown in both the analysis of job-generation and in the creation of super-quangos it is worth noting that as well as the creation of an integrated 'powerhouse' in InvestNI, the WDA also grew massively, nearly trebling its employment to over 1,000 since devolution as new functions like Finance Wales, Business Connect, and now KEF were handed to it. This can only be interpreted as the imperative that to be seen to be doing something, the Welsh and Northern Ireland Assemblies had so few powers that they could only do a few things. The first was to take over direct job-generation activity through enlarging public administration employment paid for from the block grant, which Assemblies control. This has been the default growth strategy in Wales but not in Northern Ireland or Scotland where public sector employment growth has been more modest. In fact, public sector slimdown is the order of the day in Northern Ireland, a further constraint on employment growth. The second option, practised instantly in Northern Ireland and Wales but not Scotland, was to reorganise as much as possible of the inherited administrative apparatus that became Assembly controlled. Thus, as well as the dysfunctional super-quango ELWa, now dismantled and in 2006 to disappear, the WDA was designated a 'powerhouse' by pre-devolution Secretary of State Ron Davies, was widely understood to have gone through a five-year period of Zen-like introspection that meant it lost profile and effectiveness as internal reorganisation became a permanent feature of its landscape, before it too disappears in 2006. Incidentally much the same has applied to easily Wales' biggest public body, the National Health Service, fully reorganised twice since devolution, a task of sufficient complexity that each restructuring takes two years. Thus the NHS in Wales has been in a state of permanent annual reorganisation since devolution, meanwhile performance on waiting lists is weaker than elsewhere in the UK despite per capita health budgets being higher. That such centralising aggrandisement has not ceased is testified to of course by the WAG absorption of three economic governance quangos into its Ministry of Economic Development.

In Northern Ireland, a great deal of attention was paid to administrative reform, ranging from establishment of a rating system, disposal of former military and prison real estate, closure of hospitals, modernisation of local government, introduction of the PFI, and stimulation of public–private partnerships (PPP) all at Westminster behest and, under suspension of devolution, implementation. InvestNI was Northern Ireland's version of the WDA 'powerhouse' model, and one that under circumstances of constraint and precaution where inward investment is concerned, is in danger, as shown starkly in Scotland, of receiving accusations of 'crowding out' private initiative more than stimulating enterprise and entrepreneurship, through 'over-interventionist' economic governance. But Northern Ireland, unlike Wales and Scotland, is constrained in extra ways by the New Public Management paradox. In getting away from past 'grant-dependency' business support schemes and stimulating outsourcing and private partnerships paid for with taxpayer contributions, economic governance falls foul of public accountability in the form of a particularly restrictive audit regime. This waxes and wanes according to whether Stormont or Westminster is responsible, as indeed do, in quite significant ways, public policy priorities – a factor not experienced in Scotland or Wales. It seems economic policy in Northern Ireland dances to a UK tune under suspension, with a reform agenda to the forefront while the devolved institution acts as a platform for positive policy initiative that is somewhat hamstrung by political grandstanding from the guardians of fiscal probity in the devolved administration.

Second, weak and poorly defined powers have meant there is little incentive to seek to be imaginative since it is institutionally depressing to receive bounce-backs from above. This occurred a great deal in the WAG's early years, especially concerning agricultural policy. Thus WAG boldly asserted a policy that Wales would be a 'sustainable' economy and society, whatever that meant. Wales was declared to be a Genetically Modified Organism-free zone in 2000 only to discover that the UK agriculture ministry had approved trials in three Welsh locations, and there was nothing Assembly powers could do about it. The drag-effect of Whitehall on Wales' agriculture minister's capability to respond rapidly and with good local knowledge to the foot and mouth crisis is another case, as was the uncertainty surrounding student tuition fees which Westminster wanted raised against Cardiff Bay wishes. These make for defensive or *precautionary*, even legalistic policy formulation modes from which any spark of creativity is quickly extinguished. In Northern Ireland, it is clearly hard to focus on policy implementation when devolution is in suspense and a higher level in the multi-level governance hierarchy imposes its priorities, although doing so in politically painful areas with which Stormont politicians were reluctant to be associated. In Scotland, by contrast, innovative policies have been drafted, approved and implemented, not without critique but confidently. This has led to interest in new forms of economic governance, such as the management of integration of diverse institutions, internal questioning of the effectiveness of inherited management structures such as that in Scottish Enterprise which modernising middle management sees as insufficiently 'de-layered', learning oriented and globally connected.

Finally, devolution in Wales attracted few top calibre politicians and even fewer top calibre civil servants. Because of circumscribed powers and the small majority in favour with opposition voices stressing the superfluity of 'yet another layer of bureaucracy' the style of the Assembly itself and particularly the WAG has been conditioned by a local authority mentality. There is great fear of being seen to be 'taken in' by, for example, grand architectural schemes. Most notably, the Assembly still does not have in 2005 a purpose-built Assembly building and is further from achieving it than Scotland because the Finance Minister, highly questionably, broke the contract with its designer Lord Rogers and, although he humbly re-tendered to redesign his own creation, such were the fears of cost overruns that it will inevitably cost more and be of worse quality when it is built than if it had been allowed to proceed many years ago, as planned. North Wales constituents were seriously upset at a long delay in approving £20 million in Regional Selective Assistance for the expansion of British Aerospace's Flintshire facility to build wings for the new Super-Airbus. So, even where discretion is allowed by the rules of devolution, the WAG management style can give an appearance of being highly *precautionary*, not to say timid in its decision making. Thus far, *Carpe Diem* has not become the watchword of the Welsh Assembly.

Scotland's economic governance has proceeded innovatively and smoothly, although not immune from critique, benefiting from the confidence public servants gained by the form of the settlement. Some Ministers are spoken of as of exceptional calibre – Wendy Alexander, the former Enterprise and Learning Minister who resigned unexpectedly, being the continuing subject of nostalgia among the Scottish economic governance community. Sir Reg Empey occupied a similar place in the affections of equivalents in Northern Ireland. However that cannot be said for the politicians who occupy positions on the Public Accounts Committee and to a lesser extent the professionals in the Northern Ireland Audit Commission both of whom were seen to be practising an immature 'grandstanding' posture in relation to efforts by Northern Ireland's economic governance community to forge a new set of policy instruments. Thus it is rather easy to see how 'varieties of devolution' have emerged from the differential evolution of internal and external exigencies concerned with constitutional settlements, forms of political culture, economic conditions and calibre of political incumbents. In Wales, there has developed a strong tendency for WAG to 'pull the plug' on problem projects like the National Botanic Gardens, in late 2003 effectively bankrupt, gathering more public odium for their parsimony and precaution on the way. In Northern Ireland creative politicians tend to be cowed by headline-seeking auditors thus further constraining policy manoeuvrability, while in Scotland brickbats from the far right media are not allowed to interfere with the pushing through of innovative and expensive policies. It may therefore be concluded that, if not a causal condition then at least the independent variable in this is the relative degree of autonomy provided in the UK's asymmetric devolution settlement among Northern Ireland, Scotland and Wales.

So, on a number of issues common to different elements of this book, we may ask whether devolution has helped create the kind of development coalitions that give an extra boost to a territory's economy (Goodwin *et al.*, 2002). Our answer is that, given the statistics reported earlier, this can hardly be said to be the case. Rather, New Labour Neo-Keynesianism has flooded devolved exchequers with public administration finance, which in the case of Wales, caused a throttling back on economic development expenditure. There is also the question of 'institutional fixes' whereby new institutions arrive that may cause an over-dependence of other actors and organisations upon them. It is widely argued in Wales that for the areas over which the WAG exerts authority, many interests are reliant on it and critique of WAG performance is muted accordingly. WAG senior spokesperson response that mute critique is oxymoronic belies the observation that 'absence of evidence does not equate to evidence of absence'.

Finally, what about relations between the devolved territories, especially Wales, the main focus here, and the UK government in Westminster? Briefly, it has changed from robust staking out of territory, as with the case of GMOs, to confusion as with the poorly managed Westminster response to food scares like foot-and-mouth disease, where capability on the ground was much higher locally in most cases, and over whether a decision to raise university tuition fees by Westminster meant Cardiff had to follow. In the latter case backstairs intra-New Labour discussions between the two education ministers resulted in Charles Clarke gallantly giving way to Jane Davidson. Nowadays, it is believed that Tony Blair only has to show possible indications of a sneeze for First Minister Rhodri Morgan to be at his side with the Lemsip. This contrasts with Scotland where in a classic case, DTI wanted a clear run to manage a Falkirk bus company closure threatened by the bankruptcy of its US parent 'as a global issue' only to be rebuffed by the Executive stressing that, if anything, it was 'a local issue' and Scottish Enterprise was finding a Scottish consortium willing to invest and save the company through a management buyout. Incidents such as this underline the importance to territorial identity of strong devolution settlements. The failure of devolution in North East England following a losing referendum, merely emphasises the point that the public is wary of 'expensive talking shops' but quite supportive when devolution settlements show the potential for confident envisioning strategies and solutions.

Notes

1 Martin Caton, Labour MP for Gower, Committee Member.
2 Huw Rawlings, European Affairs Division, National Assembly for Wales.
3 Welsh Affairs Committee op. cit. p. 41.

Acknowledgements

The research reported in this chapter was funded as part of the ESRC Research Programme on devolution and constitutional change in a project entitled *New Models of Development Funding in Northern Ireland, Scotland and Wales*, Grant No. L219 25 2121. The sponsors and programme partners are gratefully acknowledged. Thanks are also due to comments on the first draft by the present editors.

References

Cooke, P. (2005) 'Devolution and innovation: the financing of economic development in the UK's devolved administrations', *Scottish Affairs*, 50: 39–50.

Cooke, P. and Clifton, N. (2005) 'Visionary, precautionary and constrained "varieties of devolution" in the economic governance of the devolved UK territories', *Regional Studies*, 39: 437–451.

Department of Enterprise, Trade and Investment (2003) *Think, Create, Innovate*, Belfast: DETIN.

Department of Trade and Industry (1998) 'Our Competitive Future: Building the Knowledge Driven Economy', *DTI White Paper*, London: DTI.

Goodwin, M., Jones, M., Jones, R., Pett, K. and Simpson, G. (2002) 'Devolution and economic governance in the UK: uneven geographies, uneven capacities?' *Local Economy*, 17: 200–215.

Hall, P. and Soskice, D. (2001) *Varieties of Capitalism: The Institutional Foundations of Comparative Advantage*, Oxford: Oxford University Press.

Invest Northern Ireland (2002) *Corporate Plan 2002–2005*, Belfast: INI.

Northern Ireland Executive (2002) *Think, Create, Innovate*, Belfast: Department of Enterprise, Trade and Investment.

Osmond, J. and Jones, B. (2003) *Birth of Welsh Democracy*, Cardiff: Institute of Welsh Affairs.

Power, M. (1997) *The Audit Society: Rituals of Verification*, Oxford: Oxford University Press.

Scottish Executive (2001a) *Scotland's Science Strategy*, Edinburgh: Scottish Executive.

Scottish Executive (2001b) *A Smart, Successful Scotland*, Edinburgh: Scottish Executive.

Welsh Assembly Government (2002a) *Wales for Innovation*, Cardiff: WAG.

Welsh Assembly Government (2002b) *A Winning Wales*, Cardiff: WAG.

15 The quest for sustainable development at the Greater London Authority

David Goode and Richard Munton

Introduction

The Greater London Authority (GLA) came into being in 2000. The Authority is both an example of New Labour's modernisation programme for local government and the realisation of a political commitment made while in opposition to legislate for a successor body to the Greater London Council (GLC) which had been abolished in 1986 (see Hall's chapter in this volume). The GLA was never going to be a replica of the GLC; too much had changed since 1986. Central government had continued to accumulate authority at the expense of local government; London's 32 boroughs were never going to give back easily powers they had acquired on the abolition of the GLC; and a much more complex pattern of local governance involving a wide range of public and private bodies had taken root in a political environment where local government was expected to be more concerned with stimulating local economic competitiveness than with delivering welfare services.[1]

These changes moulded the shape of the Greater London Authority Act (1999) that severely restricted the income raising powers of the GLA and the range of services the Authority was expected to deliver. The GLA has no delivery powers in relation to housing, education, social or health services, for example. Instead, the Authority was to provide strategic direction for London's future and where it had the power to prepare strategies these had to be in accordance with central government policies and produced in consultation with a wide range of partners, most notably the Government Office for London (GOL), business interests and the London boroughs. Historically, the need for a strategic authority for London, as for other global cities, has rarely been challenged but the limited scope of the powers afforded under the GLA Act can be read, at one level, as just the most recent phase in the long-standing, ambivalent relation between central government and the political management of the capital city (Young and Garside, 1982; Travers *et al.*, 1991; Travers and Jones, 1997; Pimlott and Rao, 2002). At another, it reveals the difficulty of re-establishing a strategic authority in a terrain littered with entrenched public and private bodies with their own agendas (Newman and Thornley, 1997). Together, such observations have led to the suggestion that while the Mayor may be strong within the GLA and its component bodies, he is weak when it comes to London's wider governance (Sweeting, 2003). Travers (2004) even suggested that, 'as set out in the legislation and confirmed in practice, the powers of the mayor are largely those of patronage, persuasion and publicity' (pp. 67–8).

This chapter focuses upon one specific area of the GLA's responsibility, sustainable development, and addresses the priorities and decisions of the Mayor as chief executive, although the account cannot be understood outside the larger picture painted above. The GLA Act requires the Authority to exercise its powers 'in the way which it considers best calculated to contribute towards the achievement of sustainable development in the United Kingdom' (para. 30 (5b)). This effectively applies to all its activities. Furthermore the Act specifically requires the Mayor, in preparing his strategies for London's future to 'have regard to the effect which the proposed strategy, or revision, would have on the achievement of sustainable development in the United Kingdom' (para. 41 (4)), a potentially complex task in the context of a dynamic and highly differentiated city whose regional and global impacts extend to places well beyond the jurisdiction of the Authority. Complexity is increased by the predication of sustainable development upon an inter-generational time scale, a holistic and integrative approach to public policy, and a transparent and engaging process with the many interests involved. These requirements rarely sit comfortably with the normal timescales and practices of representative politics.

We are still in the very early days of the life of a novel form of London government. A major learning process is still going on as to its most effective institutional arrangement (see Rydin *et al.*, 2004), and the policy framework adopted by the Mayor is only now beginning to demonstrate real changes on the ground that can be attributed directly to a policy of sustainable development. We therefore concentrate our attention on the way this framework has been formulated, in particular the preparation of the Mayor's strategies, the debate over sustainability principles, the Mayor's Environment Commission, and the contribution made latterly by the London Sustainable Development Commission. Where these have already led to concrete changes on the ground such examples are included.

Background

When the GLC was abolished in 1986 some of its powers were returned to central government departments, some remained with the London Residuary Body and other pan-London bodies such as the London Planning Advisory Committee (LPAC), the London Research Centre (LRC) and the London Ecology Unit (LEU), and some passed down to the London boroughs. Few observers had much faith in this fragmented system of governance while others pointed to the democratic deficit of such arrangements and the loss of a figurehead to speak on behalf of London as a whole. Almost immediately a range of new bodies emerged with various responsibilities for all or parts of the capital to fill the vacuum, such as the business organisation London First (see Hall's discussion in this volume). As Travers (2004) suggests, these 272 bodies, separately but collectively, ensured that London's strategic functioning did not seize up completely, even if they failed to promote public confidence.

Central government responded to this concern by appointing a Minister for London in 1992 and setting up GOL in 1994, as one of ten regional government offices, to represent London's strategic interests in Whitehall. On the return of a Labour government, GOL became responsible for preparing the Green Paper *New Leadership for London* (DETR, 1997) and subsequently the White Paper *A Mayor and Assembly for London* (DETR, 1998). These policy documents served to intensify, and provide focus to, on-going debates over the future of London that had consumed the previous decade. As public concern about the state of London's environment had grown during the 1990s, the Sustainable London Trust (SLT) had initiated a major grass roots discussion about the capital's future (SLT, 1996; Jopling, 2000). It used the concept of sustainable development to outline goals and principles for the future, including an active local democracy for London, a fairer society and a healthier environment. The report highlighted the scale of London's ecological footprint whilst arguing the case for integrating economic, social and environmental concerns in a manner similar to that put forward in the UK's first sustainable development strategy (DoE, 1994). Its greatest demand, however, was for city-wide government, an elected champion for London and the establishment of a London's Citizen Forum which could bring together bottom-up demands for change (for a review of these issues see Harrison *et al.*, 2004). Some of these ideas fed into the government's Green and White Papers (see above). Among them, the Green Paper makes the first reference in a government document to sustainable development as an organising principle in policy development for London (para. 4.02), and this was reinforced in the White Paper's section on the environment (pp. 57–62) which also highlights the issues of climate change, air quality, noise, municipal waste, environmental audit and Local Agenda 21.

In varying ways many of these topics find their way into the GLA Act of 1999. The three principal purposes of the Authority, to promote economic development, social development and environmental improvement, in themselves mirror the UK Sustainable Development Strategy. The Act also states that three broad, cross-cutting themes should steer the development of policy – promoting equality of opportunity, improving the health of Londoners and contributing to the achievement of sustainable development in the UK. Guided by these themes the Mayor is required to prepare eight strategies concerned with economic development, transport, spatial development, biodiversity, noise, waste, air quality, and culture (to which the Mayor subsequently added one on energy). This menu reflects what central government intended the GLA to be responsible for rather than what the Mayor would have chosen to do.[2] At the same time, the Act calls on the GLA to ensure that each strategy is 'consistent' with all others as well as with national policies. The word 'consistent' is less powerful than the often-cited requirement that strategies for different areas of policy should be 'integrated'. The term integration has a multitude of meanings and associated practices ranging from the 'coordination' of separately derived policies cross-checked to ensure they do not contradict each other, to the positive alignment of policies in ways that make them reinforcing. Owens and Cowell (2002) launch a fierce attack on the overblown expectations implicit in the discourse of integration. They note, for example, that arguments in favour of integration often ignore 'the ways in which the sectors are themselves constructed and maintained by particular forms of knowledge and expertise, well-defined policy territories and patterns of resource allocation' (p. 69).

This warning is of particular relevance to the GLA which, following the Act, consisted of several sectors of power including a series of functional bodies, most notably in this context the London Development Agency (LDA) and Transport for London (TfL), together with the London Assembly, the Mayor's Office, and staff of the GLA itself. Moreover, these sectors were new to each other, and often had newly appointed staff who had not had much time to develop their own working cultures. Consistency in the development of policy would therefore be surprising especially as initially each sector had different expectations of their role and different objectives (see Rydin *et al.*, 2004; West *et al.*, 2003). At a minimum, time would be required for all to understand each other better, but time was at a premium. The Mayor and Assembly had no run-in time and each felt under political pressure to make a difference to the welfare of Londoners as soon as possible, and yet sustainable development, as an overarching theme, emphasises long timescales, the bringing together of a wide range of complex economic, social and environmental considerations, and the need for lengthy negotiations between the many interested parties.

In an attempt to address such potential difficulties, and following the publication of the GLA Bill in 1998, the Minister for London, Nick Raynsford, agreed to the setting up of the London Sustainable Development Forum (LSDF), chaired by the LEU, to review the area of sustainable development ahead of the establishment of the GLA and to identify the key issues which would need to be addressed by the GLA in developing a framework for sustainable development. Aware of political sensitivities, the Forum, which included major environmental agencies, related London-wide bodies and GOL, recognised that 'nothing should pre-empt decisions by the Mayor and Assembly. Rather, this report is meant as a contribution to assist the Mayor and Assembly in working towards such a framework' (LSDF, 2000: 1).[3] Nonetheless, the report outlined a number of key issues for the sustainable development of London, to which all present signed up. These included the need to reduce energy consumption and traffic congestion, to create a sustainable regional economy, and to minimise poverty and social exclusion. The report also argued that all the individual strategies should undergo a 'sustainability appraisal' and that decision making within the GLA, including its functional bodies, needed to be integrated. Integration would depend on a consultative, persuasive form of partnership politics where competing interests could be effectively managed.

Cities and sustainable development

Sustainable development is a complex process most simply understood as meeting jointly and equally the goals of economic development, social justice and environmental responsibility over inter-generational periods. It is in translating this consensual approach into action that difficulties emerge as it assumes that society can create 'win–win–win' situations between these goals in a world of inequality, disputed values and conflicting interests. Intuitively, this seems unlikely in a dynamic world of constantly changing priorities. It can, however, provide a broad overall framework for the continual reinterpretation and renegotiation of priorities that often only have meaning in particular contexts (see Owens and Cowell, 2002). But in reality, and despite the holistic rhetoric, sustainable development strategies are constantly at risk from the emergence of powerful, and often short-term, singular goals.

The UK national strategy for sustainable development, for example, has experienced such change since it was first put forward in 1994 (DoE, 1994). This first strategy focused upon economy–environment relations, but always with the need to maintain national economic competitiveness as the main priority on the assumption that only then could other goals be met (see House of Lords, 1995). A change of government led to a second strategy entitled *A Better Quality of Life* (DETR, 1999) which more fully incorporated social goals, but despite this few would dispute the suggestion that over the whole period growth in economic prosperity has been the overriding goal of government policy. This growth, whilst welcome, has conflicted with other sustainable development goals as it has been associated with greater income inequalities, increased personal mobility and only a partial decoupling of economic growth and resource use (see DEFRA, 2004). These outcomes challenge the assumption that environmental and social goals will necessarily be achieved on the back of economic growth (Blowers, 2000; Gibbs, 2000). In response to this, the current consultation paper on sustainable development priorities (DEFRA, 2004) argues for a new set of action-oriented priorities on key issues. These priorities are (i) climate change and energy, (ii) sustainable consumption, production and use of natural resources, (iii) environmental and social justice, and (iv) helping communities to help themselves (DEFRA, 2004: 9), although on this occasion the comparative lack of discussion about future macro-economic issues (there is a brief discussion of how business might make a contribution to sustainable development) leaves a sense of unreality. Even if the new agenda allows some new spaces for social and environmental considerations, it is unfortunate that their relations with economic goals are not more explicitly reviewed.

The key issue to be drawn from this discussion for this chapter is whether the GLA has the necessary powers and responsibilities to contribute significantly to the UK programme on sustainable development as spelled out in these terms. It is important that it can because the ability to assess the sustainability of cities is crucial (see Goode, 2000).[4] With about half of the world's population now living in cities, a proportion that is predicted to rise to two-thirds by 2030, city management must become the focus of local and global attempts to improve sustainability. Cities create many of the problems but this is also where the opportunities lie. Population concentrations should provide economies in the consumption and delivery of goods and services; should provide active communities with an interest in local action that can be seen to make a difference; and must provide a lead, especially in the case of global cities such as London, if more sustainable futures are to be assured. The key requirement is for effective multi-level governance, based upon a recognition of the wide range of public and private interests that have to be enrolled. Cities have to be viewed simultaneously as multiple entities in themselves but also embedded in regional, national and international relations (see Bulkeley and Betsill, 2003; Satterthwaite, 1997; Gibbs *et al.*, 2002).

As dense concentrations of production and consumption, cities are inevitably unsustainable within their boundaries. Any assessments of sustainability must take into account not only the quality of life within the city itself, but must also address the wider ecological impacts. Measures are required to see whether the resources imported from elsewhere (energy, food, water, etc.) can be used more efficiently and whether the impacts of their exported waste products can be reduced. A range of holistic accounting measures around the broad approach of ecological footprinting are now being developed, which seek to track and measure the entire resource flow of cities (or national economies) revealing the range, scale and geographical reach of the demands of the city on the global environment, rather than just describing the state of the urban environment (Goode, 2000). The first attempt to quantify London's ecological footprint (Girardet, 1996) identified key components of the metabolism and suggested that the overall footprint was 125 times the area of London. Whilst this first analysis gave an indication of the scale of the footprint it was not sufficiently detailed to provide the complete picture of London's metabolism necessary for developing policies for sustainability. In 2001, therefore, the GLA joined with the Institute of Waste Management to commission a complete resource flow and ecological footprint analysis of London. The resulting report *City Limits* (Best Foot Forward, 2002) provided a more detailed picture of London's metabolism and its wider ecological footprint. This time the total footprint was calculated to be 293 times the size of London, which is roughly the size of Spain. The per capita footprint of 6.63 gha means that Londoners are using three times more resources than their global 'earthshare' of 2.18 gha. In these terms London is far from being sustainable and the figures in this study have helped to raise greater awareness of the scale of the problem in terms of global equity.

But it was the individual components of London's metabolism that were most significant as they pinpointed key areas where Londoners could change current practices in order to become more sustainable. As well as being used by the Mayor on numerous occasions to illustrate the need for sustainability, the *City Limits* report has been followed up in greater detail by the business community in an effort to identify practical ways of reducing London's footprint (London First, 2005). It is unlikely that such a study would have been undertaken without the existence of the new London government, but it is precisely this kind of analysis which is required to ensure that sustainable development is built into policy making across both the public and private sectors.

How is sustainable development addressed in the Mayor's strategies?

Following the election in 2000, briefing was provided by officers of the GLA for the Mayor and Assembly Members on issues set out in the GLA Act. This included detailed briefing on the statutory requirements relating to the Mayor's strategies that emphasised the need for consistency across all policy areas and the requirement to address the three cross-cutting themes. On sustainable development the Mayor's briefing included the report of the London Sustainable Development Forum (LSDF, 2000) referred to earlier. This briefing recommended development of the Mayor's strategies as a holistic package, addressing sustainable development as one of the key overall objectives.

In August 2000 the Mayor set up a number of short-term Policy Commissions to address major areas of policy development. Membership of these commissions was drawn widely from the public, private and voluntary sectors, and it was intended that these commissions would make recommendations, from which the Mayor would produce an overall Prospectus forming his 'Vision for London'. Darren Johnson, the Mayor's Cabinet Advisor on Environment, who was then leader of the Green Group on the London Assembly, chaired the Policy Commission on Environment and Sustainability. Its report set the scene for development of the Mayor's environmental strategies (GLA, 2001a). It advocated that an all-embracing approach, based on sustainable development, should be imbued throughout the Mayor's vision for London, with the aim of transforming London into a sustainable world city: 'The vision should aim to improve the quality of life for Londoners and improve London's environment, but also address London's responsibilities in terms of its use of resources and its effects on the wider UK and global environment, especially climate change' (2001a: 3). The Commission welcomed the Mayor's earlier decision to carry out sustainability appraisals of all his strategies and urged him to adopt a set of sustainability principles to underpin his vision for London. The Commission also welcomed the Mayor's decision to establish a London 'Round Table' on sustainable development, recognising that this would provide a firm basis for implementing a regional framework.

A parallel commission on spatial development planning, led by the Deputy Mayor, also examined issues of sustainable development in some detail. Although no formal report was produced, many of the key issues identified were taken forward in developing early drafts of the London Plan. In January 2001 a summary of the policy proposals for the draft London Plan identified three main purposes of the plan. These were to support economic growth, to enhance sustainability, and to improve quality of life. The overall vision was 'to make London an exemplary sustainable world city'.

The Mayor's Vision was published in May 2001 in *Towards the London Plan*. This states (GLA, 2001b: 6) that:

> The Mayor's vision is to develop London as an exemplary sustainable world city. This must be based on three balanced and interlocking elements:
>
> - Strong and diverse economic growth
> - Social inclusivity to allow all Londoners to share in London's future success
> - Fundamental improvements in environmental management and use of resources.

The vision was amended slightly in the published strategies but its main thrust remained much the same.

Development of the Mayor's strategies

The GLA Act (1999) requires the Mayor to produce a series of strategies that together provide the basis for London's future development (H.M. Government, 1999). These eight strategies include the overarching Spatial Development Strategy or 'London Plan' that sets the scene for the next 20 to 30 years, together with detailed policies for transport and economic development. Alongside these are four environmental strategies: biodiversity, municipal waste management, air quality, and ambient noise. The Mayor is also required to produce a culture strategy. Shortly after his election, The Mayor Ken Livingstone decided that he would also produce an energy strategy, recognising that this was an essential element of sustainable development.

We have already referred to the requirements of the Act for consistency across all the Mayor's strategies, and the need to address sustainable development as one of three cross-cutting themes. It is fair to say that these requirements posed difficulties

from the outset. Although considerable effort was made between strategy development teams at officer level to ensure consistency, this could not always be achieved
for several reasons. The Mayor's political agenda required that his transport and
economic development strategies be produced with minimum delay. Both of these
were published by July 2001 (GLA, 2001c; LDA, 2001). This urgency was particularly acute in the case of the Transport Strategy that required a substantial lead time
for implementation of the Mayor's proposed Congestion Charge programme. In
contrast the timetable for production of the Spatial Development Strategy was
much more extended owing to detailed requirements of the GLA Act, which
included formal inspection in public. Production of the suite of five environmental
strategies also followed a longer timetable. Whilst the first was published in July
2002, the last was not published until after the London Plan in 2004. This disparity
in timing meant that the transport and economic development strategies could not
benefit from the detailed work on environmental sustainability that followed in
later strategies.

But it was not simply a matter of timing that prevented consistency being
achieved. There was a tension between the urgency to achieve specific political
goals and the need to consider wider, longer-term issues, such as sustainability.
During the Mayor's first year attention was focused at the political level on a
number of priority issues particularly with the 'functional bodies' dealing with
transport, economic development, and policing. The emphasis was on delivery of
immediate objectives such as the Congestion Charge, rather than on longer-term
strategic issues. Even in the cases of economic development and transport, where
these strategies were given priority, there was relatively little consideration of
how these strategies fitted into the wider picture, particularly the effects on environment and sustainability. This early phase of strategy formulation was certainly
not the all-embracing approach advocated by the Mayor's Policy Commission.

Since the Act placed a duty on the GLA to address sustainable development
in all its policies for regional development, the Government did not require the
GLA to produce a separate sustainable development framework as required by
other regional development agencies at that time. However, it was recognised by
GLA officers that a holistic approach would require an agreed schedule of sustainable development principles for use by the GLA and its functional bodies in
the development of the Mayor's strategies. A draft schedule was produced by
August 2000 and, following informal consultation with key stakeholders such as
GOL and the Association of London Government (ALG), this was submitted in
September to the Assembly's Environment Committee, which at that time sug-
gested only minor changes. This draft schedule provided an interim basis for
sustainable development input to the Mayor's strategies, as in the Draft Economic
Development Strategy published in November 2000 (see LDA, 2000: 63).
Several months and several revisions later, a final schedule was agreed by the
Mayor in April 2001. This provided the basis for sustainability appraisals for all
the emerging strategies, in particular for the draft London Plan, for which such an
appraisal was mandatory (Entec, 2002). It also provided the underlying principles
of sustainability for use by the Mayor until such time as a more detailed regional
framework could be produced. Somewhat perversely, the adopted schedule was
not accepted by the Assembly's Environment Committee that argued that it
should be more closely linked to the main themes and indicators set out at
national level in the UK Sustainable Development Strategy (DETR, 1999). In

fact, the schedule developed for London reflected the main themes of the national strategy in terms of economic, social and environmental issues, but since the detailed list of principles was designed to provide a consistent basis for addressing sustainable development across all the Mayor's strategies it was specifically attuned to the needs of London.

Scrutiny by the Assembly did, however, have a significant part to play in ensuring that sustainability issues were addressed more effectively, especially in the public consultation phase of the emerging strategies. The GLA Act required that each of the Mayor's draft strategies be subject to scrutiny by the London Assembly prior to public consultation. In the case of the five environmental strategies this proved to be a lengthy process, with two to three months devoted to each strategy. The Assembly appointed consultants to assess each draft strategy and took evidence from a wide range of specialists and interest groups. As well as making numerous detailed recommendations, the Assembly's Environment Committee pressed the Mayor in several cases to set more ambitious targets, and to make concrete proposals to achieve these. In the light of the Assembly's comments each draft strategy was amended to some extent by the Mayor prior to public consultation, but he resisted pressure to set unrealistic targets. However, the detailed consideration of these strategies by the Assembly undoubtedly assisted in raising both environmental issues and sustainable development on the political agenda. Preparation of the five environmental strategies represented a very substantial part of the Mayor's programme during his first term. As a result the time devoted to environmental issues and sustainability by the London Assembly proved very considerable. It is unlikely that such detailed and lengthy consideration would have been given to sustainable development had it not been for the specific requirements for scrutiny of the Mayor's strategies set out in the Act, in conjunction with the mandatory requirement to address sustainable development.

Sustainability in the Mayor's strategies

The London Plan sets the scene for London's strategic development for the next 20 to 30 years. Its policies are set within an overarching framework for sustainable development, which is a powerful strand running throughout the plan. The main elements of the five environmental strategies are reflected in this overall plan and to some extent also in the Transport and Economic Development Strategies. This series of strategic plans together provide the basis for an integrated framework. The environmental components of sustainability are achieved largely through the Mayor's environmental strategies. Whilst improvement of London's immediate environment, by reducing pollution and improving the quality of life for Londoners, is the main purpose of these strategies, this is not their sole objective. The strategies also take account of London's wider impacts on the global environment and identify action to reduce damaging or unsustainable processes in the longer term. The contribution made by these strategies to environmental improvement and sustainability is described in detail by Goode (2005).

In developing these strategies the GLA recognised the need to understand the way that London functions in terms of its daily processes and to be aware of the wider ecological footprint. We referred earlier to the detailed analysis of London's footprint, published in 2002, which quantified the energy and materials used or wasted by current practices. This was summarised in the Mayor's *State of the Environment Report for London* published in May 2003 (GLA, 2003a). It demonstrates unsustainable levels of resource use, and a metabolism characterised by linear rather than cyclical processes where vast quantities of material are imported daily for human use and waste products are discharged as unwanted residues. The GLA recognised that a detailed analysis of the components of London's functional metabolism, such as waste or energy, is necessary to identify action that can be taken to improve London's environmental performance and reduce damaging impacts elsewhere (see GLA, 2004a: iii).

The Mayor's London Plan makes it clear that 'to become an exemplary, sustainable world city, London must use natural resources more efficiently, increase its reuse of resources and reduce levels of waste and environmental degradation' (GLA, 2004b: 155). It is argued that as London grows, these objectives will become even more important. The shift towards a compact city, which is inherent in the plan, will contribute towards these objectives. It will enable more efficient use of resources such as land and energy and will also enable the 'proximity principle' to be applied to promote greater self-sufficiency. London 'should become a more sustainable and self-sufficient city, healthier to live in and more efficient in its use of resources. It should also be a better neighbour to its surrounding regions by consuming more of its own waste and producing less pollution' (2004b: 155).

The extent to which the draft London Plan addressed matters of sustainability was considered in some detail during the Examination in Public in 2003. Both the London Assembly and the London Sustainable Development Commission stressed the need for the London Plan to include a specific target for reducing CO_2 emissions and this was supported in the Panel Report (EIP, 2003) which recognised that the plan would need to deliver on policies for sustainable construction, increased use of renewable energy and reducing the need to travel. The Panel Report also commented on the need for a baseline assessment and a set of indicators against which to monitor the achievement of sustainable development in the longer term. The Panel was aware that action was underway to provide these through the Mayor's *State of the Environment Report*, and through the Indicators being developed by the LSDC and concluded that the Mayor should provide for these to be used as a means of gauging the future performance of the plan. In broad terms the Panel appeared satisfied that the Plan met statutory requirements regarding sustainable development and that the developing policy framework for sustainable development would further assist this process.

The Mayor's Energy Strategy (GLA, 2004a: iv) makes it clear that energy use is fundamental to long-term sustainability:

> If London is to make a significant contribution to the reduction of greenhouse gas emissions we need to restrain our use of fossil-fuels, encourage greater energy efficiency, and promote renewable energy. Implementation of the Mayor's Energy Strategy will help to mitigate climate change by reducing carbon dioxide emissions. This Strategy has wide implications, promoting new kinds of fuel for transport and encouraging high performance buildings with less demand for energy. It promotes good practice in new developments and supports examples such as the Beddington Zero Energy Development. Although one of the principal objectives of the strategy is to reduce our dependence on fossil fuels, it also addresses the vital social issue of energy poverty.

The Mayor launched the London Energy Partnership in 2004, which was intended to provide the main mechanism for implementing the strategy. The key objective was to increase substantially London's use of renewable energy. Following an initial assessment of renewable energy potential, which suggested that London could meet only 2 per cent of its electricity needs from renewable energy sources within its boundary by 2010 (ETSU, 2001), the Mayor instigated London Renewables, a major programme to promote renewable energy, funded largely by central government. The London Plan includes a performance target of 945 GWh of energy produced from renewable sources by 2010, including at least six large wind turbines.

In considering major new development proposals referred to him, the Mayor now expects to see a significant proportion of renewable energy incorporated in the design, and boroughs are expected to take similar action. To facilitate such changes the London Energy Partnership published a toolkit for planners, developers and consultants for integrating renewable energy into new developments (LEP, 2004). This was in addition to Draft Supplementary Planning Guidance on Sustainable Design and Construction issued for public consultation by the Mayor (GLA, 2005). During his first term the Mayor also took action through the Government's 'Heat Streets Programme' to implement a range of energy efficiency and renewable energy projects. He also established a London Hydrogen Partnership jointly with business and industry to promote development of the hydrogen economy.

The GLA also took the lead in organising a London Climate Change Partnership comprising public and private sector organisations and specialist agencies. The partnership commissioned a detailed study of the likely impacts of climate change on London. Its report *London's warming* was published in October 2002. The key messages of increased flood risk, significantly higher summer temperatures, and increased frequency of intense storms, all have implications for the future planning and management of London. The report identified impacts affecting water supply, wildlife, the built environment, transport, businesses, tourism, and health, as well as possible lifestyle changes. Possible consequences for Londoners and options for adapting to new conditions were identified, and the report provided a timely input to the emerging London Plan.

Waste is another area where significant improvements need to be made in environmental performance. The Mayor's strategy recognises that it is not simply a matter of improving levels of recycling, which is how the problem is often perceived. If London is to become sustainable, a more fundamental long-term change is required to establish a secondary materials economy:

> We need to develop a new business culture, where components of the waste stream are automatically considered as potential products for new industries. The policies contained in the Mayor's Waste Strategy set the framework for such a change. Substantial progress has already been made through the London Remade Programme, funded by the LDA, and this approach is now being promoted as a component of economic development. The Mayor's Green Procurement Code is another key initiative which provides the necessary link between environmental improvement and business performance.
>
> (GLA, 2003b)

Poor air quality is one of London's most severe environmental problems and has direct consequences for human health. The main causes are emissions from road traffic in the form of nitrogen oxides and airborne particles. London currently fails to meet EU and national targets for air quality because of the size of the conurbation and because of the density of road traffic. The Mayor's Air Quality Strategy (GLA, 2002a) makes proposals for meeting the legal targets, and for longer-term solutions through the introduction of cleaner vehicles. One of the proposals was for detailed consideration of the need for a Low Emission Zone, which led to the Mayor making the implementation of an LEZ a key priority during his second term. Other action taken to implement the strategy includes a programme to improve London's bus fleet to meet higher emission standards, and reducing taxi emissions through imposing licensing conditions. The Mayor also published guidance jointly with the Energy Savings Trust (EST, 2003) for fleet operators to encourage businesses to reduce exhaust emissions from their vehicle fleets.

Strategic policies to deal with noise have until recently been far less advanced than other areas of environmental concern. However, the requirement for the Mayor to produce the UK's first city-wide strategy for tackling noise has resulted in much progress over the past three years. His Noise Strategy (GLA, 2004c) sets out the main steps that need to be taken, including quieter road surfaces, smoother traffic flow, rail infrastructure improvements, aircraft noise measures, and improved design for new developments.

Conservation of biodiversity is addressed in detail in the Mayor's Biodiversity Strategy (GLA, 2002b) and in the London Plan. The subtitle of the strategy *Connecting with London's Nature* emphasises the social context, since one of the main objectives is to ensure the conservation of London's natural heritage for people to enjoy. The Mayor has adopted the well-established procedures for identification of important habitats in London as the basis for his Biodiversity Strategy. At present, London is the only part of Britain where there is a statutory requirement for a biodiversity strategy as part of regional planning. The value of this has already been demonstrated in London through the detailed incorporation of biodiversity in the London Plan and through publication by the LDA of guidance for development planners on design for biodiversity (LDA, 2004a). This provides a useful model for RDAs to address biodiversity conservation in urban areas elsewhere in the UK. The strategy also has an international dimension by making proposals to clamp down on the illegal international trade in globally endangered species, for which London's airports are one of the main points of entry to Europe.

The Mayor's vision for improving London's environment was clearly set out in his speech to the London Environment Conference in September 2002 (Livingstone, 2005) which demonstrates his commitment to sustainable development. It is argued that the overall effect of his strategies over the next twenty years will be to make significant improvements in London's local environment as well as reducing wider global impacts, and that together the strategies provide many of the essential ingredients to make London a truly sustainable world city. The Mayor states that, 'although the GLA has a duty to promote sustainable development, we will need to work closely with our partners and stakeholders if we are to achieve real changes'. He emphasises the role of individuals as crucial to success and concludes that 'influencing lifestyle changes and helping to forge new value systems will be fundamental in helping London to lead the way to achieving real improvements in the environment and hence quality of life for Londoners'.

London Sustainable Development Commission

The Mayor set up this Commission in May 2002 as an independent advisory body to assist him in achieving his objective of making London an exemplary, sustainable world city. During its first year the Commission concentrated on three key areas: the production of a sustainability framework for London; reviewing the sustainability content of the draft London Plan; and considering appropriate targets for CO_2 emissions and for renewable energy. Using its new framework the Commission also undertook sustainability appraisals of the Mayor's draft strategies for waste, energy, noise and culture. Its work in developing acceptable targets for CO_2 emissions was particularly important. The Commission recommended (LSDC, 2003) that the Mayor should include a target of 20 per cent reduction from 1990 levels of CO_2 emissions by 2010 in the draft Energy Strategy. This was accepted by the Mayor as the crucial first step on a long-term path to a 60 per cent reduction from the 2000 level by 2050. The Commission played a vital role in gaining a consensus on this issue.

Other work included development of London's Quality of Life Indicators (LSDC, 2004a) through a process of public consultation, and a review of the impacts of air transport on London (LSDC, 2004b). The Commission and the London Development Agency jointly commissioned Environmental Resources Management (ERM) consultants to undertake a sustainable development appraisal of the revised Economic Development Strategy. ERM's overall conclusions were published by the Commission (see LSDC, 2004c) and were also included in the public consultation draft (LDA, 2004b). This has resulted in much more effective and transparent consideration of sustainability issues than was the case with the first strategy in 2001.

The Commission also undertook a detailed assessment of sustainability issues involved in London's bid to host the Olympic Games in 2012, arguing that the Games should aspire to the principles of 'zero carbon' and 'zero waste'. This was summarised in its Second Annual Report (LSDC, 2004c), which also highlighted topics that the Commission expected to address in future, including the revised Transport Strategy and a London Food Strategy. The Commission was by now well established and has achieved a solid base with its London Framework and Quality of Life Indicators, enabling it to advise the Mayor effectively on sustainability issues.

In view of all these developments it can be expected that sustainable development would be given greater prominence in policy development than it was during the initial period of the GLA. This will be assisted by the fact that during 2004 the GLA put in place procedures for mainstreaming sustainable development across all areas of its work. That is what the GLA Act really required, but it took four years to get there mainly because the Mayor's Office did not have a champion for sustainable development on a par with those pressing for economic development and equalities issues.

The Mayor's second term

Politics had a more direct influence on the Mayor's environmental programme in his second term. On being readmitted to the Labour Party, prior to the 2004 election, Ken Livingstone made a deal with Nicky Gavron, who he replaced as official Labour candidate. He agreed that as Deputy Mayor she would take the lead in developing a London Climate Change Agency. This was to be a major new initiative, which she had proposed, to generate investment in new sustainable energy technologies and to promote their widespread application in London, building on the model successfully developed on a much smaller scale by Woking District Council over the past 10 years. This was in addition to significant environmental measures in the Mayor's own manifesto, including a large westward extension of the congestion charge zone and a commitment to a Low Emission Zone covering the whole of Greater London as the only sure way to reduce air pollution.

Environmental projects and sustainability have also benefited during the second term from the political balance of the London Assembly, since the Mayor is now dependent on support from the Green Party to gain the necessary approval for his annual budget. In setting his 2005 budget the Mayor was forced to commit substantial additional funding to environmental projects, including £20 million for protection of green space and for development of a green grid in the Thames Gateway.

We are now beginning to see real changes as the Mayor's strategic policies start to bite. Some of this is through influence, but increasingly it is the result of the Mayor's direct powers. An example is the requirement for all London taxicabs to be upgraded to meet higher emission standards over the next three years, funded since April 2005 through an environmental surcharge on fares. New built developments are also now incorporating renewable energy technologies, examples being the public transport interchange developed by Transport for London (TfL) at Vauxhall Cross, with its highly visible solar energy panels; and Willow Lane industrial estate in Merton where the developer is required to incorporate measures to deliver at least 10 per cent of predicted energy requirements from renewable sources on site.

A year after the election, the London Sustainable Development Commission produced its first audit of London's Quality of Life Indicators (LSDC, 2005). The Commission intends to publish an annual audit and thereby hold the Mayor to account regarding the success of his policies for sustainable development. London's successful 2012 Olympic bid, with its emphasis on a 'One-Planet' Olympics, provides another indication of the way that sustainable development principles are becoming more generally embedded in current thinking.

So indications are that sustainability is being addressed more directly in the Mayor's second term, with the Mayor's Office leading on implementation of a Low Emission Zone and, jointly with the Deputy Mayor, developing a Climate Change Agency for London. However, the most significant development has been the recognition of the need for stronger policy direction on sustainability within the Mayor's Office, and the establishment in September 2005 of a Policy Team for Sustainable Development on a par with other key policy areas.

Conclusions

The GLA is the most recent attempt to create a strategic authority for London as a whole but once again its powers and responsibilities can be said to represent what has been politically acceptable to central government, and to London's boroughs, rather than what might be agreed as optimal for the good governance of the capital. The outcome is an Authority good on strategic influence and leadership opportunities, but somewhat restricted in its range of direct delivery capabilities. The Mayor's independence of action, despite what might be read into his role as an 'executive' mayor, is hemmed in by powerful public and private interests encouraging him, as a political actor, to give priority to specific initiatives which can provide quick, observable results. This does not assist the achievement of sustainable development goals that largely depend upon the integration of policy areas and working to long time horizons. That said, it can be argued that after a somewhat slow start the GLA has progressively promoted a future for London that emphasises sustainable solutions within a balanced approach to economic, social and environmental objectives, as required by the GLA Act. It also has to be said that the Mayor has increasingly capitalised on his relatively limited direct powers of delivery to promote sustainable solutions across all the functions of the GLA.

In doing this, the Mayor has been assisted by the detailed requirements of the Act which demands consistency across all aspects of policy making, and includes a specific requirement to achieve sustainable development as one of three cross-cutting themes. This is further assisted by the very specific issues to be addressed by the Mayor in a range of environmental strategies. The legislation certainly provides a firm basis for achieving sustainable development, but given the tension referred to above there was a delay before sustainability issues were fully addressed. During the first year of the Authority's life the framework for sustainability was only gradually being developed, whilst the Mayor was already finalising the first two of his strategies. Inevitably there was insufficient coordination, despite the efforts of officials, between those responsible for the economic, transport and environmental strategies. Some of these problems also stemmed from the initial compartmentalisation of the functional bodies, and the wish of the Mayor to give transport issues, and congestion charging in particular, priority in policy making. To some degree, all the later strategies were out of phase in relation to this priority, a priority with its own merits but not necessarily the most obvious (nor only) starting point for a more sustainable London, but they have become progressively more consistent with each other as they have come up for revision.

What is welcome is that, despite those delays, by the end of the Mayor's first term sustainability was firmly established as part of policy formulation. Consistency was eventually achieved between the environmental strategies and the London Plan, and as the economic and transport strategies came up for review not only were they much better coordinated with the London Plan and the environmental strategies, but sustainability principles were also much more fully debated and incorporated into their goals. Examples include the emphasis on new environmental technologies, especially renewable energy (including the hydrogen economy), in economic development; greater attention to alternative less-polluting fuels in transport; and the emphasis on sustainable design and construction in planning. In addition the Mayor has increasingly used his direct planning powers to promote sustainable development and has also taken action to ensure that the functional bodies that he controls address it effectively, especially Transport for London.

At the same time, the growing influence of London's Sustainable Development Commission, and the active scrutiny of the Mayor's decisions and strategies by the Environment Committee of the London Assembly, provide support for the view that sustainable development principles are becoming more generally embedded in the thinking of the Authority. But, from experience so far, this thinking will only be routinely converted into actions and initiatives for the whole GLA when the Mayor's Office itself promotes sustainable development as one of its key policy drivers, as it has now started to do through the formation of a Policy Team for Sustainable Development.

Notes

1   This is aptly summed up by Pimlott and Rao (2002) who observe that 'for the first time, the proposal was put forward for a Greater London Authority (GLA) – not Council – which would take an area-wide view but which would not be directly responsible for the provision of services. Instead, the GLA would promote economic, transport, planning, environmental, and political strategies as well as inward investment' (p.59).
2   The Mayor made it clear on numerous occasions that his political agenda during his first term, which reflects his personal interests, the areas in which he feels he can make a difference and his electoral needs, led him to focus on transport issues, and congestion charging in particular, but was inadequate in the social sphere (see West *et al.*, 2003).
3   The London Planning Advisory Committee (LPAC) similarly prepared a strategy document based on integrated thinking which it gave to the Mayor on his election. LPAC hoped that it would act as the basis of the GLA's Spatial Development Strategy (West *et al.*, 2003).
4   We are often short of baseline data. In the case of the GLA an excellent start has been made on collating environmental information in the Mayor's state of the environment report entitled *Green Capital* (GLA, 2003a). The GLA has focused on 36 environmental indicators that relate directly to the Authority's strategic responsibilities in planning, transport and the environment, and subsequent reports will contribute to assessments of whether the targets set in the Mayor's strategies are being met.

References

Best Foot Forward (2002) *City Limits: A Resource Flow and Ecological Footprint Analysis of Greater London*, Oxford: Best Foot Forward.

Blowers, A. (2000) 'Ecological and political modernization: A challenge for planning', *Town Planning Review*, 71: 371–93.

Bulkeley, H. and Betsill, M.M. (2003) *Cities and Climate Change*, London: Routledge.

DEFRA (2004) *Taking it on: Developing UK sustainable development strategy together*, A consultation paper. London: DEFRA.

DETR (1997) *New Leadership for London*, Cm 3724, London: HMSO.

DETR (1998) *A Mayor and Assembly for London*, Cm 3897, London: HMSO.

DETR (1999) *A Better Quality of Life: A Strategy for Sustainable Development for the UK*, Cm 4345, London: HMSO.

DoE (1994) *Sustainable Development: The UK Strategy*, Cm 2426, London: HMSO.

EIP (2003) *Panel Report, Draft London Plan*, London: Examination in Public, July 2003.

Entec (2002) *Sustainability Appraisal of the Draft London Plan*, Entec UK Ltd.

EST (2003) *The Fleet Operator's Guide to Cleaner Fuelled Vehicles*, London: Energy Savings Trust.

ETSU (2001) *Development of a renewable energy assessment and targets for London*, A joint report of the Government Office for London, Mayor of London and Association of London Government, London: ETSU.

Gibbs, D. (2000) 'Ecological modernisation, regional economic development and Regional Development Agencies', *Geoforum*, 31: 9–19.

Gibbs, D., Jonas, A. and White, A. (2002) 'Changing governance structures and the environment: Economy-environment relations at the local and regional scales', *Journal of Environmental Policy and Planning*, 4: 123–38.

Girardet, H. (1996) *Getting London in Shape for 2000*, London: London First.

GLA (2001a) *Environment and Sustainability*, Report of Mayor's Policy Commission on the Environment, London: Greater London Authority.

GLA (2001b) *Towards the London Plan*; Initial proposals for the Mayor's Spatial Development Strategy, London: Greater London Authority.

GLA (2001c) *The Mayor's Transport Strategy*, London: Greater London Authority.

GLA (2002a) *Cleaning London's Air*, The Mayor's Air Quality Strategy, London: Greater London Authority.

GLA (2002b) *Connecting with London's Nature*, The Mayor's Biodiversity Strategy, London: Greater London Authority.

GLA (2003a) *Green Capital*, The Mayor's State of the Environment Report for London, London: Greater London Authority.

GLA (2003b) *Rethinking Rubbish in London*, The Mayor's Municipal Waste Management Strategy, London: Greater London Authority.

GLA (2004a) *Green Light to Clean Power*, The Mayor's Energy Strategy, London: Greater London Authority.

GLA (2004b) *The London Plan: Spatial Development Strategy for Greater London*, London: Greater London Authority.

GLA (2004c) *Sounder City*, The Mayor's Ambient Noise Strategy, London: Greater London Authority.

GLA (2005) *Sustainable Design and Construction*, The London Plan (Spatial Development Strategy for Greater London) Draft Supplementary Planning Guidance, London: Greater London Authority.

Goode, D. (2000) 'Cities as a key to sustainable development', in D. Poore (ed.) *Where Next?: Reflections on the Human Future*, London: Royal Botanic Gardens, Kew.

Goode, D. (2005) 'Environmental strategies for London', in J. Hunt (ed.) *London's Environment: Prospects for a Sustainable World City*, London: Imperial College Press.

Harrison, C.M., Munton, R.J.C. and Collins, K. (2004) 'Experimental discursive spaces: policy processes, public participation and the Greater London Authority', *Urban Studies*, 41: 903–17.

H.M. Government (1999) *Greater London Authority Act 1999*, London: The Stationery Office Ltd.

House of Lords (1995) *Report from the Select Committee on Sustainable Development*, HL Paper 72, London: HMSO.

Jopling, J. (2000) *London: Pathways to the future*, London: Sustainable London Trust.

LDA (2000) *Draft Economic Development Strategy*, London: London Development Agency.

LDA (2001) *Success Through Diversity: London's Economic Development Strategy*, London: London Development Agency.

LDA (2004a) *Design for Biodiversity: A Guidance Document for Development in London*. London: LDA.

LDA (2004b) *Sustaining Success: Developing London's Economy*, Draft Strategy, London: LDA.

LEP (2004) *Integrating Renewable Energy into New Developments: Toolkit for Planners, Developers and Consultants*, Report of London Energy Partnership and London Renewables, London: Greater London Authority.

Livingstone, K. (2005) 'Mayor of London's Vision of the Future of London's Environment', in J. Hunt (ed.) *London's Environment: Prospects for a Sustainable World City*, London: Imperial College Press.

London Climate Change Partnership (2002) *London's Warming: The Impacts of Climate Change on London*, London: LCCP.

London First (2005) *Making London a Sustainable City*, London: London First.

LSDC (2003) *London Sustainable Development Commission, First Annual Report*, London: Greater London Authority.

LSDC (2004a) *2004 Report on London's Quality of Life Indicators*, London: London Sustainable Development Commission, Greater London Authority.

LSDC (2004b) *The Impacts of Air Transport on London*, London: London Sustainable Development Commission, Greater London Authority.

LSDC (2004c) *London Sustainable Development Commission, Second Annual Report*, London: Greater London Authority.

LSDC (2005) *2005 Report on London's Quality of Life Indicators*. London: London Sustainable Development Commission, Greater London Authority.

LSDF (2000) *Towards a Framework for Delivering Sustainable Development in London*, Report of London Sustainable Development Forum, London: London Ecology Unit.

Newman, P. and Thornley, A. (1997) 'Fragmentation and centralisation in the governance of London: Influencing the urban policy and planning agenda', *Urban Studies*, 34: 967–88.

Owens, S. and Cowell, R. (2002) *Land and Limits: Interpreting Sustainability in the Planning Process*, London: Routledge.

Pimlott, B and Rao, N. (2002) *Governing London*, Oxford: OUP.

Rydin, Y., Thornley, A., Scanlon, K. and West, K. (2004) 'The Greater London Authority – a case of conflict of cultures? Evidence from the planning and environmental policy domains', *Environment and Planning C*, 22: 55–76.

Satterthwaite, D (1997) 'Sustainable cities or cities that contribute to sustainable development?' *Urban Studies*, 34: 1667–91.

SLT (1996) *Creating a sustainable London*, London: Sustainable London Trust.

Sweeting, D. (2003) 'How strong is the Mayor of London?' *Policy and Politics*, 31: 465–78.

Travers, T. (2004) *The Politics of London: Governing an Ungovernable City*, Basingstoke: Palgrave/Macmillan.

Travers, T. and Jones, G. (1997) *The New Government of London*, York: Joseph Rowntree Trust.

Travers, T., Jones, G., Hebbert, M. and Burnham, J. (1991) *The Government of London*, York: Joseph Rowntree Trust.

West, K., Scanlon, K., Thornley, A. and Rydin, Y. (2003) 'The Greater London Authority: Problems of strategy integration', *Policy and Politics*, 31: 479–96.

Young, K. and Garside, P.L. (1982) *Metropolitan London: Politics of Urban Change, 1831–1981*, London: Edward Arnold.

16 We'll have more please, but not now and not like that

Public attitudes to the Assembly in Northern Ireland

Roger MacGinty

Introduction

As Part I of this book describes, New Labour's devolution experiment offered opportunities to the Parliament in Scotland and the Assemblies in Wales and Northern Ireland to carve out their own niches and identities via policy divergence and the development of distinctive political cultures (Keating, 2003). Although the political centre placed limits on the power of the sub-units, devolution was something of a live experiment. There was at least the possibility that it would create the space for dissent and deviation from the centre, and allow innovations in policies and political style. Devolution also offered the possibility for the redress of the democratic deficit and policy inefficiency thought to stem from a large central government far removed from the citizen. Moreover, devolution held the possibility of satisfying some of the identity related aspirations and grievances in Scotland, Wales and Northern Ireland. According to nationalist proponents of devolution, the creation of elected devolved institutions would produce a more authentic form of government that would better reflect local concerns.

This chapter concentrates on just one aspect of devolution in relation to one of the devolved territories: public perceptions of devolution, utilising the Assembly in Northern Ireland as an illustration. Using data from a survey of public attitudes, the chapter attempts to gauge how inhabitants in Northern Ireland regarded the new Assembly as an instrument of governance and as an alternative to Westminster or other tiers of government. The survey findings support the somewhat counter-intuitive argument that public opinion in Northern Ireland was largely positive towards the Assembly and devolution despite the actual experience of devolution. The survey reveals surprising public support (or at least neutral rather than outright negative positions) for the Assembly despite the patent failure of the Assembly and devolution to become sustainable or deliver effective government. The chapter also contains a sub-argument that complicates the main argument of a generally positive public evaluation of the Assembly: the survey finds a distinct sectarian differential in attitudes, with Catholic–nationalists generally holding more positive opinions towards the Assembly than Protestant–unionists. This is most easily explained with reference to the wider peace process and peace accord that led to the establishment of the Assembly (see also the chapter by Ellis and Neill elsewhere in this book). In general, Catholic–nationalists believed that the process and accord were working towards their constitutional and grievance goals, while many Protestant–unionists regarded the peace process and peace accord as antithetical to the Union and their position within it.

The chapter begins by providing the context into which the survey findings of public support, or at least neutrality, towards the Assembly can be placed. It briefly outlines the 1998–2002 devolved experience of four periods of suspension, sectarian rancour and a failure of the Assembly to transcend constitutional issues and take on a life as a non-constitutional institution. But any coverage of context must also acknowledge the historical legacy of previous eras of devolution and how they augured poorly for contemporary public attitudes to the newly devolved Assembly. The historical and contemporary experiences of devolution make the relatively positive attitudes to the Assembly found in the survey all the more remarkable. The chapter then outlines the survey methodology, noting that the interruptions to the Assembly posed problems for the survey, particularly its time series ambitions. Subsequently, the chapter reports survey findings on public attitudes to three aspects of the Assembly: the extent of its powers; its position with regard to other tiers of government; and efficiency.

The legacy of earlier devolution and the context of contemporary devolution

The chief contextual point to make is that devolution in Northern Ireland, either as a contemporary or historical experience, has been extremely troubled. It is reasonable to expect this context to influence public attitudes on devolution as a form of government. Before sketching the fraught background of contemporary and historical devolution, it may be useful to stress Northern Ireland's exceptionalism with regard to similar devolution projects in Wales and Scotland. The establishment of a devolved Assembly in 1990s Northern Ireland was not a straightforward matter of the redistribution of power from the centre to the periphery. Devolution in Northern Ireland occurred within the wider context of a peace process and peace accord that attempted to manage a deeply entrenched ethno-national conflict. It was an attempt to use an enhanced form of consociationalism to put constitutional divisions on hold and encourage inter-group cooperation on day-to-day functional matters (MacGinty, 2003). Thus devolution in Northern Ireland was immediately of a different character to that in Scotland and Wales: it was accompanied by a wider package of constitutional and security changes, and the establishment of institutions to oversee human rights and equality. It was also designed with the assistance of another sovereign state, the Republic of Ireland, which would have formal linkages with devolved Northern Ireland via the newly established North–South Ministerial Council.

Most fundamentally, devolution mattered in Northern Ireland in ways not apparent in other regions of the United Kingdom. While devolution was broadly welcomed in Scotland, and greeted with widespread indifference in Wales, the opportunities and costs offered by the success or failure of devolution in Northern Ireland could mean life or death, or at least serious political instability. Devolution mattered to the extent that 81 per cent of voters turned out in the May 1998 referendum on the Belfast or Good Friday Agreement and 70 per cent in the election to the Assembly a month later. In Scotland and Wales, turnout in the 1997 referendums on devolution was 60 per cent and 50 per cent respectively. Indeed, turnout for the second election for the Welsh Assembly in 2003 was 38 per cent, as against a turnout of 63 per cent in Northern Ireland where the Assembly was suspended and there was little chance of a swift resumption. Not only were Northern Ireland's political parties passionate in their defence or rejection of the 1998 peace accord and the devolution that came with it, but civil society, the United States Presidency, the European Union and Northern Ireland's paramilitary groups all became involved in the referendum campaign. Public awareness of the establishment of the new Assembly was high (Wilford *et al.*, 2003), and Northern Ireland's four main political parties were committed to participating in the new institution.

The hangover of history is rarely far away in Northern Ireland, and it is worth noting devolution's contested history when examining attitudes to the latest version of a subordinate legislature. Although the Belfast Agreement billed itself as a 'fresh start' (*The Agreement*, 1998: 2), anticipation of any new devolved institution often made reference to previous versions. Northern Ireland had its own Parliament between 1921 and 1972, with this half century of devolution ending when direct rule was imposed from London amid a worsening security situation. Northern Ireland's first Prime Minister, James Craig, referred to the first devolved era as 'a Protestant parliament and a Protestant state' (Phoenix, 2003: 252) and presided over a regime that structurally excluded Catholic–nationalists. The latter repaid the compliment by withholding legitimacy from the Parliament, often boycotting already gerrymandered elections or electing abstentionist Members of Parliament. For many Catholic–nationalists, the Parliament building at Stormont became a symbol of an era of exclusion and discrimination (Elliot, 2000: 399–400). Yet for many unionists, '... the building remains synonymous with "Protestant Ulster" and unionist power ... emblematic of the northern state as it once was' (Officer, 1996: 131).

Stormont had also been the site for an ill-fated attempt to establish a power-sharing Assembly in 1973–4 and an Assembly attended only by pro-Union parties from 1982 to 1986. The early 1970s attempt to establish a powersharing government was largely derailed by a massive campaign of civil disobedience by Protestant–unionists that included a large rally outside the Parliament building. For Catholic–nationalists, and particularly radical republicans, Stormont in its historical incarnations represented attempts at an 'internal settlement' to the Northern Ireland conflict that copper-fastened Northern Ireland's place within the Union and excluded all-island constitutional options. For much of the Troubles, Sinn Féin and their political cousins in the Irish Republican Army refused point blank to countenance revised versions of an 'internal settlement'.

In short, the building was something of a historical comfort zone for many Protestant–unionists and reminded them of their political 'ascendancy' (Porter, 1996: 124), yet it was deeply contentious for many Catholic–nationalists. Despite its history and significance, Stormont was to be the site of the post-1998 devolved administration in Northern Ireland. The new Assembly was to be deliberately designed so that no one group could be in the ascendant. The Belfast Agreement envisaged '... a democratically elected Assembly in Northern Ireland which is inclusive in its membership, capable of exercising executive and legislative authority, and subject to safeguards to protect the rights and interests of all sides of the community' (*The Agreement*, 1998). This was something of a tall order given the regularity with which Northern Ireland's politicians mined history (real and imagined) in pursuit of modern political agendas. History augured poorly for contemporary attitudes to the newly devolved version of the Northern Ireland Assembly.

Space does not permit a full catalogue of the trials of the infant years of the new powersharing Assembly, but a brief sketch is required before outlining public attitudes to the institution. The key problem lay in the chronic mistrust between nationalists and republicans on the one hand and unionists and loyalists on the other. The Belfast Agreement had attempted to put to one side their essential conflict (a clash of incompatible nationalisms) and hoped that the Assembly and its companion institutions and mechanisms would provide avenues for harmonious day-to-day inter-group cooperation and the management of grievances. But the nationalist and unionist constitutional projects continued to pollute an Assembly that was only mandated to look after local governance matters. 'Outbidding' from intra-group competition (Sinn Féin versus the Social Democratic and Labour Party, and the Democratic Unionists versus the Ulster Unionists) meant that there was little space for centre ground accommodation. The contested interpretation and implementation of the Belfast Agreement, especially with regard to the decommissioning of republican arms stocks, provided an unpromising terrain for the new powersharing administration.

Although elections for the Assembly were held in June 1998, nationalist versus unionist wrangling meant that the Assembly did not meet and its powersharing Executive was not nominated until November 1999. Disagreement over the decommissioning of republican arms led to the first suspension of the Assembly between February and May 2000. The Assembly was again suspended in August 2001, for 24 hours, and between September and November 2001. The fourth, and longest lasting, suspension came in October 2002 following allegations that the IRA was using the Assembly as a base for spying. Since then, little of substance has occurred. Elections to the moribund Assembly were held in November 2003 and saw Sinn Féin and the DUP make large gains (later confirmed by the May 2005 general election). The DUP rejects the 'forlorn hope that Sinn Féin can eventually be sanitised' and proposes a 'voluntary coalition' to form a devolved government (DUP, 2005). This contradicts the principle of inclusion that underpinned the peace process and would require a major policy shift from the nationalist SDLP if the voluntary coalition was to have a cross-community dimension. British and Irish government attempts to broker nationalist–unionist deals have foundered amid further claims of republican involvement in violence and unionist dissatisfaction with the Belfast Agreement.

Three examples provide a flavour of the extraordinarily dysfunctional nature of relations between nationalists and unionists when the Assembly was in operation and help provide the context amid which survey respondents were questioned. First, the two Democratic Unionist ministers refused to attend meetings of the Executive because of the presence of Sinn Féin. Instead, they relied on civil servants to relay messages while Executive meetings were in progress. Second, politicians from a number of parties took part in a televised shoving match in the foyer of the Assembly, popularly daubed the 'brawl in the hall'. Third, Sinn Féin took legal action against the First Minister (David Trimble) because of his refusal to invite its ministers to meetings of the North–South Ministerial Council. Such shenanigans were symptomatic of the chronic malaise and mistrust between nationalists and unionists.

The key point is that the Northern Ireland Assembly was rarely in the news for policy initiatives and humdrum governance. Instead, it reflected a still live and keenly contested ethno-national conflict. Despite this, as the survey evidence will show, there was widespread hope that the Assembly could be something more than that.

Survey methodology and timing

The Northern Ireland Life and Times survey was launched in 1998 as a survey of social attitudes. It has an annual political attitudes module, funded in 2000, 2001 and 2003 by the Economic and Social Research Council (ESRC) under its Devolution and Constitutional Change Programme.[1] The survey is a joint initiative from the University of Ulster and the Queen's University of Belfast, and the fieldwork is conducted in the autumn of each year. Eighteen hundred adults are interviewed face-to-face and issued with an additional self-completion questionnaire. Response rates have averaged at 67 per cent in the 1998–2003 period, a figure that compares well with similar attitudinal surveys in England, Scotland and Wales. Addresses (a simple random sample) are selected from the Postcode Address File. Interviewers select one adult for interview at each address via a Kish Grid method. Interviews are carried out using computer-assisted personal interviewing. A pilot survey is conducted prior to the main survey to assist questionnaire design.[2]

The independent variables used throughout this chapter are 'Protestant' and 'Catholic' rather than unionist or nationalist. The survey shows a strong equivalence between self-identification as Protestant and unionist (over 70 per cent), and Catholic and nationalist (over 60 per cent), although a substantial number of Catholic and Protestant respondents refuse to identify themselves as nationalists or unionists. The use of the religious labels thus yields a higher sample. Indeed the use of religious identity as a virtual proxy for political identity is further legitimised as the survey repeatedly reveals the salience of Northern Ireland's sectarian differential as the key fault-line in society.

Special ethical and practical problems attend research on sensitive issues in deeply divided societies (Smyth and Robinson, 2001). In the case of this research, additional challenges stemmed from the initial delay in the establishment of the Northern Ireland Assembly followed by its repeated suspension. In order to construct a time series of public attitudes to devolution, it was hoped that the same questions could be asked annually. Since the Assembly was suspended on a number of occasions when the survey was in the field, or there was confusion as to whether it was about to be suspended, this was not possible and some questions had to be modified or omitted depending on the status of devolution. In addition, the Assembly had so much 'downtime' over the course of the survey time series that it was legitimate to ask if the institution had left enough of a legislative footprint to register as an organ of governance among survey respondents. When in session, much time was taken with nationalist versus unionist constitutional posturing (an issue over which the Assembly had no authority). The new Assembly also spent much time putting in place new systems of governance and information dissemination, and conducting extensive public consultation exercises, all of which delayed the drafting of legislation. As a result, the corpus of initiatives undertaken by the Assembly was quite limited. Nevertheless, by the time of the fourth, and longest lasting, suspension in October 2002, the Executive of the Assembly had announced its second Programme for

Government and budget, and could point to a modest range of initiatives such as free public transport for pensioners and the announcement of spending plans for the health service. In keeping with those in government in Wales and Scotland, members of the powersharing Executive in Northern Ireland were adamant that they were 'making a difference' (Trimble in Wilford, 2002). The survey attempts to ascertain if this confidence was shared by Northern Ireland's citizens.

Assembly powers

Survey evidence suggests support for the extension of the powers granted to Northern Ireland under devolution. In the first instance devolution was to be limited with some powers reserved to London. When asked in 1999 and 2000 if they thought that the Northern Ireland Assembly should have the power to raise taxation in a manner similar to the Scottish Parliament, majorities (57 per cent in both years) agreed with the proposition (Table 16.1 shows the figures for 2000). Less than a third (27 per cent in 2000) disagreed with the proposition. Noticeable is the sectarian differential that is common across a range of political attitudes to the Northern Ireland Assembly, with Catholic survey respondents (68 per cent) more positive to the idea of a tax-raising Assembly than their Protestant counterparts (52 per cent).

It is possible to argue that a preparedness to grant the Assembly additional power in the form of the ability to raise taxation is indicative of wider trust for the institution. In the first instance it suggests a faith in the probity and efficiency of the Assembly to collect and administer public funds. Second it suggests a desire to see a more active and empowered institution. This inference is reinforced by the responses to another survey question conducted later in the time series cycle when survey respondents had more experience of devolution in action. The question sought to ascertain which constitutional permutation (ranging from Irish unification to Northern Ireland's incorporation in the United Kingdom without devolution) attracted most support. Two devolved options were offered to respondents: that Northern Ireland remain part of the United Kingdom but with its own devolved Assembly, or that Northern Ireland remain part of the United Kingdom, but with its own devolved Parliament. The first of these devolved options described the situation following the 1998 peace accord, while the second described an enhanced form of devolution.

Table 16.1   Attitudes to the extension of tax-raising powers for the Northern Ireland Assembly (2000)

	Catholics	Protestants	No religion	All
Yes	68	52	47	57
No	13	33	37	27
(Don't know)	18	15	16	16

Question: 'Do you think the Northern Ireland Assembly should have the power to raise or lower income tax like the Scottish Parliament?' (all figures in percentages).

Table 16.2 Attitudes to how Northern Ireland should be governed (2003)

	Catholics	Protestants	No religion	All
Northern Ireland should become independent, separate from the UK and the EU	2	1	2	2
Northern Ireland should become independent, separate from the UK but part of the EU	10	5	9	7
Northern Ireland should remain part of the UK, with its own elected parliament	18	37	34	30
Northern Ireland should remain part of the UK, with its own elected assembly	9	31	2	22
Northern Ireland should remain part of the UK without an elected Assembly	7	17	4	12
Northern Ireland should unify with the Republic of Ireland	38	2	19	17
(Don't know)	15	7	10	11

Question: 'Which of these statements comes closest to your view?'

As Table 16.2 shows, no single constitutional option received majority cross-community support (hence the intractability of the Northern Ireland conflict), but the devolved options received most overall support. Taken together, 52 per cent of survey respondents wanted either an Assembly or Parliament devolved to Northern Ireland, with the latter option being most popular and suggesting a desire on behalf of a plurality of respondents for an extension of the powers already devolved to Northern Ireland. When analysed according to religious identification, it becomes clear that a strong majority of Protestants (68 per cent) are in favour of the devolved options; doubtless their unionist aspirations being satisfied by Northern Ireland remaining within the United Kingdom. The story is somewhat different among Catholic survey respondents, with the largest proportion (38 per cent) opting for the traditional nationalist goal of Irish unity. Yet support for this aspiration is by no means overwhelming and it is clear that the experience of devolution, and the opportunities offered by it, has had some effect. Nine per cent of Catholics favour the devolved status quo (an Assembly), but double that figure are in favour of an enhanced form of devolution (a Parliament).

Perception of the Assembly in relation to other tiers of government

Other results from the survey point to positive attitudes towards the Northern Ireland Assembly and a desire to see it as the preferred tier of government. Again these findings seem counter-intuitive when compared with the actual experience of devolution. Public faith in Assembly did not seem to be supported by the institution's actual performance. Survey questions on day-to-day governance under the Assembly and direct rule from London revealed a greater public regard for the Assembly. In 2002, survey respondents were asked if they thought the Assembly and Executive did a good job in the day-to-day running of Northern Ireland (Table 16.3). The question was deliberately framed in this way to encourage respondents to think of the Assembly and Executive as institutions responsible for public service provision rather than institutions concerned with the constitutional battle between unionism and nationalism. Overall, 41 per cent of survey respondents thought that the Assembly and Executive performed well, with 42 per cent holding a neutral position that the Assembly and Executive performed 'neither a good nor bad job' in day-to-day governance. Despite the fact that devolution collapsed amid rancour during the time that the survey was in the field, and that unionist disaffection with the Belfast Agreement had been steadily rising since 1998, only 11 per cent of survey respondents (including 15 per cent of Protestants) thought that the Assembly and Executive had performed poorly. Given the tenor of statements from some unionist politicians regarding the folly of entering into devolved government with still armed republicans the figure seems low.

Notwithstanding the relatively low proportion of Protestant respondents who thought that the Assembly had performed poorly, a distinct sectarian differential was visible among respondents who awarded the Assembly a positive or neutral rating. Catholic survey respondents were more positive than their Protestant counterparts, perhaps reflecting greater Catholic–nationalist confidence in the wider peace process. While a majority of Catholic respondents (56 per cent) thought that the Assembly did a good job, less than a third of Protestants (29 per cent) shared this view. Three times as many Protestants said that the Assembly did 'neither a good nor bad job' (50 per cent) than held a negative view. This is an interesting finding in that it suggests that many Protestants seemed to be adopting a 'wait and see' position that stopped short of a fully negative stance.

The findings on public perceptions of the Assembly make interesting reading when juxtaposed with attitudes towards direct rule, or the resumption of government via London and the Northern Ireland Office when devolution was suspended (Table 16.4). In 2003, after devolution had been suspended for a year, survey respondents were asked to rate the performance of central government in the day-to-day running of Northern Ireland. Overall, only 12 per cent thought that central government was doing a good job (as against 41 per cent who had given a positive rating to the devolved administration in the previous year). Sixty per cent were indifferent to direct rule, thinking that it did 'neither a good or bad

Table 16.3 Perception of the Assembly on day-to-day running of Northern Ireland (2002)

	Catholics	Protestants	No religion	All
A good job	56	29	33	41
Neither a good nor bad job	33	50	42	42
A bad job	6	15	13	11
(Don't know)	6	6	13	7

Question: 'How good a job did you think the Assembly and Executive did in the ordinary day-to-day running of Northern Ireland? Would you say ...'

job' in the running of Northern Ireland. While 11 per cent felt that the Assembly had performed badly, 19 per cent felt likewise about direct rule.

The sectarian differential apparent in attitudes towards the Assembly was much less evident in relation to attitudes towards direct rule. This is interesting in that it suggests that the devolved Assembly was still very much regarded by the two main politico-sectarian communities as central to the conflict. A key aim of the British and Irish governments, as co-sponsors of the peace process, had been to establish a local chamber for day-to-day governance and inter-group cooperation on functional matters, while putting constitutional differences aside. The sectarian differential in public attitudes to the Northern Ireland Assembly suggests that the Assembly failed to convince many respondents that it was a site for non-constitutional politics.

At the same time, the absence of a significant sectarian differential in relation to attitudes to day-to-day governance by the British government is noteworthy in that it challenges stereotypes associated with nationalism and unionism. Traditionally, Irish nationalists had distrusted British governments ('perfidious Albion'), while unionists regarded British governments as (not wholly reliable) bulwarks against Irish unity. These traditional views (never pure caricatures) have been open to revision, with many nationalists regarding the British government as adopting a more even-handed approach in its dealings with the Northern Ireland conflict and many unionists seeing the British government as too ready to make concessions to the Irish government and nationalists. Yet, the survey does not detect partisan attitudes towards the British government in perceptions of the British government as a day-to-day administrator. A strategic aim of the British government peace process strategy from the late 1980s onwards was to portray itself as a 'neutral' third party willing to mediate between the quarrelsome natives. With regard to the day-to-day running of Northern Ireland at least, the absence of a significant sectarian differential in public attitudes suggests some measure of success.

Another set of questions also reveals a generally positive view of the Assembly when presented alongside alternative sources of government (Table 16.5). Between 1999 and 2001 survey respondents were asked first to identify the tier of government they thought currently had most power in the running of Northern Ireland, and then to identify the tier that they thought *ought* to have most power.[3] The question was phrased to encourage survey respondents to see the power source in terms of day-to-day governance rather than as a constitutional

Table 16.4 Perception of central government on day-to-day running of Northern Ireland (2003)

	Catholics	*Protestants*	*No religion*	*All*
A good job	12	12	23	12
Neither a good nor bad job	60	62	44	60
Or a bad job	17	20	20	19
(Don't know)	12	7	14	9

Question: 'How good a job did you think the present government under direct rule is doing in the ordinary day-to-day running of Northern Ireland? Would you say ...'

Table 16.5 Attitudes to actual and desired primary tier of government, 1999–2001

	1999 Cath.	1999 Prot.	1999 All	2000 Cath.	2000 Prot.	2000 All	2001 Cath.	2001 Prot.	2001 All
British Government has most power	71	71	70	72	64	66	53	50	51
British Government ought to have most power	6	33	22	9	50	36	7	24	17
Northern Ireland Assembly has most power	7	7	7	7	12	11	27	29	28
Northern Ireland Assembly ought to have most power	40	27	31	57	35	42	74	61	65

Question: 'Which of the following do you think has most influence/ought to have most influence over the way Northern Ireland is run?'

authority. Survey respondents were presented with a wide range of tiers of government or sources of power: British Government, Irish Government, European Union, Local Councils, Northern Ireland Assembly, North–South Ministerial Council, the People of Northern Ireland, East–West Council of the Isles, and other.[4] For ease of presentation, only attitudes towards the British Government and Northern Ireland Assembly are shown. The results are revealing in that they suggest firstly a steady increase in those who thought that the Assembly wielded most power, but also a startling increase in those who thought that the Assembly *ought* to have most power. Again this challenges the observed evidence of the Assembly in operation.

In 1999, 71 per cent of both Catholic and Protestant respondents thought that most power over the way Northern Ireland was run lay with the British Government. Over the next two years, this figure fell to an overall bare majority of 51 per cent of respondents believing that the British Government held most power over the way Northern Ireland was run. So in the period that devolution was in operation (or, if in suspension, was a possible form of government) belief that the British government held primary power over the running of Northern Ireland fell by 20 per cent. The proportion of survey respondents who felt that the British Government *should* have primacy fell from 22 per cent in 1999 to 17 per cent in 2001 (though the 36 per cent figure in 2000 suggests that opinion on this issue is prone to considerable fluctuation). Unsurprisingly, and in keeping with the basic tenets of unionism, the proportion of Protestants believing that the British Government should hold most power over Northern Ireland consistently outstripped Catholics of a similar view. By 2001 though, when faced with the alternative provided by a devolved Assembly, less than a quarter (24 per cent) of Protestants believed that the British Government ought to have most power over the day-to-day running of Northern Ireland.

In tandem with the declining perception of Westminster as both the actual and desired primary day-to-day ruler of Northern Ireland, faith in the Northern Ireland Assembly increased. In 1999, 7 per cent of both Catholics and Protestants thought that the Assembly had most power over the way Northern Ireland was run, with this figure quadrupling to 28 per cent in 2001. This suggests that there was a 'devolution effect', with the Stormont administration convincing a sizeable number of respondents of its capacity as an actual unit of government. More impressive though was the increase in support for the Assembly as the desired primary tier of government. This figure more than doubled from 31 per cent in 1999 to 65 per cent in 2001. By 2001 majorities of both Catholics (74 per cent) and Protestants (61 per cent) favoured the Assembly as the primary power source.

Assembly efficiency

By way of caution to survey results showing a generally positive attitude towards the Northern Ireland Assembly, the survey also reveals public concerns on the efficiency of the Assembly. These views, held simultaneously with a desire to see Assembly powers extended, point to a contradiction in public attitudes. Opponents of devolution have argued that it creates an extra tier of government, duplicating rather than complementing the work of local and central government. The Confederation of British Industry, for example, has maintained a sceptical position on the ability of devolved institutions to control public spending and govern efficiently. The organisation warned that the costs of projected English Regional Assemblies could 'spiral out of control', and pointed to the case of Wales where the Assembly running costs were twice those of the Welsh Office and rose from £77 million in 1999 to £148 million in 2002 (CBI, 2002). To investigate public perceptions of the Northern Ireland Assembly as an efficient organ of government, the survey posed questions concerning the number of elected politicians at various levels of government and whether the Assembly represented value for money.

Post-devolution Northern Ireland had three Members of the European Parliament, 18 Members of Parliament, 108 Members of the Legislative Assembly, and 585 local government councillors. The 2003 survey asked if Northern Ireland had too many political representatives, with 46 per cent (40 per cent of Catholics and 51 per cent of Protestants) agreeing with the proposition (see Table 16.6). Twenty per cent said that the number was about right and 2 per cent said there were too few. The immediate comment to make is that there was not an overwhelming public sense of political over-representation. Respondents who thought that the region had too many elected representatives were then asked to identify which cadre of politicians should be reduced. The Assembly was identified as the level of government most eligible for a cull in the scale of representation. Twenty-eight per cent of those who thought that Northern Ireland was served by too many political representatives thought that the number of Westminster MPs should be reduced, while 25 per cent supported a reduction in the number of MEPs. A majority (57 per cent) supported a reduction in the number of local government councillors, but three-quarters of respondents (75 per cent) identified the Assembly as the most politically over-staffed chamber. Catholics and Protestants who were in favour of a reduction in the number of elected representatives were relatively united in their belief that they had a surfeit of MLAs (71 per cent of Catholics and 79 per cent of Protestants).

What explains the public identification of the Assembly as the tier of government most in need of a reduction in size? When this question was asked (autumn

Table 16.6 Percentage in favour of reduction in number of types of elected representative (2003)

Type of Representative	MLAs	MPs	MEPs	Local Councillors
Percentage in favour of reduction	75	28	25	57

Question: To those that agreed that Northern Ireland had too many elected representatives: 'Which elected representatives should be reduced in number?'

2003), the Assembly had been suspended for a year and, with direct rule from London ensuring that people experienced no interruption in the provision of public services, many respondents may have concluded that Assembly members were extraneous. Media reports that Assembly members continued to receive (albeit reduced) remuneration despite the suspension of the Assembly, and reports that the Office of the First and Deputy First Minister had twice the staff of 10 Downing Street, may not have helped create a widespread public perception of the efficiency of devolution. By the time of the November 2003 Assembly election, the Democratic Unionist Party made the 'ever growing bureaucracy' of the Assembly a campaign issue, promising to 'reduce the number of Assembly members from 108 to 72' (DUP, 2003: 22).

In terms of the ratio of Assembly/Parliament members to eligible voters, there is no question that Northern Ireland was comparatively over-represented at the devolved level. As Table 16.7 shows, in 2003 Northern Ireland had one Assembly member per 10,162 eligible voters. This compares with 30,058 eligible voters per Member of the Scottish Parliament and 31,839 per Assembly member in Wales.[5] In other words, Northern Ireland had three times the number of elected representatives at the devolved level (in proportion to eligible electorate) than the other devolved regions of the United Kingdom.

Northern Ireland's comparative over-representation at the devolved tier of government reflected the peculiar genesis of devolution amid the peace process. The architects of the peace process, the British and Irish governments, were anxious that the Assembly have as wide a representation as possible so as to include those traditionally on the political margins, particularly within loyalism. While republicans had managed to establish a sophisticated and increasingly popular political machine in the shape of Sinn Féin, loyalist political parties were less well established and had less electoral appeal. Political loyalism, with its links to paramilitary organisations, had the potential of derailing the peace process from without, and so Northern Ireland's Assembly was enlarged as a way of keeping

Table 16.7 Ratio of electors to representatives in devolved institutions (2003)

	Eligible electors	Number of representatives	Electors per representative
Scotland	3,877,460	129	30,058
Wales	2,228,697	70	31,839
Northern Ireland	1,097,526	108	10,162

Table 16.8 Perception of the Assembly as value for money (2001)

	Catholics	Protestants	No religion	All
Yes, definitely	9	5	2	6
Yes, probably	35	23	27	29
No, probably not	25	29	23	26
No, definitely not	11	24	15	18
(Don't know)	20	19	34	21

Question: 'On balance, do you think that the Northern Ireland Assembly is good value for money in Northern Ireland?'

potentially dangerous constituencies on board. In the 1998 Assembly election the Progressive Unionist Party, with links to the militant Ulster Volunteer Force, was able to secure two seats in the Assembly with 2.5 per cent of the vote. The price of this inclusion for the sake of the peace process was an enlarged Assembly in which Assembly seats were essentially a form of patronage.

In 2001 the survey asked respondents if they thought that the Assembly represented good value for money. Overall, 35 per cent thought that the Assembly probably or definitely represented good value for money, with 44 per cent perceiving the Assembly to be probably or definitely poor value for money, and quite a high proportion (21 per cent) registering as don't know. The initial comment to make is that a plurality rather than a majority thought that the Assembly represented bad value for money. But when these results are broken down according to religious identification, a distinct sectarian differential becomes apparent. A majority of Protestants (53 per cent) believed that the Assembly was either definitely or probably bad value for money, with 36 per cent of Catholics sharing this view. While 44 per cent of Catholics thought that the Assembly represented good value for money, 28 per cent of Protestants shared this view. Interestingly, most respondents from both religions cluster around the 'probably' rather than 'definitely' answers, pointing to reticence in choosing completely negative responses.

Conclusions

The survey suggests a considerable dissonance between the observed evidence of the Northern Ireland Assembly in action and public attitudes towards the Assembly. A surprising proportion of survey respondents were willing to invest faith in the Assembly as their primary provider of public goods, despite the institution's failure to live up to these aspirations. Yet this public faith in devolution was not enough to prevent the repeated collapse of the Assembly. This gap between public aspirations and political reality may be explained by a public perceptual separation between the Assembly as an institution for the day-to-day running of Northern Ireland on the one hand, and as a site for the continuation of the nationalist versus unionist constitutional conflict on the other. Ultimately, the willingness to invest faith in the Assembly as an institution of day-to-day governance was unable to outweigh the forces that made devolution unsustainable.

Given the repeated suspensions, it is difficult for the observer to compare the Northern Ireland experience of policy innovation or divergence with the experiences of Scotland and Wales. The Assembly was not an arena of innovation (indeed with all four main parties represented on the powersharing Executive, there was no 'opposition' in the classic Westminster sense and few sources of policy criticism). The default option of direct rule from London guaranteed that Northern Ireland followed the centre to the letter. It was even suggested that the local political parties were happy to let direct rule ministers deal with potentially unpopular issues (such as the introduction of water rates) (*Financial Times*, 2004). Despite the patent failure of devolution in Northern Ireland, the survey suggests that there was a perceptual re-territorialisation: or the internalisation among many survey respondents of the ability of the devolved Assembly to deliver better governance than that available under direct rule. Yet while there may have been a perceptual re-territorialisation with regard to the ability of Northern Ireland's devolved institutions to govern more effectively, decidedly old notions of territory (pro-Union versus pro-united Ireland) prevailed.

When in operation, members of the Northern Ireland Assembly were prepared to use the Welsh Assembly and Scottish Parliament as frames of reference in policy discussions, suggesting some degree of re-territorialisation. Indeed, the British–Irish Council – with its interest in various functional 'sectors' such as the environment – provided an institutional route for such comparison. The Official Record of the Northern Ireland Assembly also shows regular reference to devolution in Scotland and Wales, and reveals that Northern Ireland's politicians were aware of the opportunities devolution offered in terms of policy transfer and innovation. With the suspension of the Assembly, however, reference to Scotland and Wales has ceased in Northern Ireland's political discourse. There has been little evidence of either the public or politicians casting envious glances eastwards to see progress (or otherwise) in devolved Scotland or Wales. Perhaps Northern Ireland provides one enduring lesson for constitutional engineers: the limitations of 'institutional fixes' if political constituencies and their attendant mindsets are not prepared to retreat from entrenched positions and embrace the new institutions. Northern Ireland was the recipient of an incredibly sophisticated consociational political dispensation that was replete with a raft of safeguards to ensure pluralism and thwart majoritarianism. Yet, technocratic approaches can only go so far and, in the Northern Ireland case, failed to confront or circumvent chronic inter-group mistrust.

Notes

1   ESRC projects 'Political attitudes to devolution and institutional change in Northern Ireland' (L327253045) and 'Public attitudes to devolution and national identity in Northern Ireland' (L219252024).
2   Discussion of the sensitivities of conducting public attitudes research in deeply divided societies can be found at R. MacGinty, R. Wilford, L. Dowds, G. Robinson, 'Consenting Adults: The principle of consent and Northern Ireland's constitutional future', *Government and Opposition*, Vol. 36, No. 4 (Autumn 2001) pp. 472–92 at pp. 482–4; K. Brown & R. MacGinty, 'Public attitudes towards partisan and neutral symbols in post-Agreement Northern Ireland', *Identities: Global studies in culture and power*, Vol. 10 (2003) pp. 83–108 at pp. 87–8.

3   The question changed somewhat during the time series, with the formulation 'has most power over everyday life in Northern Ireland' being used in 1999 and 2000, and the phrase 'has most influence over the way Northern Ireland is run' being used in 2001.
4   The list was not consistent from year to year. In 1999 the choices were British Government, Irish Government, European Union, Local Councils, Northern Ireland Assembly, North–South Ministerial Council, and People of Northern Ireland. In 2000 the choices were British Government, Irish Government, European Union, Local Councils, Northern Ireland Assembly, North–South Ministerial Council, and East–West Council of the Isles. In 2001 the choices were Northern Ireland Assembly, British Government, Local Councils and European Union. In all years respondents were allowed to opt for can't choose/don't know and to suggest alternative answers.
5   Figures according to the Electoral Commission. See 'National Assembly for Wales elections – results' and 'Scottish Parliament elections – results' at www.electoralcommission.org

References

CBI News Release (2002) 'Grassroots CBI gives "vote of no confidence" in government plans for English Regions', 20 September, available at www.cbi.org.uk

Democratic Unionist Party (2003) *Fair Deal: Assembly Election Manifesto 2003*, Belfast: DUP.

Democratic Unionist Party (2005) 'Moving On', Belfast: DUP.

Elliott, M., (2000) *The Catholics of Ulster: A History*, London: Penguin.

Financial Times (2004) 'Households in Ulster set to pay water rates', 21 August.

Keating, M. (2003) 'Policy emergence and divergence in Scotland under devolution' in *Economic Governance Post-Devolution: Differentiation or Convergence? Conference Proceedings of the Regional Studies Association Annual Conference, November 2003*, Birmingham: ESRC.

MacGinty, R. (2003) 'Constitutional referendums and ethnonational conflict: Northern Ireland', *Nationalism and Ethnic Politics*, 9(2): 1–22.

Officer, D. (1996) 'In search of order, permanence and stability: Building Stormont, 1921–32' in R. English and G. Walker (eds) *Unionism in Modern Ireland*, Basingstoke: Macmillan, 130–47.

Phoenix, E. (2003) 'Lord Craigavon' in B. Lalor (ed.) *The Encyclopedia of Ireland* , Dublin: Gill & Macmillan, 252.

Porter, N. (1996) *Rethinking Unionism: An Alternative Vision for Northern Ireland*, Belfast: The Blackstaff Press.

Smyth, M. and Robinson, G. (eds) (2001) *Researching Violently Divided Societies: Ethical and Methodological Issues*, Tokyo/London: United Nations University Press/Pluto.

The Agreement (1998) Belfast: HMSO.

Wilford, R. (2002) 'The Assembly' in *Devolution Monitoring Programme Northern Ireland*, Report 10, February, available at http://www.democraticdialogue.org/devolution.htm

Wilford, R., MacGinty, R., Dowds, L. and Robinson, G. (2003) 'NI's devolved institutions: A triumph of hope over experience', *Regional & Federal Studies*, 13(1): 31–54.

Part 4

Complexities and interdependencies in spatial governance

Each individual is a component part of numerous groups, he is bound by ties of identification in many directions, and he has built up his ego ideal upon the most various models. Each individual therefore has a share in numerous groups' minds – those of his race, of his class, of his creed, of his nationality, etc.

(Sigmund Freud)

Like all other modern peoples, the English are in process of being numbered, labelled, conscripted, 'co-ordinated'. But the pull of their impulses is in the other direction, and the kind of regimentation that can be imposed on them will be modified in consequence.

(George Orwell)

17 Government, governance and decentralisation

Steven Musson, Adam Tickell and Peter John

Introduction

The UK state has long been among the most centralised in the capitalist world, as power and control have historically been concentrated in Whitehall. This reflects the wider and mutually reinforcing economic geography of the UK, in which London and the South East have experienced higher levels of growth and prosperity than other parts of the country. There have been attempts to address this centralisation over the last 40 years, for example by relocating administrative civil service activities or the more ambitious creation of Government Offices for the English regions in 1994. However, such policies did little to address the underlying geography of power and control. The 1997 general election presaged a renewed commitment to decentralised, elected government in Scotland, Wales and Northern Ireland, whilst also offering the prospect of greater decentralisation to the English regions.

In this chapter, we argue that in its existing form, English regional government has failed to realise its decentralising objectives. We argue that the structure of regional government in England, even in London where an elected assembly has been introduced, re-enforces the power of the centre. The role of Government Offices for the English regions, the enduring importance of political networks in local government and the direct intervention of central government departments on issues of national importance, are identified in this respect. Drawing on the analyses of Jessop (1999) and Peck (2001), we argue that the national state is enduringly important and that its relative demise has been overstated (Amin, 2002). As Jessop (1999: 394–5) notes:

> Just as the 'hollow corporation' retains its core command, control, and communication functions within the home economy even as it transfers various production activities abroad, so the hollowed-out national state retains crucial general political functions despite the transfer of other activities to other levels of political organisation.

Peck (2001) also argues that the transformed national state remains a powerful political institution, and suggests: 'the appropriate question to ask ... is not the extent to which the national state has somehow become "less" powerful in the process, but how it has become *differently* powerful' (2001: 447, original emphasis). We conclude that regional government in England currently fails to challenge the centralisation of power in London and the South East. Instead, a different but nevertheless powerful national state has emerged. Its interests are articulated through an increasingly complex institutional system, which serves to re-inscribe underlying and long-standing geographies of power.

Decentralisation and the state

The centralised nature of the UK, in both political and geographical terms, can be seen as something of an exception in contemporary Europe (Harvie, 1994). While comparably large states such as Germany, Italy, Spain and France have all introduced decentralised systems of government, the UK remains politically and economically dominated by London and the South East (John *et al.*, 2002; Amin *et al.*, 2003). In some respects, this reflects the long-term economic history of the UK, in which this area has experienced disproportionate levels of growth for at least 100 years (Mohan, 1999). The legacy of London as a Victorian centre of empire may also re-enforce these centralist tendencies (Weiner, 1981), while as one of the foremost financial centres in the global economy, it is also suggested that London will inevitably overshadow other regional economies in the UK (Smith, 1989; 1993, but see also Allen *et al.*, 1998).

Despite the seemingly inevitable dominance of the South East, the transformation of the UK and other advanced capitalist economies during the late 1970s and 1980s qualitatively changed the nature of uneven development. While northern England, Scotland and Wales experienced deindustrialisation and economic decline, the south and east of England became increasingly prosperous and economically successful (Massey, 1984; Lewis and Townsend, 1989). Critically, it was argued that the response of Margaret Thatcher's Conservative governments to these changing economic circumstances, in which fiscal and monetary policies seemed designed to promote more successful regional economies, led to a structural bias that favoured south eastern England (Peck and Tickell, 1995; Hamnett, 1997). Furthermore, it was argued that these problems would be difficult to address without attention to the underlying nature of political and economic power in the UK (Dunford, 1997).

Earlier attempts to address regional disparities had tended to focus on interventions by the national state, including civil service administrative relocation including the Driving and Vehicle Licensing Agency to Swansea and the National Health Service to Leeds (Marshall, J. *et al.*, 1997; Marshall, N. *et al.*, 2003), and, with a different focus, regional policies such as Industrial Development Certificates, which regulated the industrial geography of the postwar UK economy. While such policies brought temporary relief, for some, from the effects of wider economic transitions, they did little to address the underlying geography of power and control in the state. Furthermore, the geography of electoral support for the Conservative Party, which was overwhelmingly dominated by southern English votes, served to emphasise the political marginalisation of Scotland, Wales or some parts of northern England. At the 1992 general election, the Conservative Party returned 6 Welsh and 11 Scottish MPs. It failed to win a single seat outside England in 1997 and returned only one MP from Scotland and none from Wales in 2001 (Boothroyd, 2001).

Governments throughout the 1980s appeared to have little sympathy for such regional issues, promoting instead a nation-statist agenda that re-inscribed the power of the centre and emphasised the overall economic performance of the

UK. This style of government led some to question the implications of centralised control for local government (Cochrane, 1993; Stewart and Stoker, 1995). A programme of devolution and constitutional change seemed a remote possibility. In spite of, or even in response to, this prevailing political climate, by the end of the 1980s national devolution campaigns in Scotland and Wales had found voice, not only in the nationalist politics of the Scottish National Party and Plaid Cymru, but also in the Labour Party. As such, the Labour manifesto for the 1992 general election stated:

> We will move immediately to establish an elected Scottish Parliament. It will have powers to legislate for and administer Scotland's domestic affairs and modernise Scotland's economy and the ability to represent Scotland within the United Kingdom and Europe ... We will establish, in the lifetime of a full Parliament, an elected Welsh Assembly in Cardiff with powers and functions that reflect the existing administrative structure ... A regional tier of government in the English regions will take over many powers now exercised nationally, such as regional economic planning and transport. These new administrations will later form the basis for elected regional governments.
>
> (Labour Party, 1992)

By the time Tony Blair became Labour leader in 1994, devolution in some form for Scotland and Wales was established party policy. Almost immediately following the general election of 1997, the new Labour government announced referenda on elected government in these national territories. In England, a more modest but nevertheless expanded system of regional governance was introduced.

The current programme of regional government in England began in 1994 when the Conservative government established a Government Office (or GO) in each of the nine English regions. The GOs initially brought together the existing regional activities of three central departments: Employment, Environment, and Trade and Industry. However it was not until the election of the new Labour government in 1997 that regional government began to gather momentum. Two new sets of regional institutions were introduced: Regional Development Agencies (or RDAs) in 1997 and Regional Assemblies in 1999 (Tomaney, 2002). The RDAs were conceived as business-led bodies that would promote the regional agenda on economic development, skills and training and sustainable development. In contrast, the Regional Assemblies were envisaged as inclusive organisations with members from local government, community and voluntary sector groups and other economic and social partner organisations. They were to provide a forum in which regional strategic planning could be discussed and from which a 'voice of the region' could emerge (Sandford, 2002).

Parallel with these developments, the work of GOs was expanded to include nine central departments by 2004, including those without regional administrative traditions such as the Home Office. In line with these additional activities, GO budgets rose steadily. In 2004, 81 per cent of all expenditure by regional institutions was managed by the GOs, or £7,324 million compared with £1,756 million by the RDAs and just £15.3 million by the Regional Assemblies outside London (Musson *et al.*, 2004). Furthermore, in 2003, the Regional Co-ordination Unit (RCU) was established to provide a single point of contact between the GOs and central government. In part, the creation of the RCU reflects the increasing complexity of the GOs' work and the growing number of departments with which they deal. Significantly, it also offers central government departments a dedicated channel of communication with their regional representatives, and creates a clear line of accountability between the regions and the centre.

While each of this 'triad' of regional institutions in England has distinct areas of responsibility, each can only be fully understood with reference to the other two. In effect, the three institutions form an overlapping network of regional governance. For example, the Regional Assembly is responsible for scrutinising the RDAs' regional economic strategy. In April 2001, the East of England failed to endorse the East of England Development Agency's economic strategy, demonstrating that this function is more than a mere administrative formality (Tomaney, 2002). The GOs promote co-ordinated and well informed decision making within the region, communicate where necessary between the regional triad and local government, act as the 'eyes and ears' of central government in the region and provide expert information on their region to central government departments (Roche, 2001; RCU, 2003). The RDAs work with other regional partners to ensure that learning and skills policies are relevant to employment, and to monitor the sustainability of regional development plans.

The triad system became established across England over the course of the first new Labour government, although it was greeted with variable levels of public and political enthusiasm across the country (see, for example, Curtice, 2003). In London, with its history of exceptionalism (John *et al.*, 2005; Travers, 2004), the Labour Party moved more quickly on its long-term expectation that regional government would lead to elected assemblies. As such, the Greater London Assembly (GLA) and the Mayor of London were elected in May 2000. While the GLA and the Mayor have relatively limited powers, even in comparison with the Greater London Council (Sweeting, 2002), the effect of these exceptional arrangements for London has been to complicate further an already variegated system of regional government.

Although institution building in England has been predominantly at the regional scale, the implications for local government have been significant. Regional Assemblies provide a forum for local government involvement in the region, but have relatively limited funding and few statutory powers except as a regional strategic planning body (Sandford, 2002). The expansion of the English regions has been more focused on the GOs and to a lesser extent the RDAs, in both of which local government is a relatively minor partner. The GOs and RDAs have sought to establish new working arrangements in their regions, in collaboration with other new and existing regional and sub-regional partners such as local Learning and Skills Councils, county economic development partnerships, and regional employer, business and trade union organisations (see also Cochrane, 1993). However the role of local government in these new arrangements, and the extent to which the local and the regional tiers have integrated, are less clear.

In this chapter, we argue that, although moves to decentralise power to the regions changed the way in which government works in England, the centre retains a range of key strategic powers and controls (Peck, 1995). Through the triad system of regional government, the centre is able to monitor and directly influence the regional decision-making process. Furthermore, the majority of government expenditure in the regions is administered by the GOs, which represent central government in the regions. Despite the introduction of elected regional government in London, devolution in England is more modest than in Scotland or Wales (Tomaney, 1999). The current round of English regionalism may represent a further decentralisation of administration and promote new forms of partnership working at the regional scale, but we argue that it falls well short of a decentralisation of genuine control. The result of a referendum in the North East region during November 2004, in which 77.9 per cent of voters said 'no' to an elected regional assembly, indicates the enduring strength of the existing, centralised system of power (Rallings and Thrasher, 2005). Furthermore, the effectiveness of the 'no' campaign, which focused on the potential cost of the assembly and the creation of a new tier of political office, emphasised the difficulty of winning public support for changes to the political status quo in England (Tickell *et al.*, 2005).

Conceptualising regional government

Accounts of the expansion of regional government in England have focused in particular on the role of the RDAs, which were introduced shortly after the 1997 general election. The RDAs appeared to be emblematic of the wider approach of the new government, with an explicitly business-friendly agenda and a mandate to explore new forms of partnership working within their regions; they seemed to symbolise much of what was 'new' about Labour. In this respect, such an intensive research focus is not surprising. The potential appeal of the new agencies to academic researchers was quickly apparent to Morgan (1999: 666), who predicted that: '[they] will doubtless generate a new cottage industry in which researchers will explore the theoretical, policy and practical significance of the new agencies'. However, emerging studies of the RDAs can be connected to a longer series of research on local economic development in England throughout the late 1990s (for example ßPeck and Jones, 1995; Deas *et al.*, 1999). Work that explored the conceptual links between RDAs and other economic development bodies emerged rapidly after their establishment (Deas and Ward, 1999). One important theme that ran though this research was the role played by coalitions and partnerships in facilitating local and regional economic development. Peck (1995) pointed to the role of key individuals – movers and shakers – in shaping urban politics, while others, including Cochrane *et al.* (1996) and Tickell and Peck (1996), highlighted the role of elite local networks in enabling new and enterprising ways of working.

The implications of these new ways of working were extended by research that explored the wider transformation of the local state. In particular, this focused on the transfer of resources from local government to a range of non-governmental organisations, including centrally sponsored local institutions and a multiplicity of public–private and private sector ventures (Stoker, 1996). Jones (1998: 960, emphasis in original) located this process: 'within the discourse of a shift from structured govern*ment* to fluid govern*ance*'. For MacLeod and Goodwin (1999: 702), the development of new forms of governing capacity involved: 'The collective endeavour of those actors who have access to, and the power to deliver, the resources (financial, physical, human and political) of key public and private institutions'. These analyses placed the emergence of the RDAs, and indeed of regional government more generally, firmly in the context of a wider transformation of the local state. The new working practices that they adopted, and the range of private-sector resources that they incorporated, were as conceptually significant as they were politically symbolic. However, in the emerging network of regional governance, the RDAs did not command as powerful a role as their academic prominence suggests (Musson *et al.*, 2004).

Research on the transformation of local governance emerged in the context of a wider literature on the growing importance of regional economies. Drawing on a range of now familiar cases, including the Third Italy, Silicon Valley and Baden Württemberg, a rich and influential conceptual account of regional economies was developed. This emphasised the role of regional economies both as places of innovation and production in their own right, as well as being nodes within a wider global economy (Amin and Thrift, 1994). Critically, regions were conceived as increasingly important in relation to the national state, as a site of economic activity and as a scale at which effective government interventions could be made. This led some to identify the emergence of a regionalised world economy, in which regions increasingly connected with one another and to international economic processes with limited reference to the national economic scale (Storper, 1997; Scott, 1998). More radical accounts considered this transformation to herald the 'end of the nation state' (Ohmae, 1996), or at least to signify a substantial and irreversible decline in its role and status (Reich, 1991).

In response to such analyses, it was argued that the national states retain a range of powers and controls in spite of the growth of regional economies, although their precise nature may have been transformed (Amin, 1999; Jessop, 1999; Peck, 2001). Drawing on research in London and South East England, we argue that the role of central government extends beyond the key functions, including financial and legislative controls, identified by Jessop (1999). With a focus on the role of the GOs and their place within a wider network of regional government, we argue that the national state exerts a strong influence *within*, as well as over, the English regions. The establishment of the RCU and the expanded role of GOs enable central government departments to exert influence within the triad system of regional government, not least because its structure promotes co-operation and co-dependency between institutions.

Understanding regional government in England: evidence from the South East

The economic heartland of England is also the place in which public support for regional government is weakest (John *et al.*, 2002). Irrespective of political and civic debates over the coherence of the South East region and problems of integration between new regional and other existing institutions, regional government has become well established. Although political enthusiasm in the region was less than in other parts of England, the way in which the triad institutions have begun to work together and the way in which the regional tier has connected into other forms of government within the region, speak to a wider argument about the nature of decentralisation in the state.

The first challenge faced in the South East following the expansion of regional government in 1997 related to questions over the legitimacy of regional boundaries. In some respects this is unsurprising, since the functional boundaries of the region, and its cohesiveness as a homogenous economic space have long been disputed (Allen *et al.*, 1998). The extent to which the South East region can be understood without reference to its annexed economic and cultural core, London, is also debatable (John *et al.*, 2005). However, while such claims are still voiced by political opponents of English regionalism, the growing effectiveness of existing regional institutions and a pragmatic acceptance by local government that the new system had to be made to work, led to the emergence of a functional South East region.

Local government in the South East is dominated by Conservative-led county councils. Long-standing political networks, including the South East County Council Leaders' Network (SECCL) and a parallel chief executives' group, bring together leaders and officers from across south eastern England, including Essex, Bedfordshire and Hertfordshire in the East of England region. The region lacks a major urban core, although nine towns and cities have a population greater than 100,000 people. As such, the urban Labour politics that dominate other regions such as the North West and West Midlands are less evident. One representative from the South East England Regional Assembly explained where local government power in the region lay:

> Two years ago we had a district councillor [as chair of the Assembly], a [Liberal Democrat], and I think there was a risk that the [Conservative] county councillors could well have taken their ball away. They had threatened to do that – it's sad that it could be enormously damaging.[1]

The power of the county councils in the South East is reflected in the range of political contacts available to their leaders. While the aim of SECCL is to express common points of interest amongst the counties rather than to oppose the region, one senior county council leader explained to us that in some respects, regional government was irrelevant given the existing level of political access to central government:

> Most of the major problems that we have or challenges that we have, it wouldn't occur to us to go to the region ... we would go straight to the department. If we are concerned about plans for [a regional airport], we go and see Mr Spellar [John Spellar, Minister for Transport]. If we want to extend [a motorway], it's the same thing. If we have problems with the Home Office and asylum, we go this week and we talk to the minister about those particular issues.[2]

While such existing political networks have endured, the expansion of regional government, in part at least, has led to the advent of new working practices at the local level. A wide range of new strategic partnerships have re-enforced the role of the new regional institutions. However, the proliferation of strategic partnerships in particular has led to tensions between the regions and local government. One county council leader saw this as part of a search for 'more reliable sub-contractors for central government' within the region, while the leader of a large local authority told us that:

> There are certain places where the existence of quangos can be useful, and the RDA is a good example of that in one place at one time. But when I look

at all the partnerships we have had [here], the trouble is that they are not very effective at the end of the day. They end up just being talking shops.[3]

The inclusive nature of regional government has widened involvement and capacity. However, tensions with local government have followed. In particular it is perceived by local political leaders that the new forms of governance and strategy making enabled by the region are not matched by mechanisms for actual policy delivery at the local level which, in spite of the growth of the regional tier, are still dependent on local government for implementation.

The triad institutions of regional government each have clear areas of responsibility. The role of the GOs is defined by sponsoring departments, while Regional Assemblies have tightly constrained statutory powers that mainly relate to land-use planning. The RDAs operate in a more relaxed environment, and have recently been given greater financial flexibility through the introduction of a single funding pot, under which they can choose to allocate according to their own regional priorities. However, since funding is allocated according to a measure of deprivation rather than the cost of making meaningful interventions in the regional economy, the effectiveness of RDAs in relatively prosperous regions is limited. For example, in 2002–3 per-capita funding for the South East England Development Agency was £11.48, compared to £88.80 for the North East Development Agency over the same period (National Statistics, 2003; authors' own calculations). As such, the ability of regional institutions to make effective interventions in prosperous economic areas is limited, while more powerful local politicians have incentives to use their own networks rather than work with the region.

The key economic issues faced in the South East are reflected in the strategic approach taken by the South East England Development Agency (SEEDA). These issues can broadly be characterised as problems of economic success, including shortages of skilled labour, an inadequate supply of affordable housing and an overstretched infrastructure. In addition, Counsell and Haughton (2003) identify a range of environmental sustainability problems particularly associated with water supplies, while pressure for the expansion of regional and national airport capacity, ongoing debates around the construction of Crossrail, a new suburban rail link for London, and inter-regional arguments over the disposal of household waste serve to compound existing problems.

The fragmented nature of government across south east England makes strategic intervention at the regional scale difficult. In some respects, senior regional politicians, who choose to rely on their own networks of contacts rather than working through the region, reflect this. Until the end of 2000, issues of common interest between London and the rest of the South East were discussed by SERPLAN, the South East England Regional Planning Conference. However, following the transfer of strategic planning powers to individual Regional Assemblies, such arrangements moved onto a more informal footing in spite of widespread recognition of the important role that SERPLAN played (see, for example, GLA, 2000). A new committee, including five officers from the GLA and five from adjoining regions, has by common agreement failed to operate as an effective replacement in resolving substantive issues (ALG, 2002). In part, these inter-regional arrangements reflect more general problems of communication between London and the East of England and South East regions (John *et al.*, 2005). However, they also speak to a wider debate about the role played by central government in London and the South East and the extent to which regional government there represents a genuine decentralisation of power.

Central government departments are closely involved in strategic planning and decision making in London and the South East. Such interventions are frequently made in areas of perceived national importance such as increased airport capacity at Heathrow and Gatwick and on issues critical to the London economy including London Underground and key worker housing. The Thames Gateway project exemplifies the role played by central government. The Thames Gateway is a large brownfield development site extending over 40 miles to the east of central London, incorporating three regions, London, the East of England and the South East. Over 120,000 new houses are planned for the redeveloped area by 2016 (ODPM, 2005), enhancing London's status as a world city and relieving economic development pressures on the south and west of London (Government Offices, 2001).

The Thames Gateway is institutionally complex, incorporating 15 local authority areas, two county councils and three sets of regional institutions. However, the role of local and regional government is restricted to participation in three regional partnership boards, which also include business representatives from property development companies. The role of central government was signalled by the establishment of a cabinet committee, chaired by the Prime Minister, which retained overal responsibility for the initial development of the Thames Gateway project (ODPM 2005). The committee's terms of reference were to consider the timescale development and to consider the requirements and the funding implications for transport infrastructure and other key public services. Subsequently, co-ordination of the Thames Gateway project has been absorbed into a more general cabinet committee on housing and planning. As such, although three public–private partnerships are responsible for co-ordination and implementation activities in each individual region, overall control of finance and planning in the Thames Gateway is dominated by central government.

The extent to which regional government in London and the South East can be seen as a decentralisation of power is questionable. The area is institutionally fragmented, while communication between the three regions of south eastern England is limited. The ability of regional institutions to determine their own priorities is constrained by their clearly defined areas of responsibility and by the relative lack of discretionary, 'single pot' funding available to RDAs in more prosperous regions. Additionally, central government retains a strong interest in London and the South East through a number of channels including direct interventions on issues of perceived national importance.

Conclusions

In this chapter we set out to interrogate the nature of decentralisation in England. Even in London, where an elected assembly has been introduced, regional government appears to be limited in scope. We argue that central government plays an enduringly important, but qualitatively different, role in the decentralised state. A range of key powers and controls are retained by the centre, not least relating to issues of national importance such as airport capacity and the Thames Gateway development project. The fragmented nature of government across south eastern England and the lack of effective communication between different regions limits the capacity of regional government to respond to these inter-regional issues. However, the nature of central–regional interaction is more complex than these direct interventions suggest. The role of the GOs in representing central departments in the network of regional government, and the limited amount of discretionary funding available to regions, may inhibit the emergence of independent strategies. Furthermore, the political networks of local government leaders, which have been maintained even though the nature of the local state has been transformed, provide an alternative set of connections between central and local government. We argue that, as yet, regional government in England has failed to challenge the fundamental nature of state centralism. While a limited range of responsibilities have been devolved to the regional level, the role of central government in this process and the range of powers and controls that it retains, lead us to conclude that the current round of decentralisation remains at best limited.

Acknowledgements

The authors acknowledge the support of ESRC award numbers L219252038 and RES219252001.

Notes

1   Interview conducted by Steven Musson, 14 June 2002.
2   Interview conducted by Peter John and Steven Musson, 11 June 2002.
3   Interview conducted by Steven Musson, 22 May 2002.

References

ALG (Association of London Government) (2002) *Draft London Plan – ALG views*, available from ALG, 59½ Southwark Street, London SE1 0AL.
Allen, J., Massey, D. and Cochrane, A. (1998) *Rethinking the Region*, London: Routledge.
Amin, A. (1999) 'An institutionalist perspective on regional economic development', *International Journal of Urban and Regional Research*, 23: 365–78.
Amin, A. (2002) 'Spatialities of Globalisation', Environment and Planning A 34(3): 385–399.
Amin, A., Massey, D. and Thrift, N. (2003) *Decentering the Nation: A Radical Approach to Regional Inequality*, London: Catalyst.
Amin, A. and Thrift, N. (1994) *Globalisation, Institutions and Regional Development in Europe*, Oxford: Oxford University Press.
Boothroyd, D. (2001) *The Politico's Guide to the History of British Political Parties*, Tunbridge Wells: Politico's.
Cochrane, A. (1993) *Whatever Happened to Local Government?* Buckingham: Open University Press.
Cochrane, A., Peck, J. and Tickell, A. (1996) 'Manchester plays games: exploring the local politics of globalisation', *Urban Studies*, 33: 1319–36.
Counsell, D. and Haughton, G. (2003) 'Regional planning tensions: planning for economic growth and sustainable development in two contrasting English regions', *Environment and Planning C: Government and Policy*, 21: 225–39.

Curtice, J. (2003) 'Devolution meets the voters: the prospects for 2003' in R. Hazell (ed.) *The State of the Nations 2003. The Third Year of Devolution in the United Kingdom*, London: Imprint Academic.

Deas, I., Peck, J., Tickell, A., Ward, K. and Bradford, M. (1999) 'Rescripting urban regeneration, the Mancunian way' in R. Imrie and H. Thomas (eds) *British Urban Policy*, 2nd Edition, London: Sage.

Dunford, M. (1997) 'Divergence, instability and exclusion: regional dynamics in Great Britain' in R. Lee and J. Wills (eds) *Geographies of Economies*, London: Arnold.

GLA (Greater London Assembly) (2000) *Relationships between the GLA and the wider South East*, Transport and Spatial Policy Committee report number 8, 7 November, London: Greater London Assembly.

Government Offices for the East of England, South East and London (2001) *Regional Planning Guidance for the South East (RPG9)*, London: The Stationery Office.

Hamnett, C. (1997) 'A stroke of the Chancellor's pen: The social and regional impact of the Conservatives' 1988 higher rate tax cuts', *Environment and Planning A*, 29: 129–47.

Harvie, C. (1994) *The Rise of Regional Europe*, London: Routledge.

Jessop, B. (1999) 'Narrating the future of the national economy and the national state? Remarks on remapping regulation and reinventing governance' in G. Steinmetz (ed.) *State/culture: State Formation after the Cultural Turn*, Ithica NY: Cornell University Press.

John, P., Musson, S. and Tickell, A. (2002) 'England's problem region: regionalism in the South East', *Regional Studies*, 36: 733–41.

John, P., Musson, S. and Tickell, A. (2005) 'Governing the mega-region: governance and networks across London and South East England', *New Political Economy*, 10: 79–100.

Jones, M. (1998) 'Restructuring the local state: economic governance or social regulation?' *Political Geography*, 17: 959–88.

Labour Party (1992) *It's Time to Get Britain Working Again: The Labour Party Manifesto for the 1992 General Election*, London: Labour Party.

Lewis, J. and Townsend, A. (eds) (1989) *The North-South Divide*, London: Paul Chapman.

MacLeod, G. and Goodwin, M. (1999) 'Reconstructing an urban and regional political economy: on the state, politics, scale and explanation', *Political Geography*, 18: 697–730.

Marshall, J., Hopkins, W. and Richardson, R. (1997) 'The civil service and the regions: geographical perspectives on civil service restructuring', *Regional Studies*, 31: 607–13.

Marshall, N. *et al.* (2003) *Public sector relocation from London and the South East*, Report prepared by CURDS, University of Newcastle for English Regional Development Agencies. Available from www.ncl.ac.uk/curds

Massey, D. (1984) *Spatial Divisions of Labour*, Basingstoke: Macmillan.

Mohan, J. (1999) *A United Kingdom? Economic, Social and Political Geographies*, London: Arnold.

Morgan, K. (1999) 'England's unstable equilibrium: the challenge of the RDAs', *Environment and Planning C: Government and Policy*, 17: 663–8.

Musson, S., Tickell, A. and John, P. (2004) 'A decade of decentralisation? Assessing the role of the Government Offices for the English Regions', working paper available from www.bbk.ac.uk/geog/staff/musson.html

National Statistics (2003) *Sub-regional Gross Value Added: Methods and Background*, London: Office for National Statistics.

ODPM (Office of the Deputy Prime Minister) (2005) Creating sustainable communities: delivering the Thames Gateway, available from ODPM Free Literature, PO Box 326, Weatherby, LS23 7NB, London: ODPM.

Ohmae, K. (1996) *The End of the Nation State: The Rise of Regional Economies*, London: Harper Collins.

Peck, J. (1995) 'Moving and shaking: business elites, state localism and urban privatism', *Progress in Human Geography*, 19: 16–46.

Peck, J. (2001) 'Neoliberalizing states: thin policies/hard outcomes', *Progress in Human Geography*, 25: 445–55.

Peck, J. and Jones, M. (1995) 'Training and enterprise councils: Schumpeterian workfare state, or what?', *Environment and Planning A*, 27: 1361–96.

Peck, J. and Tickell, A. (1995) 'Business goes local – dissecting the business agenda in Manchester', *International Journal of Urban and Regional Research*, 19: 55–78.

Rallings, C. and Thrasher, M. (2005) *Why the North East Said 'No':The 2004 Referendum on an Elected Regional Assembly*, Briefing, 18 February 2005, ESRC Research Programme on Devolution and Constitutional Change.

RCU (Regional Co-ordination Unit) (2003) *GO! The 2002–2003 report on the work of the Government Offices for the English regions*, London: The Stationery Office.

Reich, R. (1991) *The Work of Nations: Preparing Ourselves for 21st Century Capitalism*, London: Simon & Schuster.

Roche, B. (2001) Speech by the Minister of State Social Exclusion and Deputy Minister for Women at Solace annual conference, Ipswich, 18 October 2001, available from www.archive.cabinetoffice.gov.uk/ministers/2001/Speeches/Barbara%20Roache/Solace%20101801.html, accessed 18 January 2006.

Sandford, M. (2002) 'What place for England in an asymmetrically devolved UK?' *Regional Studies*, 36: 789–98.

Scott, A. (1998) *Regions in the World Economy*, Oxford: Oxford University Press.

Smith, D. (1989) *North and South: Britain's Economic, Social and Political Divide*, Harmondsworth: Penguin.

Smith, D. (1993) *From Bust to Boom: Trial and Error in British Economic Policy*, Harmondsworth: Penguin.

Stewart, J. and Stoker, G. (eds) (1995) *Local Government in the 1990s*, Basingstoke: Macmillan.

Stoker, G. (1996) 'Normative theories of local government and democracy' in G. Stoker and D. King (eds) *Rethinking Local Democracy*, Basingstoke: Macmillan.

Storper, M. (1997) *The Regional World: Territorial Development in a Global Economy*, New York: Guildford Press.

Sweeting, D. (2002) 'Leadership in urban governance: the mayor of London', *Local Government Studies*, 28: 3–20.

Tickell, A., John, P. and Musson, S. (2005) *The referendum campaigns: issues and turning points*, Briefing 19, February 2005, ESRC Research Programme on Devolution and Constitutional Change, Swindon: ESRC.

Tickell, A. and Peck, J. (1996) 'The return of the Manchester Men: men's words and men's deeds in the remaking of the local state', *Transactions of the Institute of British Geographers*, 21: 595–616.

Tomaney, J. (1999) 'New Labour and the English question', *Political Quarterly*, 70: 75–82.

Tomaney, J. (2002) 'The evolution of regionalism in England', *Regional Studies*, 36: 721–32.

Travers, T. (2004) *The Politics of London: Governing an Ungovernable City*, Basingstoke: Palgrave MacMillan.

Weiner, M. (1981) *English Culture and the Decline of Industrial Spirit*, 1850–1980, Harmondsworth: Penguin.

18 City-regionalism

The social reconstruction of an idea in practice

Greg Lloyd and Deborah Peel

Introduction

There is a vast international literature seeking to chart, interpret and explain the territorial transformations and changing spatialities and institutional structures of cities and metropolitan regions that are taking place across the world. The diversity of the attempts to capture this evolving landscape of ideas and practices is concerned with the pot-pourri of different dimensions that reflect the real politik of political, institutional, governance, public policy, economic, social, environmental, physical, and cultural dimensions that shape our cities. The contextualising discourses are articulated in a vocabulary that encapsulates what some describe as the contemporary 'turbulence' (Thornley, 2003). The 'transformations' in city form and institutional arrangements have drawn commentators to make reference to the leitmotif of 'post-' that runs through much of the critical literature (Allmendinger, 2002). In this context, for example, Soja (1997) coined the term 'post metropolis' to refer to the city-region, and reference is made to emerging and new forms of urban settlement (Frey, 1999). Such debates clearly alert us to shifts in thinking that may (or may not) be somehow linked with earlier paradigms and schools of thought. The important point here, however, is that the ebb and flow of ideas and the embracing of heterogeneity make it very difficult to pin down what is meant by the concept 'city-region'. On the one hand, this means that the city-region idea will be contested and open to capture by different groups seeking to construct a particular hegemony of city-regionalism for particular purposes, on the other, it can lead to creative ambiguity.

In this chapter, we examine some of the theories and practices of the city-region concept through a study of Scotland. The first section considers the roots of the city-region idea. The chapter then maps the current vogue for city-regions principally as a means of effecting modern territorial management in rapidly changing economic and technological conditions. Here, the city-region idea is promoted as the conduit to the reconciliation of a complex of economic, social and environmental policy ambitions. In Scotland, this particular spatial emphasis and redefinition is acknowledged as responding to a European emphasis on economic competitiveness and social cohesion around core areas (Glasgow City Council, 2003). Yet, this apparently self-evident reality does not fully capture the clash of ideas. We identify the contemporary advocacy of city-regionalism as being the appropriate form of integrating converging ideas associated with the re-territorialisation of – and within – national space. This draws attention to parallel debates associated with the continuing reconfiguration of the boundaries and understandings of governance and regulation (Hudson, 2005).

As a consequence of globalisation and the ascendancy of regionalism and spatiality, the city-regionalism agenda has assumed an almost iconic status in modern parlance. This form of institutional restructuring is receptive to a spectrum of allegedly innovative and hybrid ideas variously articulated and drawn from a spectrum of scalar relations associated with, for example, 'new institutionalism', 'new regionalism', 'new urbanism', and 'new localism' (Deas, 2004; Vigar et al., 2002; Deas and Ward, 2000; see all contributions to this book), and which morph with respect to the peculiarities of particular circumstances. As a distinct territorial unit, Scotland's identity as a 'stateless nation' or 'national region' is a case in point (MacLeod, 1998). As a consequence, 'new city-regionalism' serves many masters.

The parallel but disparate threads of contemporary intellectual reasoning appear to have resulted in an apparent convergence and consensus around city-regionalism as a normative concept. This belies the practicalities of the concept. This chapter explores the current (re-)casting of the city-region idea in Scotland. The evidence would appear to suggest that society needs to caution against an overly simplistic interpretation of the city-region concept which may obscure a need to critically interrogate the realities of political, governance, economic, social and environmental capacities. In particular, attention needs to be paid to the changing relationships between the state and the market, the state and civil society, and the ecology and morphology of a given city-region. Moreover, these debates occur within a dynamic articulation and assertion of particular relationships and interdependencies. Thus, in the context of the UK, for example, national can become regional, then can become local, depending on the particular perspective, prism, construction and subject in question – and vice versa.

Tracing the roots of the city-region idea

Attempts to describe and explain the form and interrelationships of urban growth have long grappled with the idea of the 'city-region'. As Dickinson cautioned, however:

> This concept of the city-region, like all concepts, is a mental construct. It is not, as some planners and scholars seem to think, an area that is presented on a platter to suit their general needs. The extent of the area they need will depend on the specific purpose for which it is required. The concept of the city-region can only be made specific and definable, as a geographic entity, by reference to the precise areal extent of particular associations with the city.
>
> (Dickinson, 1964: 95)

Such a construction of the city-region offers a geographical prism, historically and initially based on the idea of regional capitals (Hancock, 1976). More recently, Geddes (1997), for example, describes how American city-regions are 'exploding into their surrounding countryside' as the geographic growth of city-regions outpaces population increases. Elsewhere, research evidence suggests that the mismatch between the social use of space and its associated governance in the

Toronto city-region, for example, necessitated a radical reorganisation of governance and jurisdictional boundaries in order to replace a system that had simply 'stopped working' (Donald, 2002). This particular concern with city-regions presents the concept as comprising an evolving and dynamic set of interrelated systems. These ideas are often associated with the work of McLoughlin (1969), for example, in confronting complex and changing state–market relations.

The city-region idea has an important lineage. Almost a century earlier, and driven by a belief that social processes and spatial form are intimately related, Patrick Geddes was already pioneering a sociological approach to investigate processes of urbanisation, and contending that the city should be studied in the context of the region so as to better shape future developments (Meller, 1990). Here, the concept of the city-region is essentially based on an interpretation of the relationship between the city and its physical and functional hinterland as providing resources for city growth and development (Dickinson, 1964). Globalisation has disrupted this particular interface of production and consumption. Further, from the more contemporary perspective of the 'sustainable city', Ravetz (2000) asserts that one alarming global trend stems from the severe environmental problems caused by the 'runaway urbanization' in developing countries (p. 7). He suggests that the nature of urban activity now extends across city-region territories from the city centre to the relatively remote countryside. Yet, the (functional) relationships between city and its hinterland are, in practice, very different across the world. In advocating the city-region as the most suitable container and conduit for delivering and sustaining sustainable development, Ravetz (2000) argued that the city-region is the most appropriate territoriality. Nonetheless, the functional potential of a particular city-region space could be enabled or disabled in a variety of ways, including imagined and real boundaries and jurisdictions.

The zeitgeist of the city-region agenda is one that is advocated as celebrating diversity and difference. The contemporary climate of transition and fragmentation, of flows and fluidity, is reflected in a plurality of ways (Blatter, 2004). Contexts and cultures are constantly subject to revision, and in popular terms this conceptual flux is evident in references to: re-thinking (Keating, 2001); re-territorialisation (Brenner, 1999); re-scaling (MacLeod and Goodwin, 1999); re-positioning (Wilks-Heeg et al., 2003); newness (Lovering, 1999; Healey, 2004); shifts (Donald, 2002); and experimentation (Brenner, 2002). The shaping of the city-region idea is occurring against this backcloth. Indeed, political enthusiasm for the idea serves to dramatise the city-region agenda yet further. Here, for example, Simmonds and Hack (2000: 3) argue that city-regions are becoming a new political force 'out of necessity'. Taken together, a particular momentum appears to be driving a normative case for city-regions across an indiscriminate spatial hierarchy.

The variety of theoretical perspectives and descriptions of metropolitan change reflect the complexity, dynamism and evolving nature of the subject under study. There is an emphasis on the use of case study examples – which nonetheless highlight the heterogeneity of emerging metropolitan regions. Here, attention tends to be drawn to the specific conditions relating to, for example, the complexities of operating within a multi-level polity (Morgan, 2004). Reconciling these disparate perspectives and contexts remains a challenge for critical theorists seeking to develop appropriate comparative theoretical frameworks that could help to justify and to legitimise the transfer of the city-region idea between scales and contexts. The available catalogues of case study evidence run the risk of being nation specific and reductionist, particularly where they seek to generalise lessons learned from specific contextual instances (Deas and Giordano, 2003). Moreover, by redefining themselves as city-regions, metropolitan areas appear to be aspiring to global relevance. Asserting a city-region identity then becomes both an excuse and a marketing device in a global, competitive, regionalising context (Jonas and Ward, 2002; Salet *et al.*, 2003). Yet, the uncritical conflation of description and prescription, adoption and adaptation, should alert us to the need to posit critical questions with respect to the normative agenda for city-regions in a devolved UK.

A vogue for city-regionalism?

The preceding discussion has highlighted that the terminology and interpretation of city-regionalism is contested. It follows that its translation into practice will vary according to time, place, space and governance. Simmonds and Hack (2000), for example, suggest that the city-region idea captures the spatial extent of closely linked economic activity, rather than a more instrumental focus on the city and its jurisdictional and administrative boundaries. Yet, intrinsic to this line of reasoning is a recognition that the spatial realities of global city-regions do vary and therefore demand new forms of governance and management. Significantly, it challenges traditional adherence to fixed administrative boundaries in a European context, for example, these same themes have been articulated with a different emphasis, as demonstrated, for example, by Tewdwr-Jones and McNeill. Thus, their very particular definition of a city-region emphasises the broader metropolitan governance arrangements within a given spatial and political context:

> [City-regions are a] strategic and political level of administration and policy making, extending beyond the administrative boundaries of single urban local government authorities to include urban and/or semi urban hinterlands. This definition includes a range of institutions and agencies representing local and regional governance that possess an interest in urban and/or economic development matters that, together, form a strategic level of policy making intended to formulate or implement policies on a broader metropolitan scale.
>
> (Tewdwr-Jones and McNeill, 2000: 131)

This definition illustrates the complex layering of ideas associated with the term city-region. It involves notions of geographical space, territories and boundaries, cultures and identities, governance and institutional capacities, and is underpinned by assumptions of collaborative working. Yet, implicit here are the realities of power play associated with political negotiations and inter-organisational working across space. It is important not to overlook the realities of historical baggage, inherited institutional relationships across and between spaces and places, and the power of past and on-going restructuring and reconfiguring processes of socio-economic change. This is particularly sensitive, given the emphasis on place competition.

An interest in seeking to better understand and articulate the 'buzzword' of 'city-region' in the context of debates about contemporary and emerging urban and regional development is well acknowledged (Herrschel and Newman, 2002). A number of common threads can be identified in the academic and policy literature on city-regions – even if these city-regions differ in terms of scale, context, and position in the urban hierarchy, heritage, and trajectory. Whilst, on the one hand, there is a focus on understanding the materiality of metropolitan areas and the changing physicality of contemporary human settlements (Romppanen and Ujam, 2004), on the other, attempts are also being made to understand the implications of cross-border city-regions and border-less region states in a global context (Shen, 2004). Such complexity in terms of whether we are dealing with dimensions such as form and function, or process and outcome, or notions that are fixed or fluid, explain, in part, why attempts to explain and justify human settlements and governance arrangements tend to focus on particular perspectives. Thus, there is, for example, an important academic literature that seeks to contextualise, measure, debate and explain the turn to city-regionalism in terms of the processes, outcomes and demands of globalisation and its associated effects (Scott, 2001; Taylor, 2000; Knox, 1995). The all-encompassing context of globalisation has inevitably led to a particular interest in the nature and types of 'competitive city-regions' (Wilks-Heeg *et al.*, 2003; Scott, 2001; Healey, 1998). Such a Darwinian perspective naturally encourages and fosters a culture of a competitive survival of the fittest.

A parallel and associated line of intellectual enquiry concerns the renaissance of regional agendas as the appropriate arena for promoting place competition (Malecki, 2004), facilitating economic competitiveness and development, whilst, at the same time, securing social cohesion (Tewdwr-Jones and Mourato, 2005), and designing and implementing context specific regional policy interventions (MacLeod, 1998). There are, however, unresolved tensions around the ways in which the inherited dominant policy discourses of economic competitiveness and environmental conservation still prevail (Owens and Cowell, 2002). At the core of these debates is a primary concern with the appropriateness of the institutional forms of governance in place, particularly in a reconfiguring nation-state context that comprises a composite of integral regional territories (Deas and Giordano, 2003; Morgan, 1997; MacLeod and Goodwin, 1999). This fluidity of industrial, corporate, and spatial restructuring and change is reflected in the questioning of nation states as the appropriate arena for global competitiveness (Bond and McCrone, 2004), and this may then be mediated through the concept of city-regions, changing city-regional relations, and differential metropolitan spatial configurations (Deas and Ward, 2000; Herrschel and Newman, 2002). In effect, globalisation is, in many ways, setting the agenda for spatial and institutional arrangements. Therefore, reconsiderations of place, space, boundaries and identity are being asserted (Paasi, 2001). So what are the implications for our social construction of the concept of city-regions, particularly given the term's ambiguity and contested use at a spectrum of scales and in such a diversity of socio-political and spatial economic contexts? The elusiveness (and by implication the vulnerability) of the term city-region can be illustrated by its use at different scales.

International context

In the context of global city-regions, there is a concern that traditional planning and policy strategies are increasingly problematic. In the context of spatial planning and regional economic development, this has been described as being 'out of sync' (Tewdwr-Jones and Mourato, 2005). Attention is increasingly focused on devising more appropriate spatial interventions to address the needs and conditions of these important nodes and containers in the activities of the global economy. Yet, at the level of global city-regions, Simmonds and Hack (2000: 3), for example, suggest that many of the problems which contemporary settlements face call for 'region-wide' policies and co-ordinated action across jurisdictions. This intuitive argument seeks to link core cities with their natural hinterlands, and confirms that the 'spatial reality of regional cities, fuelled by global trends demands new forms of governance and management' (2000:3).

Globalisation has involved a fundamental restructuring of corporate and business activities, power, control and management and the spatial economy of the (post-) modern world. It has transformed arrangements for the management and control in the corporate sector, provision for knowledge management, technological innovation and product differentiation, and the structure of supply relations in order to respond to the competitive conditions of global and national markets. This has then had an impact on the standing of individual nation states, the constituent functional regions, and the spatial dispersal of economic activity, corporate power and regulation. Nation states do not necessarily provide the most appropriate dynamo for change. As a consequence, city-regions have emerged as a more effective spatial arrangement to facilitate the needs of global business, and to provide the efficient organisation of markets.

In this context, Scott (2001: 815) asserts that global city-regions, 'form a global mosaic that is now beginning to override the system of core periphery relationships that has hitherto characterised much of the macro geography of capitalist development'. Such changes are perceived as promoting a new politics of scale within which governments at all levels, stakeholders and social movements are attempting to adjust (Brenner, 2002). Research evidence is now exposing the nature of the restructuring of governance arrangements necessitated by the complex processes of economic globalisation and restructuring (Donald, 2002). Further, this is leading to implications for global city-regions in terms of their particular social geographies, comprising cultural and demographic heterogeneity, spatial morphology and associated public policy agendas (Scott *et al.*, 1999). Globalisation is clearly a critical and dominant canvass to any discussions about the relevance and appropriateness of governance arrangements. City-regions, emerging from the globalisation agenda, have then served to challenge nation states and the established arrangements for spatial governance, planning and territorial management.

European context

At the European level, there is considerable interest in both the term and the concept of city-region (Deas and Giordano, 2003). Research has shown that the European urban hierarchy comprises a complex layering of the spatial scales of control and management capacities in the globalising corporate sectors, and the spatial distribution of structures of production (Brenner, 2000). This complex of activity then results in a diverse hierarchy and pattern of cities and city-regions that reflect the composition of their individual economic morphologies. The origins of this interest may be traced to the emergence of the global city-regions, and their role in the global economy, and the understanding and experiences of competitive city-regions in the United States (Jonas and Ward, 2002). Indeed, it would suggest that Europe is seeking to compete on this world stage, and in terms of these global forces. Indeed, experience within the European Union would suggest that a sophisticated restatement of the city-region idea is in progress, together with a strong assertion of its appropriateness to the European context. Notwithstanding a general advocacy of the city-region idea, in practice this has resulted in a complex and differential mosaic of 'city-regions' in the various constituent parts of the European Union (Salet *et al.*, 2003). Here, structure, process and agency have been blurred in the enthusiasm for what is seen as a global panacea to promote regional competitiveness and city development. Indeed, this line of thinking has progressed to such an extent that it would appear that the terms 'region', 'nation' and 'European space' have gained popular currency as 'metaphors' for a European identity (Healey, 1998).

In practical terms, the enthusiasm for the city-region concept has developed to a point where many of the debates now centre on the appropriate city-regional forms. These distinguish between monocentric and polycentric structures and relations (Herrschel and Newman, 2002; Shaw and Sykes, 2004). This would tend to suggest that the intellectual rationale for city-regions (despite its assorted lineage and contested meanings) is now generally accepted and that it is simply a question of design and implementation in order to minimise urban sprawl. In light of the alleged advantages of the polycentric urban region form which, it is asserted, offers both regional economic competitiveness and equity environmental benefits, Bailey and Turok (2001) discuss the advocacy of this form of co-operative interactive territorial management by neighbouring cities in the Netherlands, Belgium and Germany, and question the transferability of the idea in the context of Scotland's central belt.

Furthermore, it is held that the city-region idea is intrinsic to the development of strategic spatial planning in Europe (Healey, 2004). Spatial planning has been advocated as a means of addressing a number of issues within the European project. These include addressing the changing political context within which policies are devised and executed, tackling problems of public policy co-ordination, devising ways of making urban regions more economically competitive, and securing a spatial form that is compatible with the aims of sustainable development. Thus, the European Spatial Development Perspective has asserted in its framework for spatial priorities in Europe (Faludi and Waterhout, 2002) a number of spatial concepts, including polycentric urban development around configurations of city-regions. Whilst this is advisory in nature, in a time of dynamic and sensitive change, it asserts a powerful agenda of influence and contributes to the continuing interest in the city-region idea.

The UK context

As elsewhere in Europe, the reconfiguration of state–market relations and civil society provides the context for the critical understanding and practical receptivity to ideas around spatial planning and city-regions. Here, Giddens' (1998) broad map of shifting political economy ideas in the UK illustrates the very different circumstances facing regional identities and metropolitan governance in the social democratic, neo-liberal and Third Way periods. Further, the current interest in city-regionalism in the UK may yet be interpreted as contributing to what has been described as the relatively more recent emergence of 'the congested state' (Sullivan and Skelcher, 2002). In effect, such policy and institutional congestion may be exacerbated by the promotion of the city-region idea – which would prevail over and above established institutional arrangements, political jurisdictions and government boundaries. In the absence of clear leadership and guidance as to how city-regions may operate in practice within the confines of existing organisational parameters, city-regions may serve to dilute existing land use planning practice and arrangements for economic development. Research evidence has shown that attempts to create territorial rationality through the city-region idea have often involved a degree of conflict (Jones et al., 2004). In direct contrast, however, it has been argued that the economic competitiveness of cities and regions can be enhanced by collaborative and associative forms of governance (Leibovitz, 2003). It would appear, therefore, that the parallel imperatives in train are likely to lead to confusion around the implementation of city-regions. Even this appears to be articulated in an instrumental way that obscures the cultural, historical and identity characteristics associated with Geddes' (1904) distrust of parochialism.

Clearly, then, the term city-region may be defined differently and comprises descriptive and prescriptive elements that reflect the different contexts in which the city-region idea is intended to operate. How is the nascent state of Scotland interpreting and socially constructing its city-regions – from a spatial, institutional and local perspective?

City-regionalism in practice: the case of Scotland

In Scotland, it has been asserted that there has been an explicit turn to 'spatially sensitive policy-making. Devolution has provided an opportunity to put Scotland at the forefront of modern, integrated approaches to territorial management within the UK' (Scottish Executive, 2002, p.4). In particular, it is held that the relationship between its ecological and economic bases, its democratic politics, and the nature of its civil society is shifting (Scandrett, 2003), thereby precipitating a need for appropriate governance arrangements and organisational frameworks for social and economic activities. Moreover, devolution has also provided a new stage for a clear articulation of 'Scottishness' consistent with its new standing as a putative nation state (McCrone, 1994; cf. Allmendinger in this volume). Subsequent political science research has shown the continuing importance of this sense of identity (Bond and Rosie, 2002), and this insight provides a critical prism with respect to understanding place and space in Scotland.

Context, however, is all-important. In terms of size, density and function, the morphology of Scotland's urban centres does not rest on a single metropolitan centre, but rather a system of what may be described as regional centres (Rodger, 1996). When further contextualised with reference to historical change in industrial, corporate and economic activities, this land poses a particular cultural setting of diversity and difference. What observations may be made about the implicit or organic approach to city-regional planning in the light of the contemporary emphasis on territorial management and spatial planning to inform local policy and intervention?

The explicit attempt at territorial management and spatial planning finds its current expression in the National Planning Framework (Scottish Executive, 2004a). This is intended to assert a

> broad recognition of the need for a more nuanced understanding of the geographies and scales at which policies are played out on the ground in our neighbourhoods, cities and nation state. In developing strategies, in allocating resources and in delivering services, the [Scottish] Executive needs to have both a clear sense of how geography influences the functioning of our economy, society and environment, and knowledge of where [Scottish] Executive policies will impact.
>
> (Scottish Executive, 2002: 4)

Scotland is not alone in this respect as other devolved territories are exploring such approaches to territorial management and spatial planning, as in Wales (Harris *et al.*, 2002; Harris and Hooper, 2004; and this volume). As Alden (2003) demonstrates, however, the urban settlement pattern in south east Wales is dominated by Cardiff within a broader urban metropolis. The same concept is being introduced in very different urban and cultural contexts, and seeks to embrace the construction of city-regions in broader spatial frameworks.

In effect, the Scottish Executive is seeking in a pragmatic way to capture and craft the city-region zeitgeist to meet its specific economic and social objectives, whilst also acknowledging that these must play out differentially. This must be understood, however, not only within the prevailing devolved governance arrangements in the UK, but also in the context of an enlarging Europe, an awareness of changing global relations, and within the influence of spatial planning agendas, such as those associated with the European Spatial Development Perspective. The status of the National Planning Framework is non-statutory, yet it seeks to provide the potential basis for competing in the potential allocation of future European Union funding streams. *Plus ça change, plus c'est la même chose?*

It is important to note a parallel policy convergence leading to the idea of a city-region approach in Scotland. This itself reflects a broader recasting of the urban policy agenda to promote the intended renaissance. In practice, however, urban policy change has been put into place in a 'hesitant and fragmented way' (Healey, 2004: 259). Further, urban policy has failed to address the tensions between economic competitiveness, environmental conservation and social justice in urban areas (Owens and Cowell, 2002). Yet, within this confused agenda, there is the intention of taking and developing a more holistic view of cities and city-regions as the prime engines of the nation's economic renaissance. This is captured in an alternative view of cities 'as locales imagined as complex webs of relations, traversing many social worlds and spatial scales, intertwining to promote synergy and a capacity to coexist in shared spaces' (Healey, 2004: 170). This viewpoint is supported by Meijers' (2005) argument of the importance of urban economic networks and complementarity within city-regional contexts. Indeed, urban policy is increasingly set within regional arrangements for institutional innovation and development (Ward and Jonas, 2004). Such exhortations for synergy may prove more elusive in practice.

Indeed, in Scotland, for example (Scottish Executive, 2004a: 126), in reviewing the changing character of urban policy concluded, 'it would be premature to say the [Scottish] Executive has an urban policy, or has even fully recognised the distinctive challenges and opportunities of cities'. Here, urban policy is being recast within the broader national strategic framework for city-regions. Bearing in mind these observations, it is clear that processes of city-regionalism have clearly been confined to a specific urban context and to a particular scale. Can we learn from earlier thinking and practical experiences of city-regionalism in Scotland?

City-regions: experiential learning?

In 1904, Patrick Geddes' Dunfermline Report explicitly asserted the case for the reconfiguration of the town to be achieved in the context of its wider territorial region. The following quotation, although lengthy, captures the spirit of the time and offers important insights into the perception and understanding of the then prevailing organic relations that existed between city and hinterland. This line of reasoning is clearly of importance to understanding the intuitive social construction of the city-region idea:

> Our inevitable and permanent provincialism must be accepted as one of the facts of life. Dunfermline may and will enlarge and develop, but it cannot become a Glasgow or Edinburgh. What is the vital element which must complement our provincialism? In a single word, it is regionalism – an idea and movement which is already producing in other countries great and valuable effects. It begins by recognising that while centralisation to the great capitals was inevitable, and in some measure permanent, this is no longer so completely necessary as when they practically alone possessed a monopoly of the resources of justice and of administration, a practical monopoly also of the resources of culture in almost all its higher forms. The increasing complexity of human affairs, with railway, telegraph and business organisation, has enabled the big cities to increase and retain their control; yet their continued advance is also rendering decentralisation, with local government of all kinds, increasingly possible. Similarly for culture institutions; the development of the local press has long been in progress; the history of the city library movement is in no small measure identified with that of this very town; while the adequate institution among us of other forms of higher culture is just what has been discussed in the preceding pages. We see, then, that the small city is thus in some measure escaping from the exclusive intellectual domination of the greater ones, and is tending to redevelop, not, indeed, independence, but culture individuality.
>
> (Geddes, 1904: 216)

Notwithstanding its very particular moment in time, a number of points chime with the present intellectual debates about city-regions. First, Geddes draws our attention to the existence of 'new regionalism' – as evidenced by a recognition of an increasing complexity of social relationships in terms of mobility, communications and business operations in the natural hinterland of Dunfermline. Second, the decentralisation of activities away from Edinburgh hints at what may be considered an early expression of 'new localism'. Third, and arguably most importantly, Geddes challenged the tendency to 'provincialism', even within what today may be considered the natural hinterland of Edinburgh. This brings together a number of issues around the instrumental dimensions of territory and space, with the interpretive aspects of culture and identity. Following Morley and Robins (1995), for example, the case of Dunfermline illustrates how a sense of identity is effectively constructed in relation to Glasgow and Edinburgh. It is in this context of a sense of 'otherness' that Geddes forms his prescription for the development of Dunfermline within its natural city-region. These ideas relating to the city-region template were given more sophisticated expression in Geddes' (1968) seminal work on 'cities in evolution'.

Subsequently, the development of planning practice to deal with specific metropolitan agendas demonstrated a sensitivity to adopting a city-regional canvas (Lloyd and Edgar, 1998). Glasgow is a case in point where attempts to deal with the reconstruction of the city were articulated through an explicit emphasis on its identity and relationship with its natural economic region. The exogenous structural forces for change associated with economic expansion and industrial contraction overwhelmed the city itself and its natural economic region (Rodger, 1996; Lever, 1991). Moreover, there was a very much stronger and sharper delineation of political and institutional boundaries. Metropolitan planning practice in the Glasgow city-region (and its articulation successively as West Central Scotland, Clyde Valley, Clydeside, Strathclyde Region and, today, the joint structure planning territory of Glasgow and the Clyde Valley) has been described as 'an unique experiment in regional governance; the only metropolitan regional council ever introduced in the UK and with an exceptional range of services by which to support its strategies' (Wannop, 1995: 113).

Significantly, these bottom-up attempts to capture the spirit and purpose of city-regionalism were not formalised through defined administrative arrangements (Turok, 2004b), and involved an implicit reliance on partnership working and what has been described as associational economic behaviour (Cooke and Morgan, 1998). This intuitive grasp of the importance of the metropolitan area in its city-region context to mobilise its restructuring and redevelopment may be considered prescient in the light of subsequent debates about the relationship between associative forms of governance and the economic competitiveness of cities and regions (Leibovitz, 2003). Yet, the creation of 12 regional authorities in Scotland in the mid 1970s, and which provided a strategic context to the principal cities within their broadly defined administrative hinterlands only survived for some twenty years. Urban policy, influenced by neo-liberal ideas, led to the dismantling of this regional apparatus in favour of the current fragmented unitary arrangements. It is against this latter complex tapestry of jurisdictions and boundaries, with associated financial and funding relations, and potentially parochial politics, that the city-region idea is being promoted in Scotland.

Elsewhere, an emphasis on national economic priorities has also revealed a sensitivity to putting into effect appropriate and effective territorial management around the cities (Scottish Office, 1966). Interestingly, this economic policy agenda sought to divert public and private sector investment away from Edinburgh and into its natural economic hinterlands. This informed an increasing awareness of the need for a national spatial planning framework that could provide the context to city-regional development in Scotland. Such an outcome emerged from the explicit and clearly asserted national and strategic economic imperatives. In 1970, the Select Committee on Land Resource Use in Scotland studied the anticipated social, economic and environmental consequences for the management of land resources in Scotland in an emerging and slowly recognisable global context (HMSO, 1972). Its deliberations drew particular attention to the ongoing processes of corporate and industrial restructuring, the implications of the offshore oil and gas industry, and the general slowing up of the economy with its attendant unemployment. In conclusion, the Select Committee recommended:

> There is a need to prepare an indicative plan for Scotland on a national scale which will show how it is intended to utilise the land for urban, industrial and recreational purposes. To prepare such a policy plan it will be necessary to take into account the views of planning authorities, industrialists, trade unions and many other interested parties. The structure plans of the new regional planning authorities must conform to the national indicative plan. This is an essential step if Scotland is to compete for industry on favourable terms with other European countries. Developers must be encouraged by simplifying and shortening planning procedures, and as a first step a forward looking national plan is required, setting out what areas can be designated for future use. More detailed plans would then be drawn up by regional and district authorities. Where sites have been identified as suitable for particular industrial purposes inquiries should commence prior to planning applications being received so that prospective developers could be told for certain which sites are available and in accord with the strategy for Scotland. They could then plan a programme of development knowing that one decision did not stand upon the result of a planning inquiry yet to come. Such an arrangement would not in any way prevent the local planning authority from imposing proper planning conditions to achieve the necessary environmental and pollution standards.
>
> (HMSO, 1972: 13)

The National Planning Framework (Scottish Executive, 2004a) can therefore trace its intellectual lineage from these intuitive and reasoned debates around the appropriateness of territorial management in part through city-regionalism. Further, it draws on the deliberations and diagnostic of the Cities Review (Scottish Executive, 2002) that advocated the idea of city-regions as the principal drivers of economic growth and development in Scotland. As a consequence, the nascent idea of four city-regions in Scotland is firmly grounded within the contemporary construction of national space. Yet, as Turok (2004a) demonstrates, this also takes place within a much more specific and focused trajectory of an agenda for urban policy development.

Translating the city-regional idea into effect in Scotland

The 'event' of devolution in the UK has resulted in a particular concern in Scotland with facilitating relatively more participative processes for ensuring that the intentions of the devolution project are fully realised. There is evidence of a tangled mesh of interests, opinions and power structures jockeying to influence the shape of the nation, and for reasserting Scottishness at a 'regional' level (Lloyd and Peel, 2003). Here, the advocacy of city-regions moved centre-stage. The initial Cities Review (Scottish Executive, 2002) had originally focused on the four main cities. The City Growth Fund was created to provide funding for

Scotland's cities to execute their city visions (Peel and Lloyd, 2005). On the one hand, the urban hierarchy in Scotland was enhanced by Stirling and Inverness gaining the city status, on the other, this adds another twist in the relations between cities, city-regions, regions and nation-regions in the UK.

The case for city-regions in Scotland has clearly been asserted. Now attention has turned to putting the city-regions into effect. This agenda appears to take two forms, and both approaches could trace their intellectual justification from the European emphasis on city-regions within particular planning and regulatory traditions. First, there is the relatively technical, instrumental approach associated with the physical delineation of defining the territories of Scotland's possible city-regions. This would imply an interpretation of city-regions that take a monocentric form that could further be associated with place competition. This line of reasoning could be supported by the nature of the city visioning process that encouraged individual assertions of growth trajectories. The city visions were devised within the established parameters of local government jurisdictions, boundaries and joint practices (Peel and Lloyd, 2005). Second, there is a debate around the specific relationship between Edinburgh and Glasgow in terms of realising economies of scale, establishing demographic thresholds, emphasising community planning, and promoting the notion of a polycentric urban region in the central belt (Bailey and Turok, 2001). The challenge of translating the cityregional idea into effect in Scotland is being addressed by a range of players, including central government, local government, the economic development and enterprise bodies, think tanks and non-governmental bodies. As a consequence, there is a frenzy of activity around the idea of the city-region, and competing ideas for the implementation of the concept rival for ascendancy.

City-regionalism: lines on the map?

The City-region Boundary Study (Derek Halden Consultancy, 2002) set out to identify the broad areas of influence of the four principal cities – Aberdeen, Dundee, Edinburgh and Glasgow. The city-region is here translated into a geographical understanding of the sphere of influence as measured by statistical evidence relating to housing market areas, transport links, travel to work, and retail catchment areas. Within this physically deterministic context, the study identified those local authorities that could then participate in joint working for strategic planning agendas within the delineated city-region. Indicative composite catchments were identified at a relatively crude level – essentially local authority areas. This does not entirely clarify the spheres of influence as the approach does not permit a sufficiently sensitive disaggregation of the economic and social relations associated with city-regional working. Thus, for example, the Edinburgh city-region was defined as including Fife, which was also included in the proposed Dundee city-region. Yet, Fife, for example, comprises a complex of different labour and housing markets associated with both Edinburgh and Dundee. This approach represents an attempt to define city-regions within the parameters of existing local authority jurisdictions and obscures the functional, cultural and identity considerations of city-region economic relationships.

A similar approach to informing the possible configuration of city-regions is provided in an analysis of commuting patterns across Scotland drawn from the 2001 Census Travel to Work Statistics. This research evidence shows a 'hub and spoke' pattern of commuting in the four city-regions in Scotland, with Glasgow being the most dependent in its surrounding areas (Experían, 2004). The study suggests that in-commuting to the city-regions has become more important in the last decade, suggesting again a physical dimension to the city-region idea delineated by contiguous local authority areas. Parallel arguments for a city-regional approach have also been advocated with respect to specific activities, most notably transport and housing. Each inform the case for a city-regional approach, but do not necessarily contribute to its articulation in practice. Thus, Begg and Docherty (2002) have argued that a city-regionalist structure is the most appropriate means of providing for an effective transport network within the cities and between them. In the specific context of Edinburgh, MacKay (2004) has raised serious questions about the ability of conventional land use planning to deliver the required housing needs for the capital. This clearly suggests the need to promote a more functional view of housing sectors in the Edinburgh city-region.

City-regionalism: spatial relationships?

Reflecting the importance of localities and local interests within the internal organisation and operating frameworks of regional economies, polycentric arrangements are perceived as an effective way of establishing a common regional agenda with better integration between local interests for a broader regional purpose (Herrschel and Newman, 2002). Here, polycentric regional strategies are held to offer the potential to realise broader spatial economies of scale and, through co-operative behaviour, facilitate greater regional competitiveness. City-regions can likewise provide the conditions to secure competitive advantages, either through the benefits of scale and diversity arising from the physical proximity and concentration of economic activity, or the synergies arising from inter-organisational co-operation (Turok, 2004a).

The focus of attention has also turned on the potential of the central Scotland urban axis performing as a polycentric urban region (Bailey and Turok, 2001). This is based on the perceived economic, physical and connectivity relationships between Edinburgh and Glasgow (Bailey et al., 1999). Reflecting intuitive arguments that an Edinburgh–Glasgow economic relationship, with associated infrastructural and transportation links, would perform as a city cluster and generate further economies of scale (Glaeser, 2004), there is a case for establishing a customised strategic spatial framework to inform key investment decisions in the two cities and to promote closer collaboration in their governance arrangements. Notwithstanding the geographical proximity of the two urban areas, the dynamism of intercity movement, and the evident synergies in the industrial, commercial and retailing sectors, it is argued that a polycentric urban region is not necessarily the model to be adopted because it may be too broadly specified to guide the types of public and private decision making required to realise the potential of the city-regional idea (Bailey and Turok, 2001).

Conclusions: city-regionalism – towards a new mercantilism?

This chapter has highlighted the ambiguous and contested concept of the city-region in contemporary global and glocal circumstances (Deas, 2004). Importantly, Simmonds and Hack (2000) demonstrate the heterogeneity of the city-region idea and form at the global scale. Such a celebration of diversity at the global level could usefully inform the translation of the city-regional idea in Europe. Indeed, Herrschel and Newman (2002) articulate a case for a city-regional agenda within Europe that is sensitive to both monocentric and polycentric spatial forms. Further, as Healey (2004) asserts, the context to city-regions in Europe is based in the resurgence of interest in strategic spatial planning. From this it follows that the social construction of the city-region can be legitimated from a number of perspectives. Clearly, what Lovering (1999) expressed as the 'new regionalist picture of the world' is very relevant for socially constructing the city-region concept in the particular circumstances of a devolved UK, with differential institutional arrangements for governance, economic management, and trajectories of institutional leadership, thickness and capacity. Hence, the city-region in Scotland is being articulated within a national spatial context – here, The Cities Review, city-visions, the Growth Fund, and the National Planning Framework reinforce this particular city-region discourse. In this context, the city-region concept is being deployed in order to capture the perceived advantages of European spatial planning agendas from a clearly articulated economically deterministic perspective:

> Sustainable cities need thriving regions and, in turn, the success of the national economy depends on the economic competitiveness of our city regions. Our city regions do not exist in isolation. They are diverse – geographically, culturally and socially – but, in a global setting, they are not large in scale. The connections between city regions must be exploited and their strengths combined to enable Scotland to compete effectively in a global environment.
>
> (Scottish Executive, 2004b)

The city-region concept also sits comfortably within debates on the role of regional economies as the most appropriate level for efficient market activity, economic management and the design of state–market relations (Ethier, 1998). Yet, successful city-regions in Europe vary considerably in scale. Moreover, very different social and economic divisions inform particular city-region visions. Here, the policy objectives of social justice and economic competitiveness collide head on. Indeed, the concept clearly invites 'bullishness' and concerns for equity, as the following observations from the Scenarios for the Edinburgh City Region illustrate:

> In twenty years time if all goes well the Edinburgh City Region would economically be outperforming Scottish and British levels, and ideally, other European cities of a similar size and type.

> My vision is Edinburgh and Lothian being one of six city-regions with a very specific place in the Scottish economy that is recognised, allowed to get on with things, supported and there is no confusion.

> Will the future regional economy be spread wider or focused on the centre – which argument wins?
>
> (McKiernan, 2004)

Here, it is evident that the exhortation of city-regions is seeking to cultivate a cat-alogue of innovations, inspiration, and utopian visions and is not afraid to make reference to potentially apocalyptic outcomes. Here, power relations are assumed to be neutral. Gaffikin and Morrisey (2001), for example, highlight the complexities of managing city-regional change in conditions where inherited and prevailing socio-cultural traditions are not in harmony, and where the spatial segregation which results from ethno-national conflict can confound an intended city-region dynamic. Further, an attractive characteristic of the city-region idea is that it has a powerful appeal to a wide range of constituencies – environmentalists, politicians, policy makers, multinationals, corporate interests, ad infinitum. Yet, this is a prob-lem – whilst these different interests appear to be able to identify with the city-region ideal, defining the actors and institutional arrangements for carrying the vision through can prove to be much more elusive – and potentially divisive.

Significantly, the city-region approach invokes a duality of place-competition alongside co-operative working in shared spaces and across boundaries. In this context, the trading of ideas about the importance and potential effectiveness of city-regionalism is clearly important; indeed the ambiguity of the term is such that it may be used creatively across a scalar spectrum of world cities, metropolitan areas, and small to medium cities. This may be considered as introducing a new mercantilism of spatiality where ideas about appropriate territorial managements are being traded across different spaces and places and metropolitan contexts. Here, space and place are the new prism for addressing inherited problems. Yet, this new spatial framework directly challenges the established institutional struc-tures, actors, ways of working and geographical boundaries.

Experiential learning and visions from the past should not be neglected in devising city-regional arrangements today. Yet, the current reality of transferring the city-region idea in practice has so far demonstrated a relative poverty of criti-cal reflection. Here, we can draw useful insights from the policy learning literature that cautions against uncritical mimetic behaviour (Rose, 1993). In a world, then, where ideas can flow uncritically, the ease with which the city-region idea is being advocated as a solution to metropolitan territorial development has developed its own momentum. Yet, the city-region idea will of necessity have to be tailored to particular scale, state–market relations, economic and social context.

References

Alden, J. (2003) 'The experience of Cardiff and Wales' in W. Salet, A. Thornley and A. Kreukels (eds) *Metropolitan Governance and Spatial Planning: Comparative Case Studies of European City-regions*, London: Spon Press: 77–90.

Allmendinger, P. (2002) 'The post-positivist landscape of planning theory' in P. Allmendinger and M. Tewdwr-Jones (eds) *Planning Futures: New Directions for Planning Theory*, London: Routledge: 3–18.

Bailey, N., Turok, I., and Docherty, I. (1999) *Edinburgh and Glasgow: Contrasts in Competitiveness and Cohesion*, Glasgow: University of Glasgow Department of Urban Studies.

Bailey, N. and Turok, I. (2001) 'Central Scotland as a polycentric urban region: useful planning concept or chimera?', *Urban Studies*, 38(4): 697–715.

Begg, D. and Docherty, I. (2002) *The Future of Strategic Transport Planning: Rediscovering the City-Regional Approach*, Aberdeen: Robert Gordon University.

Blatter, J. (2004) '"From spaces of place" to "spaces of flows"? Territorial and functional governance in cross-border regions in Europe and North America', *International Journal of Urban and Regional Research*, 28(3): 530–48.

Bond, R. and McCrone, D. (2004) 'The growth of English regionalism? Institutions and identity', *Regional and Federal Studies*, 14(1): 1–25.

Bond, R. and Rosie, M. (2002) 'National identities in post-devolution Scotland', *Scottish Affairs*, 40: 34–53.

Brenner, N. (1999) 'Globalisation as reterritorialisation: the re-scaling of urban governance in the European Union', *Urban Studies*, 36(3): 431–51.

Brenner, N. (2000) *Entrepreneurial cities, glocalising states and the new politics of scale: Rethinking the political geographies of urban governance in Western Europe*, Cambridge: Harvard University Centre for European Studies Working Papers 76A and 76B.

Brenner, N. (2002) 'Decoding the newest "metropolitan regionalism" in the USA: A critical overview', *Cities*, 19(1): 3–21.

Cooke, P. and Morgan, K. (1998) *The Associational Economy*, Oxford: Oxford University Press.

Deas, I. (2004) 'From a new regionalism to an unusual regionalism? Mapping the emergence of non-standard regional configurations in Europe', Paper presented to the Association of American Geographers, 100th Annual Meeting, Philadelphia, March.

Deas, I. and Giordano, B. (2003) 'Regions, city-regions, identity and institution building: contemporary experiences of the scalar turn in Italy and England', *Journal of Urban Affairs*, 25(2): 225–46.

Deas, I. and Ward, K.G. (2000) 'From the "new localism" to the "new regionalism"? The implications of regional development agencies for city-regional relations', *Political Geography*, 19: 273–92.

Derek Halden Consultancy (2002) *City-region Boundary Study*, Edinburgh: Scottish Executive.

Dickinson, R. (1964) *City and Region: A Geographical Interpretation*, London: Routledge and Kegan Paul.

Donald, B. (2002) 'The permeable city: Toronto's spatial shift at the turn of the millennium', *The Professional Geographer*, 54(2): 190–203.

Ethier, W.J. (1998) 'The New Regionalism', *The Economic Journal*, 108(449): 1149–61.

Experían (2004) *Commuting in Scotland*, Edinburgh: Scottish Futures Forum.

Faludi, A. and Waterhout, B. (2002) *The Making of the European Spatial Development Perspective: No Masterplan*, London: Routledge.

Frey, H. (1999) *Designing the City: Towards a More Sustainable Form*, London: Spon Press.

Gaffikin, F. and Morrisey, M. (2001) 'Regional development: an integrated approach?' *Local Economy* 16(1): 63–71.

Geddes, P. (1904) *City Development: A Study of Parks, Gardens, and Culture Institutes. A Report to the Carnegie Dunfermline Trust*, Bourneville: The St George Press.

Geddes, P. (1968) *Cities in Evolution: An Introduction to the Town Planning Movement and to the Study of Civics*, London: Ernest Benn Ltd.

Geddes, R. (1997) 'Metropolis Unbound' *The American Prospect*, 8(35) November 1–December 1.

Giddens, A (1998) *The Third Way*, Cambridge: Polity Press.

Glaeser, E.L. (2004) *Four Challenges for Scotland's Cities*, Glasgow: University of Strathclyde (The Allander Series).

Glasgow City Council (2003) *Metropolitan Glasgow: Our Vision for the Glasgow City-region*, Glasgow: Glasgow City Council.

Hancock, T. (1976) 'The city region and the changing basis of planning' in T. Hancock (ed.) *Growth & Change in the Future City Region*, London: Leonard Hill: 1–14.

Harris, N., Hooper, A. and Bishop, K.D. (2002) 'Constructing the practice of "spatial planning": a national spatial planning framework for Wales', *Environment and Planning C: Government and Policy*, 20: 555–72.

Harris, N. and Hooper, A. (2004) 'Rediscovering the "spatial" in public policy and planning: an examination of the spatial content of sectoral policy documents', *Planning Theory and Practice*, 5(2): 147–69.

Healey, P. (1998) 'The place of 'Europe' in contemporary spatial strategy making', *European Urban and Regional Studies*, 5(2): 139–53.

Healey, P. (2004) 'The treatment of space and place in the new strategic spatial planning in Europe', *International Journal of Urban and Regional Research*, 28(1): 45–67.

Herrschel, T. and Newman, P. (2002) *Governance of Europe's City-regions: Planning, Policy and Politics*, London: Routledge.

HMSO (1972) *Land Resource Use in Scotland, Volume 1: Report and Proceedings*. Select Committee on Scottish Affairs, House of Commons Papers 511-i, Session 1971–2, London: HMSO.

Hudson, R. (2005) *Economic Geographies*, London: Sage.

Jonas, A.E.G. and Ward, K. (2002) 'A world of regionalisms? Towards a US-UK urban and regional policy framework comparison', *Journal of Urban Affairs*, 24(4): 377–401.

Jones, R., Goodwin, M., Jones, M. and Simpson, G. (2004) 'Devolution, state personnel, and the production of new territories of governance in the United Kingdom', *Environment and Planning A*, 36(1): 89–109.

Keating, M. (2001) 'Rethinking the region: culture, institutions and economic development in Catalonia and Galicia', *European Urban and Regional Studies*, 8(3): 217–34.

Knox, P.L. (1995) 'World cities and the organization of global space' in R.J. Johnston, P.J. Taylor and M.J. Watts (eds) *Geographies of Global Change: Remapping the World in the Late Twentieth Century*, Oxford: Blackwell: 232–47.

Leibovitz, J. (2003) 'Institutional barriers to associative city-region governance: the politics of institution-building and economic governance in "Canada's Technology Triangle"', *Urban Studies*, 40(13): 2613–42.

Lever, W.F. (1991) 'Deindustrialisation and the reality of the post industrial city', *Urban Studies*, 28(6): 983–99.

Lloyd, M.G. and Edgar, W.M. (1998) 'A place apart? Metropolitan planning in West Central Scotland', in P.W. Roberts, K. Thomas and G. Williams (eds) *Development Planning in Metropolitan Britain*, London: Jessica Kingsley: 193–210.

Lloyd, M.G. and Peel, D. (2003) 'Shaping national space in Scotland', *Town and Country Planning*, August 72(7): 224–5.

Lovering, J. (1999) 'Theory led by policy: the inadequacies of the 'New Regionalism' (Illustrated from the case of Wales)', *International Journal of Urban and Regional Research*, 23(2): 379–95.

Mackay, D. (2004) *Planning Famine*, Edinburgh: The Policy Institute.

MacLeod, G. (1998) 'In what sense a region? Place hybridity, symbolic shape, and institutional formation in (post-) modern Scotland', *Political Geography*, 17(7): 833–63.

MacLeod, G. and Goodwin, M. (1999) 'Space, scale and state strategy: rethinking urban and regional governance', *Progress in Human Geography*, 23(4): 503–27.

McCrone, D. (1994) *Understanding Scotland: The Sociology of a Stateless Nation*, London: Routledge.

McKiernan, P (2004) Scenarios for the Edinburgh City Region: One-to-one and group workshops detailed interview notes: Workbook, Appendix A, St Andrews: University of St Andrews Scenario Team.

McLoughlin, J.B. (1969) *Urban and Regional Planning: A Systems Approach*, London: Faber & Faber.

Malecki, E.J. (2004) 'Jockeying for position: What it means and why it matters to regional development policy when places compete', *Regional Studies*, 38(9): 1101–20.

Meijers, E. (2005) 'Polycentric urban regions and the quest for synergy: is a network of cities more than the sum of the parts?', *Urban Studies*, 42(4): 765–81.

Meller, H. (1990) *Patrick Geddes: Social Evolutionist and City Planner*, London: Routledge.

Morgan, K. (1997) 'The learning region: institutional innovation and regional renewal', *Regional Studies*, 31: 491–503.

Morgan, K. (2004) 'Sustainable regions: governance, innovation and scale', *European Planning Studies*, 12(6): 871–89.

Morley, D. and Robins, K. (1995) *Spaces of Identity*, London: Routledge.

Owens, S. and Cowell, R. (2002) *Land and Limits: Interpreting Sustainability in the Planning Process*, London: Routledge.

Paasi, A. (2001) 'Europe as a social process and discourse considerations of place, boundaries and identity', *European Urban and Regional Studies*, 8(1): 7–28.

Peel, D and Lloyd, M.G. (2005) 'City-visions: visioning and delivering Scotland's economic future', *Local Economy*, 20(1): 40–52.

Ravetz, J. (2000) *City-region 2020: Integrated Planning for a Sustainable Environment*, London: Earthscan.

Rodger, R. (1996) 'Urbanisation in twentieth century Scotland' in T.M. Devine and R.G. Finlay (eds) *Scotland in the 20th Century*, Edinburgh: Edinburgh University Press: 122–52.

Romppanen, M. and Ujam, F. (2004) 'Monitoring the physical structure of the Helsinki Metropolitan Area', *Journal of Urban Design*, 9(3): 367–77.

Rose, R. (1993) *Lesson Drawing in Public Policy: A Guide to Learning across Time and Space*, New Jersey: Chatham House Publishers.

Salet, W., Thornley, A. and Kreukels, A. (eds) (2003) *Metropolitan Governance and Spatial Planning: Comparative Case Studies of European City-regions*, London: Spon Press.

Scandrett, E. (2003) 'Towards a new agenda for a sustainable economy in Scotland' in E. Scandrett (ed.) *Scotlands of the Future: Sustainability in a Small Nation*, Edinburgh: Luath Press Ltd.

Scott, A. J. (2001) 'Globalization and the rise of city-regions', *European Planning Studies*, 9(7): 813–26.

Scott, A.J., Agnew, J., Soja, E.W. and Storper, M. (1999) 'Global city-regions', Conference theme paper, available at: www.sppsr.ucla.edu/globalcityregions/ (accessed 09.10.2004).

Scottish Executive (2002) *Review of Scotland's Cities – The Analysis*, Edinburgh: Scottish Executive.

Scottish Executive (2004a) *National Planning Framework*, Edinburgh: Scottish Executive.

Scottish Executive (2004b) *A Smart, Successful Scotland*, Edinburgh: Scottish Executive.

Scottish Office (1966) *The Scottish Economy 1965–1970: A Plan for Expansion*, Edinburgh: HMSO. Cmnd. 2864.

Shaw, D. and Sykes, O. (2004) 'The Concept of Polycentricity in European Spatial Planning: Reflections on its Interpretation and Application in the Practice of Spatial Planning', *International Planning Studies*, 9(4): 283–306.

Shen, J. (2004) 'Cross-border urban governance in Hong Kong: The role of state in a globalizing city-region', *The Professional Geographer*, 56(4): 530–43.

Simmonds, R. and Hack, G. (eds) (2000) *Global City-regions: Their Emerging Forms*, London: Spon Press.

Soja, E.W. (1997) 'Six discourses on the post metropolis', in S. Westwood and J. Williams (eds) *Imagining Cities*, London: Routledge.

Sullivan, H. and Skelcher, C. (2002) *Working Across Boundaries: Collaboration in Public Services*, London: Palgrave Macmillan.

Taylor, P.J. (2000) 'World cities and territorial states under conditions of contemporary globalization', *Political Geography*, 19: 5–32.

Tewdwr-Jones, M. and McNeill, D. (2000) 'The politics of city-region planning and governance: reconciling the national, regional and urban in the competing voices of institutional restructuring', *European Urban and Regional Studies*, 7(2): 119–34.

Tewdwr-Jones, M. and Mourato, J. (2005) 'Territorial cohesion, economic growth and the desire for European "balanced competitiveness"', *Town Planning Review*, 76(1): 69–80.

Thornley, A. (2003) 'London: Institutional turbulence but enduring nation-state control' in W. Salet, A. Thornley and A. Kreukels (eds) *Metropolitan Governance and Spatial Planning: Comparative Case Studies of European City-regions*, London: Spon Press.

Turok, I. (2004a) 'Scottish urban policy: continuity, change and uncertainty post-devolution' in C. Johnstone and M. Whitehead (eds) *New Horizons in British Urban Policy. Perspectives on New Labour's Urban Renaissance*, Aldershot: Ashgate.

Turok, I. (2004b) 'Cities, regions, and competitiveness', *Regional Studies*, 38(9): 1069–83.

Vigar, G., Healey, P., Hull, A. and Davoudi, S. (2002) *Planning, Governance and Spatial Strategy in Britain*, Basingstoke: Macmillan.

Wannop, U. (1995) *The Regional Imperative: Regional Planning and Governance in Britain, Europe and the United States*, London: Jessica Kingsley Publishers.

Ward, K. and Jonas, A.E.G. (2004) 'Competitive city-regionalism as a politics of space: a critical reinterpretation of the new regionalism', *Environment and Planning A*, 36(12): 2119–39.

Wilks-Heeg, S., Perry, B. and Harding, A. (2003) 'Metropolitan regions in the face of the European dimensions: regimes, re-scaling or repositioning?' in W. Salet, A. Thornley and A. Kreukels (eds) *Metropolitan Governance and Spatial Planning: Comparative Case Studies of European City-regions*, London, Spon Press.

19 Global localism

Interpreting and implementing new localism in the UK

Janice Morphet

Introduction

The notion of 'new localism' has been reverberating in England since 2002 and, although it is seen as primarily an English issue, it is located in both European and US debates on the future shape, size and span of governance at national level. In the international context, the principles of 'smaller government' generated by Osborne and Gaebler (1992) continues to generate contested space, leading to a current debate about both the most effective size of nations from an economic perspective (Alesina and Spolaore 2003) to a further developed form of 'Reinventing Government' from Osborne and Hutchinson (2004), which relates efficiency, effectiveness and performance to size and social capital. In the EU, the concerns relate to the more rigorous application of subsidiarity which in part is related to a churn in the recognised pan EU layers of governance which is yet to be resolved but is strengthening at all sub-national levels, i.e. mega-regional, regional, sub-regional, local and neighbourhood. EU member state sub-national government structures may look very similar at the end of this process and are likely to be reinforced by other emergent EU measures on public administration data and information sharing at all levels.

In England, the emergence of new localism has started a considerable debate on appropriate levels of governance (Morphet, 2004; Morphet, forthcoming). Fuelled by low voter turnout figures and loss of the democratic mandate, an examination of new localism has been associated with the continuing growth of central government in scale and power over the last thirty years, being seen as a potential antidote. Since 2000, there has been a reversal of this political position in the advocacy of more local working which has been manifest in a variety of formats including the ascendancy of local authorities, neighbourhood renewal areas, Business Improvement Districts, the generation of new parish councils and the emergence of small area boards to manage health and education. Some of this debate has focussed on genuine attempts to introduce new governance formats whilst others may be seen as attempts to reinvent the concept to support the status quo. This has been reinforced through a ten year review of local government (ODPM, 2004a) and the promotion of initiatives for devolved decision making (HMT, 2004a,b). A particular form of new localism that is gaining ground, particularly in England, is the establishment of Local Public Service Boards that are emerging from Local Strategic Partnerships (LSPs), a form of voluntary multi-agency local partnerships established in 2000. These new 'LSPs with teeth' and Local Public Service Boards are being promoted by Government with the specific support of the London Borough of Hammersmith and Fulham, which is to the west of London's core. The Government announced the next iteration of this approach through the publication of a prospectus on the development of Local Area Agreements (ODPM, 2004b). This chapter will establish these emergent trends for new localism within a changing context of the relationship between territory and governance that is emerging in England.

The international and European context on new localism

The notion of 'new localism' is one that has been at the heart of a growing debate in central and local government since 2002. The origins of 'new localism' derive from pressures on government to consider whether it should divest more central power to the local level. The first set of pressures emanates from the US and can be seen to be economic in the underlying drive. In studies such as that by Alesina and Spolaore (2003), which examine the relationship of the size of a country to its economic success, the premise has been that since 1945, a great number of smaller states have been created which, given their size, are becoming more homogenous in their population characteristics. This homogeneity is leading to a more culturally concentrated market that becomes easier to serve in economic terms. This is a wider version of Putnam's notion of social capital (2000), where a smaller state creates the community unity which he indicates leads to happier, more socially engaged and stable communities. Alesina and Spolaore also indicate that smaller states on the whole have smaller government structures, once their size is considered, which are also more economic and less onerous on their citizens in terms of cost, and also in the more responsive ways in which they can act. Thus an economically efficient future could be achieved through the adoption of smaller states as an organising principle. These states are autonomous in terms of government but join with other states for those activities that are most appropriate. This also supports an argument for regionalism within states. In effect it also generates a reason for the existence of the larger world trading blocks such as the EU, at the same time as supporting the *raison d'être* of the United States, a principle also supported by Bobbitt (2002).

This interest in the size of government and the 'weight' it places on economic efficiency has also been a consistent theme in the United States since the Gore initiative to 'reinvent government' (Osborne and Gaebler, 1992) as something smaller, more transparent and customer led. This approach also takes the view that 'central' government has become too large, and is now hampering the economic health of the state. This notion has since developed further into a greater interest in the relationship between the scale of government and its performance. Better performance management could be a necessary prerequisite for 'letting go' and downsizing. On the other hand, performance management may displace other more traditional central government tasks such as policy making but not lead to smaller government, just the same number of people undertaking different tasks. This debate, on the relationship between performance and scale, has been taken forward by Osborne, but this time with Hutchinson. In *The Price of Government* (2004), Osborne and Hutchinson argue that government activity should be more focussed on key priorities and results rather than maintenance. In this way, energies can be focussed on what is important and costs can be reduced. Their views are fuelled by the scale of the fiscal crisis in local and central government in the

United States. However, their ten principles are influencing other governments to consider the size and efficiency of their own central administrations and may have the same degree of influence as 'Reinventing Government' around the world. This approach is generating a radical review of the size of government everywhere and it is useful to consider these principles to be contributing drivers to what is being translated as the movement to 'new localism in England. Osborne and Hutchinson set out these principles as ten operational activities to create efficient government:

1 Strategic reviews – divesting to invest by combing through programmes and identifying redundancy
2 Consolidation – rather than concentrating on merging organisations, which take time and reduce delivery effectiveness through the confusion they cause, rather merge budgets and put them in the hands of 'steering' organisations which purchase from 'rowing' or provider organisations which may be from any sector
3 Rightsizing – understanding that the right size is critical for the success of some activities, but this does not mean 'one size fits all'
4 Buying services competitively – making public institutions compete with other sectors can save money
5 Rewarding performance not good intentions – in this way those who improve their outcomes receive higher rewards
6 Smarter customer service – putting customers in the driving seat introducing more choice and more appropriate means of delivery to suit the service and the customer
7 Don't buy mistrust – eliminate it – win voluntary compliance rather than generate rules which people will attempt to evade or cheat
8 Using flexibility to get accountability – encourage more freedoms and flexibilities for those who accept more performance based structures
9 Making administrative systems allies not enemies – organisations are prisoners of their internal systems and these have to be modernised and streamlined which can generate major savings; and
10 Smarter work processes: tools from industry – organisations need to change the way in which they work using a variety of tools including Total Quality Management (TQM), Business Process Reengineering (BPR) and Team Workouts or small problem solving groups brought together for fixed periods.

(Osborne and Hutchinson, 2004: 13–17)

These transatlantic economic drivers for smaller central government have an international resonance and are beginning to lead to a more fundamental questioning of the nature of the 'best' arrangements for effective governance and economic growth. There is also a further element to the economic arguments for smaller governance for smaller places coming from Europe, where there is a further economic imperative through the demographic problem facing all EU states, i.e. of a falling population of working age needing to support a higher proportion of dependent, retired adults. If this demographic situation continues without any further action, it could generate higher labour costs to match a scarce supply, act as a disincentive for people to enter higher education or encourage greater inward migration. A number of measures are being taken to address this concern including the encouragement and the subsequent enforced extension of working life so that a greater proportion of the population remains in the labour market. Another area to consider is greater productivity to generate labour shake out, and some EU member states such as France and Germany are commencing this approach. In England, much of the Labour downsizing has occurred in the private sector and in formerly owned public sector companies. However, the move to 'smaller' government could lead to both lower initial recruitment and a reshaping of the public sector labour market over time to make people available to enter other sectors and to reduce the risk of the economy overheating. This approach is mirrored over the whole of Europe, although the EU also has the advantage of the increased labour market available from the new member states to support a wider ageing population. These then create the economic principles for the pursuance of new localism.

There are also other drivers for new localism within the EU that are political in character. Within the EU, concerns about the low levels of electoral turnout in European Parliamentary elections and the continuing debates about the nature of accountability, its 'democratic deficit', and of its indirect political structures have encouraged a greater interest in focussing on the relationship between the EU and sub-national tiers of government in particular regions and neighbourhoods, through the means of specific programmes which fund or support at these levels. The application of the principle of subsidiarity in the EU has more generally been regarded as a means of ensuring that the Council of Ministers and Commission do not seek to remove powers from member state governments. At the same time, there has been an increasing drive to ensure that the principle is applied at sub-national level (Loughlin, 2001a). Since 1994, this has had a specific voice in the EU since the establishment of the Committee of the Regions (Morphet, 1994). Loughlin (2004b) also argues that the emergence of a difference relationship between the state and its sub-national levels, through policies such as privatisation, has also helped to build a more competitive approach between regions needing support both within states and within the EU. Within the EU, it is also clear regional economic performance is seen to be closely linked to regional identity, bringing the political and economic drivers more closely together. It is also now understood that the performance of any state in Europe can no longer be driven from the centre of each state but has to be led from within the region and it is through this growth from the bottom up that the state and European economy is expected to flourish (ODPM *et al.*, 2004). The application of the principle of subsidiarity is also having other effects much as the review of devolved decision making that is taking place in the UK (HMT, 2004a,b).

In all, locally based governance initiatives are seen to be desirable across the world for a number of reasons including:

- wider ownership of decision making
- increasing interest in voting thus improving the democratic accountability of decision making
- efficiency by reducing layers of bureaucracy
- effectiveness in delivery – smaller areas for delivery can provide more targeted approaches
- sustainability – local provisioning can reduce environmental and economic costs.

New localism in the UK: an example of the fugue of devolved policy delivery

Within the United Kingdom, the twin economic and political pressures have lead firstly to the devolution of central powers to Scotland, Wales and Northern Ireland in 1999, with proposals for devolution within England following in 2002 (Cabinet Office/DETR, 2002). The extent to which separate policy making for England emerges as an identified consequence of UK devolution has yet to be debated. The levels of devolution within each part of the UK vary, with Scotland having the greatest degree of autonomy. The implementation of devolution has allowed different approaches to emerge. Taking the example of spatial planning it is possible to see these variations but also the extent to which an underlying trend for similarity in delivery is occurring. In Northern Ireland, the Regional Strategy commenced work in 1996, before the subsequent initiatives to restore the role of Stormont in governance (see Ellis and Neill's contribution in this volume). In Wales, the publication of *'People, Places, Futures'* – *the Wales Spatial Plan* in 2003 was the occasion for the Minister, Sue Essex, to state that 'Devolution has provided us with an opportunity to do things differently in Wales' (Welsh Assembly Government 2003:5). In the Wales Spatial Plan, there is a strong focus on the ability to bring geography into wider policy making and bring the spatial policies together with the Welsh Assembly Government's economic development strategy, 'A Winning Wales' (see Harris and Hooper's contribution in this volume). An analysis of the drivers of change that the spatial plan for Wales needs to address is also included and is significant. This analysis reflects European factors such as enlargement, constitutional change and EU structural funds as being key issues together with the competition that was expected following the establishment of English Regional Assemblies. The Wales Spatial Plan also includes the key issues to be dealt with in relation to neighbouring regions, i.e. the South West, the West Midlands, the North West and Ireland, but these issues are seen to be primarily at the UK level – that is in terms of overarching transport infrastructure. Thus, in the case of Wales, integration is seen to be necessary only at the level defined in the devolution settlement. The remaining relationships are seen to be essentially competitive and beyond the borders of the UK.

In Scotland, the National Planning Framework is in preparation following a series of participative seminars in 2003. The National Planning Framework is seen as a spatial plan for Scotland, covering a variety of issues starting with the economy and sustainability of Scotland's key cities, European links, transport infrastructure and Scotland's natural resources (Scottish Executive, 2004). However, within Scotland more emphasis has been given to the Review of Strategic Planning (Scottish Executive, 2003a) at the local level and the role of participation within these processes with the publication of 'Your Place, Your Plan', a White Paper on public involvement in planning (Scottish Executive, 2003b).

Thus, differences have been accommodated within parallel initiatives and timescales but without this necessarily being formally acknowledged as part of the same approach (Allmendinger, Morphet and Tewdwr-Jones, 2005). The implementation of devolution in 1999 has led to the emergence of differing planning priorities in Scotland and Wales that continue to change. However, it is difficult to estimate how far these different priorities in policy development will lead to fundamentally different planning systems in the longer term. What may appear is a difference in ordering and language but not necessarily a difference in function and role. This is already apparent in other areas of policy making since devolution such as community plans and strategies. There is also the overarching role of the European Spatial Development Perspective (ESDP) (European Commission, 1999). A longer term issue for consideration will be the extent to which difference is engineered to obtain a more competitive edge within the UK and with Europe, which is already a preoccupation of the spatial plans in preparation. Devolution thus is a major contributor to the 'new localism' agenda. It provides more immediate control over policy and delivery, it allows more local homogeneity of approach within cultural norms (thus meeting one of Alesina and Spoloare's conditions), it encourages some competition within the state enabling wider growth objectives and it creates more local accountability and transparency to encourage more political engagement and ownership. There are some who might argue that more integration is required at the UK level, not least for national issues which have already been defined or that there is likely to be some overall loss to the UK as a result of following these emerging trends to more localised decision making.

New localism in England: destination agreed – routes various

The implications of a more devolved governance structure for England have yet to be more fully considered. The immediate responses have been the generation of the debate about the implementation of regional government and numerous sub-regional and local experiments in different forms of neighbourhood governance. The approach to regional government within England has been implemented through variable geometry as in other countries such as Spain. Three regions were offered the preliminary opportunity to take this forward, with one, the North East, ultimately going to a referendum on 4 November 2004. The result of the referendum was to decline the proposed offer of regional government that, in turn, has led to a rethinking of the regional governance approach. This is now more likely to be similar across all English regions, with the exception of London, where a directly elected mayor has executive powers over a number of functional areas. This may become a model for the future, as city–regions become more prominent, using the regional mayor model as in Spain.

All English regions have a Regional Assembly and greater devolved decision making within the region through the Chapter 2 proposals of the White Paper (Cabinet Office/DETR, 2002). This multi-speed approach is also set within a set of mega-regional initiatives within England through the Northern Way, a combination of the North East, North West, and Yorkshire and Humber regions (The Northern Way Steering Group, 2004), and followed by the Midlands and the South West, both of which are at earlier stages of development. It is also noticeable that within all the regions of England, the Regional Development Agencies established in 1999 have also been actively encouraging the adoption of pragmatic sub-regional structures as a means of promoting initiatives. These sub-regions are apparent now across most of England although they have not had any formal governance structures that are common to each sub-region. This opportunity to relate the future constituencies of elected members of the Regional Assemblies to these sub-regional areas thus bringing them into a common democratic framework has now gone but may emerge in other ways such as European Parliamentary constituencies or sub-regional mayors, again as in Spain (Morphet, 2005; Morphet, forthcoming).

At regional level, the internal reforms within the English state continue as existing democratic governance structures and their relationship to funding is continually reviewed. Within regions, recently established Regional Housing Boards and Regional Planning Boards will be merged. Regional finance is now being brought together into Regional Funding Allocations with single pots that give increasing flexibility for regional programmes to be developed and implemented. The Regional Assemblies continue to have representatives from all constituent councils, although these are not seen to have a significant role at present. However, Robinson (2005) argues that this role could be enhanced through the constitutions of each local authority including their councillor representative of the regional Assembly as a member of their local cabinet. At the same time, the powers of the regions, to negotiate Local Area Agreements with individual local authorities, are being enhanced, creating some specific hierarchical and networked relationships for delivery (Bevir and Rhodes, 2003).

Alongside these changes in the relationship between governance and scale of territory, there have also been moves to consider the effectiveness of 'smaller' government that can be created by internal reform of government (Blair, 2004). This has been manifest by a variety of policy papers that include proposals to move central government departments to the regions in part as a means to encourage regional growth and in line with similar developments in Scotland, Ireland and France, but also to impose more fundamental restructuring on the scale of the centre restructuring (Lyons, 2004). Relocating can provide an opportunity to downsize. In other initiatives, it is proposed to join together central government departments such as Inland Revenue and Customs and Excise or to review the delivery structure of government as a whole (HMT, 2004a). These approaches, which are accompanied by a move away from a more generalised to a more professionally qualified civil service, lead to the expectation of a much smaller governance machinery. Taken with the impact of devolution, they reflect a series of actions that are changing the structure of governance. These changes are also being reflected more broadly on the way in which government and governance are considered in more theoretical terms (Bevir and Rhodes, 2003).

At the same time as these initiatives are being developed in central government to alter its own structure, there have also been proposals to generate a mood to create 'new localism' at the most local levels. 'New Localism' first appeared as a concept in 2002 (Corry and Stoker, 2002) in a pamphlet published by the New Local Government Network. In some ways, it was seen as a response to the adoption of centrally set targets, which were viewed as a strongly defining feature of the post-1997 government and which are in effect versions of Osborne and Hutchinson's approach to better performance management (2004). However, again as indicated by Osborne and Hutchinson, the process of collecting better performance information had been seen to become an industry in its own right and was threatening the freedoms and flexibilities offered to local authorities (DETR, 2001). These performance and outcome targets were often individually conceived albeit now more frequently set within Public Service Agreements (PSAs) between the Treasury and individual government departments. Individual local authorities have also entered into Local Public Service Agreements that have set stretch targets at the local level, but which have also been centrally set until January 2004; however, these have had a far smaller effect than the implications on the Central Departments requirements to deliver their targets to the centre. The ability for local flexibility is almost nil when central targets for literacy, numeracy or reduction in domestic burglaries are set. Where central departments are dependent on local delivery to ensure the achievement of their national targets in education, crime and health for example, the failure of these targets to be integrated has a fundamental effect on what happens at the local level. It is this issue which has dominated the debate about public service reform in the second term of the Labour government from 2001 and fuelled some of the debate on 'new localism', where the need to provide the freedom to respond to the 'local' has come to the fore. Central targetry has local unintended consequences, which can be just as difficult to deal with.

Thus the principle of new localism has been accepted as a positive concept worth debating and implementing, but beneath the headlines, there is a contest, with much more at stake than the future of local government. Any change which brings something like parity between local and central governance is fundamental. The recognition that effective economic and social delivery needs different forms of governance is seen as critical. At the same time there will be many who do not welcome the structural and cultural changes that such approaches would bring. As all those involved in change management appreciate, the ability of the status quo to exert the great force of inertia is often underestimated. At best what happens is partial reform and implementation. What is interesting about the package of reforms which support the broad notion of new localism, is their scale and the number of institutions which are involved.

There are a variety of newly emergent policy experiments that are seeking to take forward new localism in a practical way. The emergence of the Local Public Service Boards (Kent CC, 2003; Corry *et al.*, 2004) and Local Area Agreements (Lorimer, 2004; ODPM, 2004b) command support as a means of moving forward, as not only do they provide means of overcoming the problems generated by silo public services for the individual, community, locality or company, but also they can create significant savings through back office systems, staff and premises. As described by Corry *et al.*, this version seeks to generate benefits from competition between organisations that can also hold each other to account. This may be a negative outcome and could in some ways represent what is already the status quo. However, used more positively, a group of localities with a wider variety of strengths and specialisms on offer, say in health and education, could provide a more enriched offer for communities and their people within their area.

A further version of these new local arrangements is the 'pluralist model' proposed by Corry *et al.* (2004), said to provide the benefits whilst minimising the problems of delivery between agencies at the local level, which they set against the opposites of silo and municipal options. The pluralist approach seems difficult to understand unless perhaps taken to mean adopting differing approaches at each level – the Local Public Service Board at local authority level, with more single service based approaches at the neighbourhood level, although the authors do not make this distinction in their proposals. The pluralist model may be considered a 'fudge' not to upset either central or local government but it may have lost its sense of purpose. Corry *et al.* support more sub-authority devolution although not with the clarity which may be needed for successful implementation.

In addition to the New Local Government Network other agencies have also been reviewing ways in which more localised working would be more beneficial. The Audit Commission has also been reviewing ways in which more locally joined up working could be more successfully pursued. In their study *People, Places and Prosperity* (2004), the Audit Commission demonstrates that the continuing separation of streams of funding and objectives serve to undermine joined up working at the local level and that this is still having a deleterious effect on communities. They found that the promotion of economic, social and environmental wellbeing was more likely to be successful where national and local priorities are 'fully aligned and where local partners achieve coherence in establishing their priorities and targets' (p.2). The continuing numbers of partnership arrangements and funding streams that still exist are still seen to hamper delivery at the local level, which is supported by other research (Buck et al., 2002). These partnerships seem to be primarily engaged in attracting short-term project funding which renders mainstreaming and long-term planning for change difficult as it is seen to be undertaken as a separate activity from day to day delivery. However, the Audit Commission does not suggest that the partnerships, including Local Strategic Partnerships that cover every local authority area in England, should be replaced by another vehicle but rather they should be strengthened to become the focus of Local Area Agreements (LAAs) which should be the basis of a contract with central government through the Local Public Service Agreement.

The government has responded to the Audit Commission's proposals with two approaches. The first is the publication of *Local Area Agreements: A Prospectus* (ODPM, 2004b) and the second is an invitation to contribute to the consideration of *The Future of Local Government: Developing a 10 Year Vision* (ODPM, 2004a). In the Ministerial foreword to the prospectus, Local Area Agreements are described as representing 'a radical new approach to improve coordination between central government and local authorities and their partners, working through the Local Strategic Partnership' (p.5). The key components remain much the same as in earlier partnership working and are stated as:

- Simplified funding for Safer and Stronger communities
- Strengthened Local Public Service agreements
- Strengthened national strategy for Neighbourhood renewal
- A stronger role for Government Offices; and
- Pilot Local Area agreements.

(ODPM, 2004b, p.7)

The LAAs are proposed to have a series of key themes around specific groups and communities that are already identified in the jointly agreed priorities for local and central government (CLP, 2000). These will be included in three 'blocks' – children and young people, safer and stronger communities, and healthier communities and older people – which are to be negotiated separately and brought together into a single LAA. The more developmental features proposed also include a 'single pot' for funding in a locality with no barriers between different public sector budgets and funding streams, with greater local flexibility in order to address local needs. The second is to further rationalise the separate locally based funding initiatives that run in parallel to bring these into the LAA. The proposals for LAA implementation are that a few LAAs would be introduced as pilots to be followed with more mainstream implementation soon after. The effectiveness of delivery is to be assessed in a number of ways including proposed Local Area Profiles developed by the Audit Commission and other forms of local performance management.

New localism in practice: neighbourhoods as the repository of trust?

Local government is therefore the major beneficiary of the application of the principle of new localism. It may not receive all that it would wish in terms of local freedom – concerns about postcode services remain as a brake on this freedom (Walker, 2002) but at the same time, the new degrees of freedom may be as great as those expected by local government in 1997. At the sub-local authority level, area based improvement is critical and elections to neighbourhood development groups have attracted much higher turnouts than local elections. This drive for representation at neighbourhood, or parish level, driven by international comparison, is one which local authorities may not entirely welcome but is probably set to develop. The means to achieve this are already there whether from the powers for sub-authority working in the Local Government Act 2000, to the implementation of Business Improvement Districts (ODPM, 2004c) and the application of new town and parish councils. Directly elected single purpose boards might have a role in this kind of structure, although better local accountability may also be able to provide for their role as well (Blears, 2003). The developing governance structures may also be an amalgam of all the approaches proposed.

These more recent approaches sit alongside some that have been promoted since the introduction of the 'modernised local government' programme. This has been as part of the post-1997 approach to community leadership and locally focussed working, which has included a number of different initiatives to encourage local governance at a sub-local authority level. These initiatives have built on existing activities but have been enhanced over time. The first is the opportunity for town and parish councils to further develop in terms of their roles and responsibilities. The second has been the enhanced role that can be played by neighbourhood renewal and Neighbourhood Development Companies (NDCs), that can operate in areas which are most deprived whether they be in urban or rural areas, although their work is primarily focussed on urban areas where deprivation is seen to be at the most concentrated. The third initiative has emerged from the Local Government Act 2000 where councils are now able to deliver services in areas which are within their areas and to give responsibility for this delivery to the councillors and community representatives of each of these areas. Although these initiatives arise from separate provenances and have a differing focus, one of the key issues for the future is how they may be brought together into a more coherent sub-local authority model, which would be similar to the arrangement of local government in France for example. With new localism, all of these approaches provide a legitimising basis for working at the lowest appropriate level – in some ways a rebranded principle of subsidiarity.

The issue of trust has also been seen to be a factor in a failure of people to engage in the political process. In looking at the issue of trust in public institutions, the Audit Commission and MORI (2003) found in a joint study that trust is a difficult notion to define but in situations where citizens use public services, where they have no choice of provider, trust is an important issue for them. In communities, local TV and radio stations were seen to have the greatest 'trust' at 70 per cent in their communications, compared with 22 per cent for local councillors and 11 per cent for national politicians. There was seen to be a declining deference to experts and yet a strong link between experts and accountability in the public's mind. Improving trust was seen to be a function of a number of factors including more information and the independence of those supplying it. Further, and critically here for the development of neighbourhood policy, trust was seen to be higher in those with whom there was some personal contact – the local decision maker was seen to be more likely to be trusted than someone who was more distant. Thus, 'trust is strengthened by the visibility of service delivery', (2003, p.40). The effectiveness of locality-based approaches has also been confirmed by other studies such as that by the Audit Commission on connecting with citizens and users (2002b), where handing over control to the users was seen to be a key success factor in project and programme delivery.

The contribution of the neighbourhood in generating social capital, trust and engagement are seen to be important factors in delivering the quality of life and liveability factors which people are seeking. Giving the community more control and more choice is seen to be an important factor in the way in which people feel about their community and themselves. The introduction of Anti-Social Behaviour Orders (ASBOs), where communities can deal with 'drug dens' or the 'neighbours from hell' are seen to be an important component for communities exerting control of their own localities. It is at the neighbourhood level that local services which make up the important components of the perceived quality of life are delivered including health, education, housing, shopping and public transport (Audit Commission, 2002a).

There is no doubt that the development of the role of the neighbourhood governance body in the management of communities and localities is emerging as one of the most significant in the next ten years. In the period before 1997, local authorities were concerned about their own role and rights for independence. Many of the reforms in the period since 1997 have been aimed at helping to achieve this, although local authorities may argue that these need greater local authority independence from targets than is still permissible. The period from 2005–2010, and possibly longer to 2025, is one where the role of the neighbourhood in establishing social cohesion, along the lines suggested by Putnam (2000) and increasing greater democratic engagement is on the agenda.

This approach is also part of a higher level political philosophy which is part of the move towards achieving choice in public services and more contestability for the user in terms of the provider, whether this be health, education or care (Lent and Arend, 2004). Milburn gave one of the first indications of this shift in a speech to the Social Market Foundation where he stated:

> National targets and standards are important but ultimately improvement is delivered locally not nationally by frontline staff in frontline services. And that is where power needs to be located if those services are genuinely to be responsive to local communities.
>
> (Milburn, 2003)

Milburn also identifies four key components in this process, which he has based on his own experience of public services whilst growing up in the North East. These are:

1 Better information, greater choice
2 Strengthening citizens' voice

    3   Shifting accountability outwards and onwards
    4   Devolving power to local neighbourhoods.

(Milburn, 2005: 30)

Much of this imposed delegation may be controversial to local authorities that may see their newly achieved status and role being undermined. Milburn is clear about the need for these neighbourhood bodies to be accountable, although it is not clear to whom they should be accountable – the local authority, a regional body or government departments. Some local authorities are very critical about the promotion of a new tier of governance in areas where communities are already confused about who represents them and where accountability for service delivery lies.

The reforms at neighbourhood level are significant and many of the potential problems and issues are yet to be fully examined and overcome. They also sit within a reformed local governance structure which is generated by Local Area Agreements, where the local public services can be seen to be merging together in common public programmes or through Local Public Service Boards. As the local authority becomes more managerial and programme orientated, led by its executive or directly elected mayor, the neighbourhood could be the level at which proposals are developed to meet problems, where solutions are delivered and where people and communities make choices over their priorities and provider in the application of the choice principle. What is not clear is how local decision taking sits within national targetry. Local authorities have frequently complained that local partnerships, including LSPs, cannot tackle local priorities within a framework of nationally set targets to bodies such as the police and health. If this system remains, the complexity of local delivery may be considerable. All main political parties are committed to greater local devolution and this means that there will need to be a new breed of local political leaders and managers in order to implement this emerging and increasingly important neighbourhood governance.

These newer approaches sit alongside longer-term arrangements. In 2003 there were approximately 8700 local parish and town councils in England. Local councils have a number of powers that they may choose to exercise but very few duties. There has been a strong encouragement for the development of 'local councils', i.e. town and parish councils that can have a range of functions and roles. The ability of these local councils to take on new roles was enhanced by the Local Government and Rating Act 1997, when it became possible for local councils to undertake the provision of additional services including public transport, traffic calming and community safety. This approach has also been accompanied by new powers provided by ODPM for local councils to increase the level of funding which is to be utilised to £5 per head; this can be used on projects that bring direct benefit to their area but for which they may not otherwise have the powers. There is also an intention that central government should be able to fund local councils directly in future, although this will need new legislation. The 1997 Act also made it possible for residents in an area to petition for the establishment of a parish council in their area. Parish councils are also seen to have a more general structure that would enable them to take up different issues simultaneously or over time. The parish council, which is set up on an open electoral basis, is also seen to be a potentially more accountable structure. Other forms of neighbourhood management have also been established for localities with specific needs, such as Action Zones or New Deal for Communities, which are available only to the most deprived neighbourhoods. The parish council model is appropriate to any locality wishing to engage in neighbourhood management, and they can also sit alongside other forms of locality management such as local authorities or their neighbourhood management arrangements.

Thus it is clear that the concept of 'new localism' is continuing to develop and some argue that it is being diluted as a result. There are also those who see 'new localism' as a means of helping to serve primarily central government ends through the localisation of services that are currently centrally delivered (Cross, 2004), although this does not necessarily mean that the outcome would not be beneficial. The results from the Government review of the most appropriate scale for each service to be delivered may serve the public better and be much cheaper to operate (HMT, 2004a). There are other parts of governance that are being accused of either paying lip service to 'new localism' through Local Area Agreements or establishing their own version of it through local police or health boards which in effect would create an even stronger direct relationship between the centre and the locality without necessarily divesting any central power.

Conclusions

New localism is now understood to be a main organising principle of sub-national governance in England. It is seen as a means of improving democratic accountability, providing a local mandate and producing inter-agency approaches to localities. The notion of 'new localism' is therefore, as Mulgan (2004) states, at times a confused one, or even a term that is used promiscuously (Corry *et al.*, 2004). It is possible to see how this combination of drivers is pushing a real reassessment of what should be undertaken locally, on local decision making and effective means of delivery. To be successful it is likely to take a new form, although it is unclear which of these current experiments will dominate. The Local Public Services Board, as a streamlined version of the Local Strategic Partnership, may hold responsibility for the Local Area Agreement at local authority level and has potentially a new overarching role yet without the duty of public service agencies to cooperate, as exists in Scotland. At the sub-local authority scale, it seems that place is also the dominant determinant – whether through deprived neighbourhoods, parishes or estates. The overarching principle of new localism, derived from all these international and national pressures is now becoming a determining influence on the shape of governance in England.

References

Alesina, A. and Spolaore, E. (2003) *The Size of Nations*, London: MIT Press.

Allmendinger, P., Morphet, J. and Tewdwr-Jones, M. (2005) 'Devolution and the modernisation of local government: prospects for planning', *European Planning Studies*, 13(3): 349-70.

Audit Commission (2002a) *Quality of Life Using Quality of Life Indicators*, London: Audit Commission.

Audit Commission (2002b) *Connecting with Users and Citizens*, London: Audit Commission.

Audit Commission (2004) *People, Places and Prosperity: Delivering Government Programmes at the Local Level*, London: Audit Commission.

Audit Commission and MORI (2003) *Trust in Public Institutions*, London: MORI.

Bevir, M. and Rhodes, R.A.W. (2003) *Interpreting British Governance*, London: Routledge.

Blair, T. (2004) 'Reform of the Civil Service', 24 February, available at http://www.number-10.gov.uk

Blears, H. (2003) *Communities in Control: Public Services and Local Socialism*, London: Fabian Society.

Bobbitt, P. (2002) *The Sword of Achilles: War, Peace and the Course of History*, New York: Alfred A. Knopf.

Buck, N., Gordon, I., Hall, P., Harloe, M. and Kleinman, M. (2002) *Working Capital Life and Labour in Contemporary London*, London: Routledge.

Cabinet Office/DETR (2002) *Your Region Your Choice*, London: DETR.

Corry, D. and Stoker, G. (2002) *The New Localism*, London: New Local Government Network.

Corry D., Hatter, W., Parker, I., Randle, A. and Stoker, G. (2004) *Joining-up Local Democracy: Governance Systems for New Localism*, London: New Local Government Network.

Cross, M. (2004) 'Public domain', *Guardian*, 12 August, available at http://technology.guardian.co.uk/online/story/0,3605,1280769,00.html

CLP (2000) 'Original Framework and Atrangements for the Conduct of Central Local Relations (1997)', available at http://www.lga.gov.uk

DETR (2001) Best Value and Procurement: Handling of Workforce Matters in Contracting, Local Goverment Act 1999, Section 19, London: HMSO

European Commission (1999) *Third Report on Economic and Social Cohesion*, available at the Innovations Forum at http://www.europa.eu

European Commission (2004) *A new partnership for cohesion convergence competitiveness cooperation*, available at the Innovations Forum at http://www.europa.eu

HMT (2004a) *Devolved Decision Making: 1 Delivering Better Public Services: Refining Targets and Performance Management*, London: HMT.

HMT (2004b) *Devolved Decision Making: 2 Meeting the Regional Economic Challenge: Increasing Regional and Local Flexibility*, London: HMT.

Kent CC (2003) 'LSPs with Teeth – The Kent Model', *Governance Paper* delivered at workshop October 2003, London: ODPM, available at the Innovations Forum at http://www.odpm.gov.uk

Lent, A. and Arend, N. (2004) *Making Choices: How Can Choice Improve Local Public Services?* London: New Local Government Network.

Lorimer, K. (2004) 'Bosses grab win-win opportunity', *Local Government Chronicle*, 21 May, p. 9.

Loughlin, J. (ed.) (2001) *Subnational Democracy in the European Union: Challenges and Opportunities*, Oxford: Oxford University Press.

Loughlin, J. (2001) 'Introduction: The transformation of the democratic state in Western Europe' in J. Loughlin (ed.) *Subnational Democracy in the European Union: Challenges and Opportunities*, Oxford: Oxford University Press : 1–33.

Lyons, M. (2004) *Well Placed to Deliver? Shaping the Pattern of Government Service: Independent Review of Public Sector Relocation*, London: HMT.

Milburn, A. (2003) Speech to the Social Market Foundation, 30 April, available at http://www.dh.gov.uk/NewsHome/Speeches/SpeechesList/SpeechesArticle/fs/en?CONTENT_ID=4031877&chk=1ipHDg

Milburn, A. (2005) 'A four-step plan to wrest power from the state to the citizen', *Public*, February: 30–31.

Morphet, J. (1994) 'The Committee of the Regions', *Local Government Policy Making*, 20(5): 56–60.

Morphet, J. (2005) 'The rise of the sub-region (the emergence of a sub-rgional scale of working in governance)', *Town and Country Planning*, 74(9): 268–70.

Morphet, J. (forthcoming) *Modern Local Government*, London: Sage.

Mulgan, G. (2004) 'Introduction' in D. Corry *et al.* (eds) *Joining-up Local Democracy: Governance Systems for New Localism*, London: New Local Government Network.

Northern Way Steering Group (2004) *Moving Forward: The Northern Way*, available at http://www.thenorthernway.co.uk/report_sept04.html

ODPM (2004a) *The Future of Local Government: Developing a Ten Year Vision*, London: ODPM.

ODPM (2004b) *Local Area Agreements: A Prospectus*, London: ODPM.

ODPM (2004c) *Guidance on Establishing Business Improvement Districts*, London: ODPM.

ODPM/HMT/Core Cities Group (2004) *Competitive European Cities: Where Do Core Cities Stand? Summary No. 13*, London: ODPM.

Osborne, D. and Gaebler, T. (1992) *Reinventing Government: How the Entrepreneurial Spirit is Transforming the Public Sector*, New York: Penguin.

Osborne, D. and Hutchinson, P. (2004) *The Price of Government*, New York: Basic Books.

Putnam, R. (2000) *Bowling Alone*, London: Simon and Schuster.

Robinson, E. (2005) *Living with Regions: Making Multi-level Governance Work*, London: New Local Government Network.

Scottish Executive (2003a) 'Guidance on Community Planning', *Consultation Draft* , Edinburgh: Scottish Executive.

Scottish Executive (2003b) 'Making it Work for Sotland', *Community Planning Advice Notes, March 2003*, Edinburgh: Scottish Excutive.

Scottish Executive (2004) *A Spatial Plan for Scotland*, Edinburgh: Scottish Executive.

Walker, D. (2002) *In Praise of Centralism: A Critique of New Localism*, London: Catalyst.

Welsh Assembly Government (2003) *People, Places, Futures – the Wales Spatial Plan*, Cardiff: Welsh Assembly Government.

20 Building new subjectivities

Devolution, regional identities and the re-scaling of politics

Mike Raco

Introduction

In many EU states the devolution of political and economic power from the centre to the regions has become something of a panacea for tackling spatial inequalities. Regions, it is argued, are the most appropriate scalar units for capturing and nurturing increasingly fluid global forms of investment. They also represent *places* or territories of meaning and attachment in and through which populations share collective identities, interests and imaginations. Regional empowerment can, therefore, help to encourage the evolution of stronger and more legitimate relationships between citizens and states and strengthen democratic systems. Devolution programmes can increase the *congruence* between broader, more abstract spaces of government and regional places and establish a new 'institutional fix' in which the boundaries of government and decision making have a closer fit to the outlooks and expectations of regional populations. However, this new regionalism, whatever its progressive claims, is premised on assumptions of internal homogeneity and unity within regional boundaries and how these can be mobilised to create new sustainable, democratic and empowering forms of economic development. The process of 'fixing' identities and institutional structures is fraught with difficulties, contradictions and tensions as the search for congruence necessitates political choices over how to identify and define 'authentic' regions.

This chapter examines the relationships between this recent resurgence of regionalism and the changing form and character of territorial identities. It is divided into two parts. The first assesses different conceptual approaches and begins by examining new regionalist discourses and their assumptions and implications. It then turns to recent relational critiques that highlight broader processes of identity formation and connectivity between regions. This is followed by an examination of governmentalist approaches in addressing such questions and argues that one of the strategies inherent in establishing regional devolution is that of creating new spatial institutional fixes that in turn shape the governmentalities, identities and frames of reference for regional actors. Seen in this way, one of the central rationales for devolution is to change ways of thinking about the types of problems that exist and how they can be addressed. It is more than a new regulatory project through which national state power is delivered through territorial programmes of action (cf. Jones, 2001). Instead, it seeks to create new ways of seeing the world and acts as a strategy for making regions and regional actors both the subjects and objects of policy. The second part of the chapter briefly explores some of the diverse rationales that have underpinned devolution agendas in England and argues that they have been underpinned by a governmental instrumentalism – a strategy that seeks to activate and mobilise regional identities to make regional economies work more efficiently and improve the efficiency and effectiveness of governance systems. Regional empowerment is itself a power relation that has paradoxically recentralised control at the same time as it has promoted new forms of territorial 'freedom'. Collectively, the chapter argues that a *hybridity* of conceptual approaches is required to explore and assess the rationales and practices of New Labour's agendas and that a governmentalist approach establishes new research directions in exploring devolution agendas across Western Europe.

Empowerment, identity and English devolution – conceptualising the new regionalism

Congruent regions: place–space tensions, identity and the new regionalism

Over recent decades there has been a transformation in the discourses and practices of regional development policy. At a time when meta-narratives are supposed to be out of fashion, it is ironic that the diagnoses of, and policy solutions for, regional economic problems in different countries have followed broadly similar trajectories. Thus terms such as regional sustainability, empowerment, devolution and decentralisation have become ubiquitous with their message that economic development can be more equitable, accountable, and sustainable than in the past (Meadowcroft, 1999). These discourses have been particularly attractive to nation states that are faced with a growing range of demands and problems to be addressed at the same time as their capacities for action are undergoing systematic restructuring through globalising economic, political and social forces (Giddens, 2002; Jessop, 1998). Nowhere have such ideas had greater purchase than in the emergence of the intellectual paradigm of the 'new regionalism' in which the promotion of regional economies, governed by regional agencies and executives, is put forward as the panacea for faltering regional economic growth (see Lovering, 1999; Porter, 1998). This approach is underpinned by three interrelated concepts that see the region as a focus for: the formation of common *economic strategies* in a context of enhanced globalisation (or a holding down of the global); new forms of *cultural identification*; and the mediation of *co-present social interactions* (see Amin and Thrift, 1995; Gilbert, 1988; MacLeod and Jones, 2001: 674–5). Regional government, it is argued, can link the economic, cultural and political in new and progressive ways if empowered by national governments through a radical devolution of powers and responsibilities.

It is in this context that for new regionalists, devolution represents a 'win–win–win' solution to the regional political tensions and economic inequalities of the UK. Overcentralisation has been criticised for its role in creating uneven forms of regional development, particularly since the early 1980s when the Conservative governments of Mrs Thatcher were perceived to be concerned primarily with the needs of the South East of England and unrepresentative of wider differences in regional political cultures and identities (Brown and Alexander, 1999; Gamble, 1994; Mitchell, 1996). In political terms, therefore, devolution and a wider 'politics of dispersal' has come to represent a progressive and much overdue process of decentralisation that challenges the political power of the London-dominated central state (Amin *et al.*, 2003). It provides new platforms in and through which strategically significant development partnerships and common agendas can be established, whilst forging new, closer links between citizens and the state. Devolution also enables regional actors to establish strategies and programmes that are 'in touch' with regional needs and priorities. For example, where regional economies are heavily reliant upon manufacturing industries, regional strategies could reflect this and provide the basis for new regionally organised neo-corporatist relationships to be established between employers, workers and the state. Local authorities, it is argued, are too small to carry out such a task, whereas the nation state is too detached.

Underpinning these new regionalist prescriptions for devolution are specific conceptions of what constitutes regional *identities* and attachments. The new regionalism is premised on a dualism between meaningful (regional) places and empty bureaucratic spaces or where place is seen 'as the seat of genuine meaning and global space as in consequence without meaning' (Massey, 2004: 9). For Taylor (1999), this dualism creates 'place-space tensions' in that modernist spaces of governance can only achieve 'efficiency in administrative theory … by converting messy places into calculable spaces' (p.15). Places, it is argued, are characterised by the existence of specific, historically established cultures of social interaction passed on from generation to generation, yet these do not always sit comfortably alongside the spatial boundaries of government established by modern states. It is, therefore, the extent of *place–space congruence*, or the 'fit' between place identities and institutional boundaries that are critical in determining the extent of tensions that exist in specific regions. In some regions, particularly those that are relatively homogenous in social and economic terms and have a low turnover of population, there is perceived to be a strong congruence between spaces of government and places of community imagination. In others, there may be a very different history of conflictual local politics in which regional boundaries and identities have a clear incongruence. Rounds of local government reorganisation in the 1970s and 1990s, for example, were fiercely politicised and contested at the local level with some newly created regional authorities, such as Avon and Cleveland, drawing little in the way of popular legitimacy with local populations (Barnett and Chandler, 1997). Similarly, the boundaries of the English regions were established in the mid-twentieth century and are seen by many as incongruent with regional imaginations (MacLeod and Jones, 2001).

The concept of place–space congruence is a core element in new regionalist thinking. In Massey's (2004) terms, it is characterised by 'Russian doll' assumptions of a nested hierarchy of regional identities, cultures and practices that should be both reflected in, and reproduced by, bureaucratic boundaries of governance. Ostensibly, the emphasis of devolution is on establishing a 'spatial fix' in which regional identities become more congruent with decision-making structures, thereby providing opportunities for the emergence of new, more legitimate and effective government agencies. However, the notion of place–space congruence and the rather static understanding of territorial imaginations and cultures enshrined in some of the strong regionalist discourses is fraught with difficulties and inconsistencies and it is to these and the relational critique of new regionalist thinking that the next section now turns.

Relational identities, devolution and territorial politics

The notion of place–space congruence contains a number of a priori assumptions that limit its utility as an approach. First, it 'fixes' regional identities as pre-given objects of government around which administrative boundaries can be drawn. Yet as Agnew (1991: 53) argues, regional identities are developed through particular cultures constituted by a 'set of practices, interests and ideas subject to collective revision, changing or persisting as places and their populations change or persist in response to locally and externally generated challenges'. Second, notions of congruence are built on an *internal* focus to regional identities and ways of thinking that homogenises and simplifies regional imaginations. The process of boundary drawing is always one that creates new forms of inclusivity and exclusivity and promoting devolution as a process that is inherently democratic and inclusive underplays the very real differences between groups and different interests within regions. Third, *external* links and associations, likewise, become undervalued and new boundaries are established, in both an institutional way (for example through the creation of different regulatory systems in different regions) and in terms of identities as imagined regional differences become magnified. This is exemplified by the asymmetrical nature of UK devolution with regional settlements differing in Scotland, Wales, Northern Ireland and the English regions based on assumptions about the strength and congruence of regional identities, shared histories and expectations (Elcock and Keating, 1998; see also Michael Keating's chapter earlier in this book).

Given the limitations of new regionalist conceptions of regional identities and practices, it is surprising that many critics draw on similar assumptions. For example, MacLeod and Jones (2001: 671) argue that the government's devolution proposals for England draw upon regional boundaries characterised by a 'gauche insensitivity to local civil society and the staggering lack of imagination' (MacLeod and Jones, 2001: 671). The implication is that 'local civil society' is something that exists as an object of government and that regional boundaries *could* be drawn that represented a closer place–space congruence. Similarly, other writers have focused on the strong regional identity of the North East of England even though the empirical evidence tends to be derived from erratic and limited opinion polling. As Lovering (1999) notes, the push for regionalism often comes from 'regional service classes' of professionals, business leaders and other elites for whom the discourse of a coherent, identifiable region serves their own interests. In the case of the North East, Morgan (2002) argues that this is fuelled by a politics of envy towards devolved Celtic nations and the belief that there is 'an economic dividend to political devolution' (800).

Overall, new regionalist writings fail to adequately address what has been termed the *relational construction* of regional identities and the complex processes in and through which they are formed. A relational approach sees 'cities and regions with no automatic promise of territorial or systematic integrity since they are made through the spatiality of flow, juxtaposition, porosity and relational connectivity' (Amin, 2004: 34; see also Amin and Thrift, 2002). Seen in this way, regional identities are constructed as much by external flows and connections between places as they are by internal, collective experiences and imaginations. Consequently, regional (territorial) identities must be understood as 'internally complex, essentially un-boundable in any absolute sense, and inevitably historically changing' (Massey, 2004: 5). Congruent regional identities are built on assumptions of authenticity and it is therefore, 'important to challenge the identities themselves and thus – *a fortiori* – the relations through which those identities have been established' (2004: 5). From a relational perspective devolution, as it has been promoted in the UK and elsewhere in Europe, is flawed by its grounded imaginary of the region as 'a space of intimacy, shared history or shared identity, and community of interest of fate' (Amin, 2004: 37). As Amin (2004: 42) goes on to argue, 'there is nothing to be gained from fetishizing cities and regions as particular kinds of community that lend themselves to territorially defined or spatially constrained political arrangements and choices'. Relational thinking, therefore, provides some alternative and fruitful ways of conceptualising both the form and character of the new regionalism and the types of regional imaginations that are being promoted. If regional identities are created through a mixture of flows and connections across different scales, then any attempt to 'fix' them through policy initiatives will be characterised by oversimplification and an inability to capture their dynamism and ever-changing character.

However, whilst this draws attention to the processes in and through which identities are constructed, there is also a tendency to underplay the imagined associations that groups of people may possess within particular regions. Whilst these are not fixed, it does not follow that particular subjectivities and governmentalities are not strongly influenced by boundary-drawing processes, such as those that characterise devolution. Although relational approaches rightly emphasise processes of identity formation, they do not always give adequate weight to how subjectivities are constructed through government programmes and the power dynamics and rationalities associated with devolved 'empowerment'. For example, regions may be given new roles as active, united, coherent and identifiable *subjects* who are to be empowered to take greater responsibility for themselves and their own circumstances. The effects of this on identity formation can be highly significant as the (re)creation of regional agencies covering bounded areas such as the North East and North West of England, may generate new regional subjectivities so that local actors begin to think and act in regional terms. The next section highlights the potential significance of governmentalist approaches in addressing such issues and how governmentalist conceptions open up potentially new ways of thinking about the links between territorial identities and spaces of power.

Governmentalities, devolution and empowerment

A *governmentalist* approach to regional empowerment and identity formation is taken from the work of Michel Foucault and his writings on the 'problem' of governing modern societies. Foucault examined how it was possible for governments to shape and control increasingly disparate and numerous subjects in ways that facilitated effective governance to take place. He argued that the art of modern government was to govern without governing society, or the development of reflexive government – a rationality where the ends of policy also become the means (Dean, 1999). Governmentality is the basis of political thought and action, or, as Foucault (1998: 8) defines it, 'an ensemble formed by the institutions, procedures, analyses and reflections, the calculations and tactics, that allow the exercise of this very specific albeit complex form of power'. It is characterised by particular ways 'of thinking about the kinds of problems that can and should be addressed by various authorities' (Miller and Rose, 1990: 2), that can be understood as political rationalities or discourses that seek to direct the conduct of others or ourselves. Government is concerned primarily with the 'conduct of conduct' in that it seeks to establish and build *subjectivities* in and through which government programmes and strategies can be operationalised and implemented. In Bevir's (1999: 353) terms, it is concerned with tracing 'the operation of power as it creates subjects, discourses and institutions through time'.

Such an approach draws attention to the processes in and through which *subjects are created* and the role that identities and cultural attachments, including those to places and regions, have in shaping subjectivities. It highlights the ways in which government has become associated with an increased concern for defining and shaping 'appropriate' individual and community conduct, regulation and control. Subjects are created in different places and at different times and it therefore provides an insight into the ways in which liberal states use space and place to pursue their strategies of action and ensure, what Foucault (1998: 24–5) termed, 'the ordered maximisation of collective and individual forces'. By creating boundaries and territories and granting them powers and responsibilities states create subjects with new frames of reference and new ways of perceiving the world (Rose, 1999). These are institutionalised through particular technologies and practices that turn policy agendas into knowable, calculable and administrative objects that in turn 'enables a problem to be addressed and offers certain strategies for solving and handling the problem' (Lemke, 2001: 191). In devolution or community empowerment programmes, this process of (re)subjectivisation establishes populations as the *subjects* responsible for their own destinies at the same time as they become the *objects* of policy or that which is to be assisted, worked on and changed. Responsibility becomes a mechanism of control and compliance in that the grounds and boundaries of empowerment are set by those devolving power who will do so in their own interests, often as an instrumental move in order to make their own policies more effective.

Applying a governmentalist approach to UK devolution, therefore, requires an examination of the processes through which the regional 'problem' is characterised; the technologies and mechanisms used to make these problems visible; the form and character of regional empowerment; the processes through which subjects and objects are established; and the ways in which boundaries shape identities and modes of thinking about particular 'problems' and how they can best be addressed. In this sense, it goes beyond the new regionalist focus on place–space congruence and the relationist emphasis on flows and connections to examine how and why governments establish programmes of action and how regional identities can be created through the 'fixing' of boundaries of action and the empowerment of (directed) subjects. It provides a set of insights into how governments seek to create (a governing) order out of potential chaos and why the devolution process has been fraught with politically mediated tensions and contradictions and the relationships between these agendas and regional identities. At this early stage in the devolution process, it also provides a framework through which to interpret the *rationales* of government as they stand as well as establishing a number of research questions about the changing nature of regional identities and the relationships between decision-making processes and regional associations.

The next section draws on the above to assess the rationales and discourses that underpin the government's devolution programme in relation to the English regions. It demonstrates that a *hybridity* of approaches are evident. There are strong echoes of new regionalist thinking, for example, with an emphasis on regional empowerment and the creation of new, internally focused regional projects. At the same time, however, the broader process is underpinned by a rationality of circumscribed empowerment and an attempt to create new regional subjectivities in and through which central government agendas will be implemented in more effective ways by 'empowered' regional actors. What is less evident is recognition of relational identities or the internal incoherence of the regions. This, it is argued, is a consequence of the instrumental focus of devolution and its need to simplify and categorise the regions as objects of government.

English devolution and the re-subjectivisation of the English regions

> We are not partitioning England into regions. We have set up Regional Development Agencies – such as in Yorkshire and Humberside – to help tackle local economic problems and attract new jobs.
>
> (Tony Blair, 2003)

> I view the government's [regional] officers as a local raj set out to impose Whitehall's will. They have departmental agendas.
>
> (Peter Kilfoyle, MP, quoted in Chrisafis, 2001)

As earlier chapters in this book demonstrate, the issue of English regionalism within New Labour's broader regional agendas has yet to be satisfactorily resolved. The governmental object of 'England' as a place and the processes in and through which regional and national identities have been fixed, have been highly variable and riven with contradictions, complexities and politically constructed tensions and inconsistencies (see Paxman, 1998). Englishness has long been bound up with the cultural, economic and political dominance of the capital and the South East region. Consequently, as Taylor (1993) argues, in many parts of England, English identity is at the same time a source of inclusion and exclusion, something that relates to the wider boundaries of England yet is determined by a specific set of dominant and exclusive 'southern' values and outlooks (see Amin *et al.*, 2003). Deep structural imbalances exist between the economic performance of the north and the south of the country and these have added a further dimension to debates over identity, belonging, and a just spatial politics (see Morgan, 2002; Sandford, 2002; Tomaney, 2000).

Tensions over what constitutes regional English identities and governmentalities have been reflected in the government's relative inertia on English regional devolution. Nine Regional Development Agencies were established in 1999 along with Regional Chambers to co-ordinate economic development strategies across the regions. However, it was not until 2002 (5 years into New Labour's term of office) that the government published its White Paper *Your Region, Your Choice* in which it tentatively set out plans for elected Regional Assemblies in three regions, the North East, North West, and Yorkshire and Humber (DTLR, 2002). The purpose of these agencies is to generate regional governance that 'is based on the principles of increasing prosperity, pride and democracy in the regions' (Raynsford, 2004a: column 319).

The selection of northern regions, with laggard economies, reveals much about the Labour government's priorities and rationales for action. There has been a vigorous debate within government over the form and character that devolution should take and what the outcomes of a devolution programme should be. As John Prescott, the strongest advocate of regional government in the Cabinet, argues:

> On any economic measure the North lags behind the South. Under the existing system – the status quo – the North has fallen behind the South decade after decade. And decade after decade the gap has widened ... we cannot just top-slice growth from the South and move it North. But, we can achieve real economic progress in the North. And that means doing a lot more regionally and locally. If the economic performance of the North was brought up to the national average the productivity of the three Northern regions would be a staggering £35 billion a year more than it is today.
>
> (Prescott, 2004a: 1)

For Prescott, English devolution has, therefore, become focused on resolving the economic problems of the northern regions and bringing them up to an 'average' performance for the UK economy as a whole (see Goodwin *et al.*, 2004). This is exemplified by the White Paper's direct references to the work of Michael Porter and his writings on regional competitiveness and region building. Porter (1990, 1998) explicitly calls for the empowerment of regional actors so that they may establish their own agendas and create a collective regional 'project' around which diverse actors and interests can coalesce[1]. In this way, the problems associated with uneven development can be resolved and managed in ways that do not directly impact on the competitiveness of the most successful regions. Instead, they come about through a greater utilisation of existing capacities in laggard regions that are, therefore, responsible for forging their own economic destiny.

In order to fulfil this broader economic rationality, the White Paper also adopts an explicitly political dimension in calling for the creation of new elected Regional Assemblies, arguing that:

> by taking powers from Whitehall and government quangos, assemblies can reduce bureaucracy, enhance efficiency, improve co-ordination, bring decision-making under closer democratic control and offer the regions a distinct political voice.
>
> (Prescott, 2004b: 1)

Or, as another minister, Nick Raynsford, asserts:

> an elected regional assembly would give people in each English region a distinct political voice and a real say over decisions that matter to them ... it will give power and responsibility back to the people. It will make our politics more open, more accountable and more inclusive.
>
> (Raynsford, 2004b: 1)

In essence, the proposals will 'recognise [that] the people best placed to make decisions affecting the regions are the people who live in the regions' (Raynsford,

2004c: 1). Devolution is presented as a mechanism for the establishment of more sustainable, democratic and empowering forms of government. It will enhance the congruence between spatial boundaries of government and the identities and attachments of populations that in Tomaney and Mitchell's (2000: 2) words, 'feel remote and out of touch from the centres of political power'.

At the same time, however, the new agendas are indicative of subjective and conditional forms of empowerment and seek to develop new governmentalities or ways of thinking within the regions. New Labour's rationale is to use regional spaces as territories of government that play a simultaneous role as both the policy *subjects* – in that they are left to develop their own strategies and agendas for change and become active agents in their own salvation – at the same time they represent the policy *objects*, as laggard, backward regions in need of significant investment and regeneration. For example, the White Paper states that:

> the diversity of the English regions demands a diversity of targeted responses. The regions themselves are often best placed to determine the most effective solutions to their needs. They need to have greater control over the key decisions that affect them and to be able to respond to the differing needs and desires of people in their regions. The government is, therefore, offering each region the flexibility to choose effective solutions for strengthening its performance.
>
> (DTLR, 2002: paragraph 1.6)

Paradoxically, therefore, this type of devolution reinforces the power relations between the regions and the centre, in that regions will primarily give 'added value' to the centre's policy agendas. Tailoring solutions to regional circumstances is another method of making central government policy agendas more effective and using the regions to play an active role in bringing them about. This (re)subjectification is evident in the government's argument that devolution will 'bring decision-making closer to those it affects' as the purpose is explicitly to 'make the delivery of programmes and policies more efficient and ultimately lead to better outcomes in all regions' (2002: paragraph 2.4). It is focused on allowing public sector actors in the regions 'the freedom and flexibility to deliver against national standards' (2002: paragraph 1.8), so that regional variations in approach and delivery practices are primarily concerned with increasing the efficiency of national agendas.

This bounded empowerment is evident in relation to issues such as regional planning policy where the government is introducing a new system of regional spatial planning to establish structures in and through which a broader range of regional actors will participate (see contributions in this book by Deas, Counsell and Haughton, Baker and Wong, and Vigar). This is presented as a good example of regional devolution and a mechanism for creating new, more inclusive decision-making frameworks. However, this selective empowerment is being 'given' by the centre to improve '*the quality and inclusive nature of the regional input ... [whilst] responsibility for issuing the regional spatial strategies* will remain with the government' (DTLR, 2002: paragraph 2.19: emphasis added). Similarly, in relation to strategy making, the main purpose of the new regional agencies will not be so much to create their own strategies but to 'join-up' existing strategies (most of which are developed by central government or government quangos) and work out ways in which they can be implemented at the regional level. Perhaps most significantly regional subjects are to be empowered to 'strengthen the building blocks for economic growth in all regions: enterprise, skills innovation, higher education, scientific excellence, and improving quality of life' (2002: paragraph 1.13). The purpose in devolving power is not, therefore, to strengthen the role of regional centres, but to enhance the grip of central government and force actors to make their regions more like the South East as a regional role model.

This policy emphasis is reinforced by broader government techniques and technologies that seek to make regions visible policy objects that can be compared, measured and regulated. Any regional agencies will be subject to rigorous auditing and accountability criteria with targets set by central government departments. In governmental terms, such mechanisms of accountability have a strong role in shaping how subjects act and what their 'regional' agendas and strategies should be focused on. Unlike Scottish and Welsh devolution, where the political rationale for establishing executives has been highly significant, devolution to the English regions will primarily be judged by its quantitative impacts in terms of jobs, investment levels and GDP per capita. Increased 'efficiency' in government programmes will equate with technologies and practices that identify the 'devolution effect' and the return for government from its devolved investments. There is relatively little emphasis on devolution as a political, moral or ethical imperative and more of a focus on its instrumental capacities to deliver an economic renaissance. Regional identities are repackaged as an asset that both justifies the implementation of a regional agenda and creates the conditions in and through which new ways of working and co-operating can be established. It has an instrumental value for government as a frame that can facilitate new co-operative and entrepreneurial governmentalities.

By subjectivising and objectifying the English regions in this way, a number of (new regionalist) assumptions are institutionalised. Regional 'problems' and policy 'solutions' are homogenised and internalised so that, through new ways of thinking and operating, regional actors are able to take greater responsibility for their own economic circumstances. Regions become single, identifiable economic units with internal coherence and a collective sense of identity. Internal differences between sub-regional localities, businesses, and communities are put to one side as the region becomes *the* scale that defines ways of seeing and understanding how economies work and the level at which new forms of entrepreneurialism and competitiveness can be fostered. In becoming an object of policy, it is anticipated that actors and subjects will work towards improving and re-creating 'the region' and new governmentalities and identities will be established and reproduced. There is little acknowledgment of the more relationally constructed nature of regional associations and identities, instead the new regionalist emphasis on the creation of a common regional enterprise is reinforced and expanded. Indeed, the external relations between these regions and others across the UK are almost entirely absent in the devolution proposal. Yet, as Dorling and Thomas' (2004) recent study of regional trends highlights, one of the biggest problems facing the northern regions is a movement of investment and skilled, young people to the 'successful' regions of the South East. The relationally constructed nature of regional economies and identities is not discussed in the policy proposals as it draws attention to a complexity that detracts from the wider objective of region-building and the creation of a governmental order out of chaos.

Much of this strategy derives from internal political struggles within the Labour Party and its indecision over what type of English regionalism should be promoted. As Mawson (1998) shows, the issue of English devolution is one that has gradually moved up New Labour's political agendas without ever gaining strong ground as a primary policy objective. One of the reasons for the instrumentalist focus of the subsequent agendas has been to garner political support, not least from the Prime Minister and the Chancellor of the Exchequer, both of whom have publicly expressed strong reservations about English devolution (Hetherington, 2001). There is little in the government's wider public sector reform programmes to indicate any significant devolution of power from the Treasury and Whitehall departments to regional actors. Devolution has, therefore, been linked to discourses of public sector efficiency and an 'economic' focus on improving the economic statistics of the English regions. This explains why the initial moves towards regional devolution have concentrated on the creation of Regional Development Agencies and Regional Chambers, both of which promote regional business interests in policy making and seek to engender regional ways of thinking and working within business communities. Early research in Scotland has highlighted the awakening of such subjectivities within business communities with many business associations experiencing a surge in membership over the devolution period and a re-politicisation of their activities as a response to the new political arrangements (see Raco, 2003a; 2003b).

In the English regions, where devolution has, thus far, been less well developed, there has been less of a response from voluntary organisations (Valler *et al.*, 2004). It seems likely that given the partial nature of devolution, the extent to which the new institutions will shape the governmentalities and identities of populations in the English regions will remain limited. The Scottish and Welsh experience suggests that regional empowerment has created a new focus for governmental thinking that some argue has diverted attention away from bigger questions of the break-up of the UK state (see Nairn, 2000). In this sense, devolution has been as much about seeking to maintain the cohesion of the UK (and England) as it has been concerned with empowering the regions and encouraging them to adopt and develop their own agendas. Strengthening regionalism and devolving responsibilities without powers paradoxically enhances the strength of the centre, something that optimistic new regionalist accounts fail to acknowledge. These tensions have been borne out in the first referendum for Regional Assembly that took place in the North East in November 2004. The North East was piloted since it was perceived to have the strongest sense of regional identity within England. This regionalism, it was argued, had developed from long standing imaginations of difference from the rest of England and a corporatist regional politics in which the regional brand has provided a political focus for agenda building (see Shaw, 1993). Despite this seemingly fertile ground for a new regional politics, the referendum vote only produced a turnout of 50 per cent, with a split of 78 per cent against devolution and 22 per cent in favour. As a direct consequence of this, a new vote cannot now be held for seven years and the remaining referenda have been put on hold.

This result can be interpreted in a number of ways. On the one hand, it is indicative of the oversimplifications that exist within the New Labour agenda and the thin support on which regional elites base their calls for regional empowerment. It demonstrates the incoherence of seemingly coherent English regions (cf. Amin,

2004), and that place attachments and spatial governmentalities draw on a variety of different imaginations across varying scales. On the other hand, the referendum also shows that this *type* of regional governance fails to resolve place–space tensions in the governance of England. In governmentalist terms, it is indicative of the inherent tensions within citizen–state relationships and shows that the identities and attachments of the former are not simply a functionalist result of the strategies of the latter. In terms of offering a way forward, the 'No' vote may indicate that a new, wider and outward-looking regional politics is required that rejects the notion of internal coherence and seeks to develop new, inclusive governmentalities. It also draws attention to the relationships between scale and identity formation and the extent to which place attachments and imaginations are developed in relation to perceptions of the locality, community or even nation state, rather than the regional spaces of policy makers and elite groups.

Conclusions

This chapter has looked at the relationships between identities, subjectivities, structures of governance, and the government's devolution agendas. It has argued that in order to understand the rationale and purpose of devolution, a hybridity of different theoretical and conceptual approaches is required. New regionalist approaches assume that the reform of political structures can create new types of congruence between territorial identities and state boundaries of action. These, it is argued, can represent the basis for a more democratic and progressive politics at the same time as it enables a more targeted and flexible set of policy programmes to be established. Relationists, however, suggest that this approach is fraught with assumptions concerning the homogeneity and unity of regional identities and focuses too much on the internal characteristics of populations rather than the complexities of cross-regional flows and networks. This chapter has argued that a governmentalist approach to devolution and regional identity can provide a fruitful addition to these concepts in highlighting the rationalities that underpin state practices and focusing on the ways in which boundary-drawing seeks to create policy subjects (and objects) and establish new territorial parameters in and through which identities are (re)created and embedded. Identities can become a vehicle of governance to be shaped and targeted to meet wider ends.

The chapter has briefly discussed some of these concepts in relation to English devolution and demonstrated that the government's rationale for creating regional assemblies in three northern regions is complex and multifaceted. On the one hand, it appears to represent a straightforward case of new regionalist thinking in expanding on the ideas of writers such as Michael Porter and their policy prescriptions for regional empowerment and economic regeneration. Devolution is presented as a 'good', something that will bring a new legitimacy to the political system in the UK and provide new institutional platforms in and through which policy agendas can be tailored to regional needs and aspirations. In economic, political and social terms it is portrayed as a 'win-win-win' policy agenda. On the other hand, the chapter has highlighted some of the tensions within the programme and the ways in which regional subjectivities are defined and (re)created. Regional empowerment will be highly selective and will paradoxically reinforce the powers of central government. Policy programmes are designed to improve the effectiveness of central government policy *in* the regions rather than enabling the development *of* regional policy-making autonomy. The focus is on creating new ways of thinking about economic 'problems' in the regions and mobilising subjects to focus on regional priorities in the context of national agendas. The economic implications of encouraging new forms of zero-sum, regional competition have not been thought through or the impacts that this may have on regional politics. The evidence suggests that enhanced regional competition does little to enhance the economic strength of regions or to promote forms of embedded economic investment.

This focus on the rationalities and practices of government inverts some of the more exaggerated claims of new regionalists. In so doing, it challenges the widely adopted meta-narratives of sustainability and empowerment and opens up a range of broader research questions. For example, the focus of governmentalist work tends to be on the rationalities of government and the strategies involved in producing subjectivities. Less empirical research has been undertaken on how *effective* such strategies are and what *types* of subjectivities are created, yet such questions are critical to governmentalist theoretical propositions. Does the governmental focus on 'problem' regions, for example, succeed in encouraging regional actors to think about the internal characteristics of their regions or does it, paradoxically, lead to the opposite outcome in making them more aware of the relations between regions and the processes underpinning (relational) identity formation and uneven development? Does devolution reinforce the unity of the UK state or will it generate political forces that eventually break it up? In light of the 'No' vote in the North East, it seems that the regional scale in England lacks the power to mobilise a new place-based politics or to create new regionally focused governmentalities. The future of democratic political regionalism in England therefore remains uncertain, although one possibility is that non-elected regional government may be empowered instead to implement the government's wider objective of boosting the economic competitiveness of England's regions.

Note

1   Michael Porter (1998) has also been amongst the most powerful advocates of the new regionalism and the policy prescriptions required to improve regional competitiveness. Regional competitiveness, he argues, is generated through the interactions of co-present actors and specific measures designed to boost competitive industries, particularly those in the knowledge sector. Successful regions are those that have successfully established clusters around industries such as bio-technology and computer software. Porter has been extremely influential on governments across the world and has been an advisor to the Labour government. His appearance in the White Paper is a recognition of his influence on government thinking and his wider reputation.

References

Agnew, J. (1991) 'Place and politics in post-war Italy: a cultural geography of local identity in the provinces of Lucca and Pistoia', in K. Anderson and F. Gale (eds) *Inventing Places: Studies in Cultural Geography*, London: Halsted Press: 52–79.

Amin, A. (2004) 'Regions unbound: towards a new politics of place', *Geografiska Annaler*, 86B: 33–44.

Amin, A. and Thrift, N. (eds) (1995) *Globalisation, Institutions and Regional Development in Europe*, Oxford: Oxford University Press.

Amin, A. and Thrift, N. (2002) *Cities: Rethinking Urban Theory*, Cambridge: Polity Press.

Amin, A., Massey, D. and Thrift, N. (2003) *De-centering the Nation: A Radical Approach to Regional Inequality*, London: Catalyst.

Barnett, N. and Chandler, J. (1997) 'Local government and community' in P. Hoggett (ed.) *Contested Communities: Experience, Struggles and Policies*, Bristol: Policy Press: 144–62.

Bevir, M. (1999) 'Foucault and critique: deploying agency against autonomy', *Political Theory*, 27: 65–84.

Blair, T. (2003) Statement to the CEP, http://www.thecep.org.uk/letter/colinray.htm

Brown, G. and Alexander, D. (1999) *New Scotland – New Britain*, London: The Smith Institute.

Chrisafis, A. (2001) 'Labour ignoring north-south divide, says MP', *Guardian*, March 2.

Dean, M. (1999) *Governmentality – Power and Rule in Modern Society*, London: Sage.

Department of Transport, Local Government and the Regions (2002) *Your Region, Your Choice*, London: HMSO.

Dorling, D. and Thomas, B. (2004) *People and Places: A 2001 Census Atlas of the UK*, Bristol: Policy Press.

Elcock, H. and Keating, M. (eds) (1998) *Remaking the Union – Devolution and British Politics in the 1990s*, London: Frank Cass Press.

Foucault, M. (1998) *The History of Sexuality*, London: Penguin.

Gamble, A. (1994) *Britain in Decline: Economic Policy, Political Strategy and the British State*, London: Macmillan.

Giddens, A. (2002) *Where Now for New Labour?* Cambridge: Polity Press.

Gilbert, A. (1988) 'The new regional geography in English and French-speaking countries', *Progress in Human Geography*, 12: 208–28.

Goodwin, M., Jones, M. and Jones, R. (2004) 'Economic governance in Wales, clear red water?' paper presented at *Devolution in Wales – the Scorecard Findings from the ESRC Devolution Programme*, Cardiff, 24 June.

Hetherington, P. (2001) 'Chancellor backs English devolution', *Guardian*, January 30.

Jessop, B. (1998) 'The rise of governance and the risks of failure: the case of economic development', *International Social Science Journal*, 155: 29–46.

Jones, M. (2001) The regional state and economic regulation: "Partnerships for prosperity" or new scales of state power?' *Environment and Planning A*, 33: 1185–1211.

Lemke, T. (2001) 'The birth of bio-politics: Michel Foucault's lecture at the College de France on neo-liberal governmentality', *Economy and Society*, 30: 190–207.

Lovering, J. (1999) 'Theory led by policy: The inadequacies of the "New Regionalism"', *International Journal of Urban and Regional Research*, 23: 380–95.

MacLeod, G. and Jones, M. (2001) 'Renewing the geography of regions', *Environment and Planning D: Society and Space*, 19: 669–95.

Massey, D. (2004) 'Geographies of responsibility', *Geografiska Annaler*, 86B: 5–18.

Mawson, J. (1998) 'English regionalism and New Labour', in H. Elcock and M. Keating (eds) *Remaking the Union: Devolution and British Politics in the 1990's*, London: Frank Cass Publishers: 158–76.

Meadowcroft, J. (1999) 'Planning for sustainable development: what can be learnt from the critics', in M. Kenny and J. Meadowcroft (eds) *Planning Sustainability*, London: Routledge: 12–38.

Miller, P. and Rose, N. (1990) 'Governing Economic Life', *Economy and Society*, 19(1): 1–31.

Mitchell, J. (1996) *Strategies for Self-Government*, Edinburgh: Polygon.

Morgan, K. (2002) 'The English question: Regional perspectives on a fractured nation', *Regional Studies*, 797–810.

Nairn, T. (2000) *After Britain – New Labour and the Return of Scotland*, London: Granta Press.

Paxman, J. (1998) *The English: A Portrait of a People*, London: Michael Joseph.

Porter, M. (1990) *The Competitive Advantage of Nations*, Basingstoke: Macmillan.

Porter, M. (1998) 'Clusters and the new economic competition', *Harvard Business Review*, November–December: 77–90.

Prescott, J. (2004a) 'Elected Regional Assemblies', speech to business forum, Manchester, 22 January.

Prescott, J. (2004b) Foreword, in *Your Region, Your Choice: Summary – North East*, London: Office of the Deputy Prime Minister: 1.

Raco, M. (2003a) 'Governmentality, subject-building and the discourses and practices of devolution in the UK', *Transactions of the Institute of British Geographers*, 28: 75–95.

Raco, M. (2003b) 'The social relations of business representation and devolved governance in the United Kingdom', *Environment and Planning A*, 35: 1853–76.

Raynsford, N. (2004a) *Hansard*, Column 319, London: HMSO: June 30.

Raynsford, N. (2004b) *Let the Regions Choose – Statement by Local Government and Regions Minister*, London: Office of the Deputy Prime Minister.

Raynsford, N. (2004c) *New Planning Power for Elected Regional Assemblies*, ODPM Press Release, February 12.

Rose, N. (1999) *The Powers of Freedom*, Cambridge: Cambridge University Press.

Sandford, M. (2002) 'What place for England in an asymmetrically devolved UK?' *Regional Studies*, 36: 789–96.

Shaw, K. (1993) 'The development of a new urban corporatism – the politics of urban regeneration in the North East of England', *Regional Studies*, 27: 251–68.

Taylor, P. (1993) 'The meaning of the north: England's foreign country within', *Political Geography*, 12: 136–55.

Taylor, P. (1999) 'Place, space and Macy's: Place-space tensions in the political geography of modernities', *Progress in Human Geography*, 23: 7–26.

Tomaney, J. (2000) 'End of the empire state? New Labour and devolution in the United Kingdom', *International Journal of Urban and Regional Research*, 24: 675–88.

Valler, D., Wood, A., Atkinson, I., Betteley, D., Phelps, N., Raco, M. and Shirlow, P. (2004) 'Business representation and the UK regions: mapping institutional change', *Progress in Planning*, 61: 75–135.

21 Mapping the geographies of UK devolution

Institutional legacies, territorial fixes and network topologies

Gordon MacLeod and Martin Jones

> In this world so often described as a space of flows, so much of our formal democratic politics is organized territorially.
>
> (Doreen Massey, 2004: 9)

Asymmetrical devolution and the re-institutionalization of the UK

When analyzing any aspect of contemporary UK devolution, perhaps we should be ever mindful of the view espoused by the former Welsh Secretary, Ron Davies (1999), that it is 'a process not an event' (cf. Jones *et al.*, 2005). And if there are times when some of us who live and work in this bewilderingly complex multinational state (officially designated *The United Kingdom of Great Britain and Northern Ireland*) tend to give the impression of being fixated with analyzing this very process, then perhaps we might also be excused, at least in part, by the fact that 'devolution' is a distinctly *British* term (Keating, 2005). It was coined in the nineteenth century. And while it bears some resemblance to federalism and, in turn, to the systems of regional government that have emerged in other European countries, it also exhibits three distinguishing features (2005: 19–20).

The first is that whilst certain powers and institutional capacities have been transferred from London to the nations and regions of the UK, this has not seriously affected the overall *sovereignty* of Westminster (cf. Morgan, 2002; Amin *et al.*, 2003). So, on the one hand the Prime Minister, Tony Blair, can claim that the 'government's progressive programme of constitutional reform is now moving us from a centralised Britain, where power flowed top-down, to a devolved and plural state' (Blair, 2000: 1). On the other hand, however, Blair has been at pains to articulate his ostensibly firm conviction that the devolution 'package' – which sees the granting of an elected Parliament for Scotland, a National Assembly for Wales, an Assembly for Northern Ireland, an elected London Mayor and Greater London Assembly, alongside Regional Development Agencies for the eight English regions – adds up to a 'sensible modernisation' of the country's constitutional arrangements, helping to safeguard the very *integrity* of the United Kingdom.[1]

Second, this modernized UK constitutional settlement[2] is punctuated by an asymmetrical geometry that 'runs through every clause and schedule of the devolution legislation, from the fundamentals of powers and functions down to the niceties of nomenclature' (Hazell, 2000: 268). Thus, in contrast to certain federal European states like Belgium and Germany, and as illustrated in the very naming of the different national and regional institutions mentioned above, the refurbished political arrangements for the UK assume a strikingly uneven territorial expression (and see Giordano and Roller, 2004, for a useful comparison of asymmetrical devolution in the UK and Spain). Perhaps most significantly, England – with the notable exception of the global city-region of London – remains the only nation in the UK not to be supplemented with additional elected political representation (Jeffery and Mawson, 2002; Jones and MacLeod, 2004). For the remit of Regional Development Agencies is limited to policies for (non-macro) economic development. One net outcome of this asymmetrical devolution is a political imbalance whereby Westminster doubles up as the UK *and* English government (Jeffery, 2003).

Third, this asymmetry reflects the manner in which the contemporary programme of devolution and constitutional change is being overlaid on a comparatively uneven and quite eccentric institutional inheritance (Taylor and Thomson, 1999; Marr, 2000; Jones, 2004). A primary example is the way that since the 1707 Union between the English and Scottish parliaments, the nation of Scotland has continued to carve out a distinctive public sphere and a relatively autonomous set of civic institutions (Paterson, 1994). Moreover, during the era of Thatcherism and Majorism – a period when the UK government was thoroughly unrepresentative of the Scottish nation – it was this institutional milieu that nourished the creation of the Scottish Constitutional Convention, which itself ultimately generated the political voice to establish the Scottish Parliament (MacLeod, 1998). In turn, devolution is believed to have strengthened the Scottish political arena, throwing up some distinctively Scottish policies in education and welfare alongside a bold strategy to foster a 'knowledge economy' (Cooke and Clifton, 2005; Goodwin *et al.*, 2005). Of course, the broader context for UK devolution is an integrated, albeit transforming, welfare state and there are real questions about how much policy divergence may be possible, particularly with a government that – at least outwardly – is anxious to deliver equal standards of social justice (Keating, 2005).

Even from this brief foray into the labyrinthine complexities of contemporary devolution, it is evident that we remain ever vigilant about the path-dependent institutional legacies and the evolutionary geohistory of the United Kingdom (MacLeod, 2002). Furthermore, when embarking on any given analysis of the UK's asymmetrical devolution, it also surely behoves us to appreciate the particular economic, cultural and political discourses, practices and inflections around and through which its nations and regions are, in Anssi Paasi's terms, *institutionalized*. In other words we ought to consider the socio-spatial process during which certain territorial units emerge as a part of the spatial structure of society and become 'established and clearly identified in different spheres of social action and social consciousness' (Paasi, 1986: 121). Stretching this terminology, the post-1997 institutional accomplishments enacted in the name of, and performed through the process of, devolution have led the UK to undergo a substantial *re-institutionalization*.

This chapter is primarily concerned to explore the value of some theoretical methods for researching this process of re-institutionalization. First, we examine how UK devolution has come to be conceptualized as a re-scaling of the state and of political and institutional capacity (cf. MacLeod and Goodwin, 1999; Jones *et al.*, this volume). We then discuss an alternative method, which in part has arisen out of a backlash against scalar perspectives, and which advocates a 'relational' approach to space and topological conceptualizations of spatiality (Allen *et al.*,

1998; Amin, 2004; Amin *et al.*, 2003). Following a synopsis of these two approaches, we aim to transcend what appears to be a potential theoretical impasse by demonstrating how each can lend itself towards a fruitful analysis of different expressions of devolution. This argument is presented through a discussion of: 1) the struggle to institutionalize the South West of England as a geopolitical unit during Labour's second term of office; and 2) recent endeavours to foster a new region called 'The Northern Way'. In each of these cases, we aim to highlight how territorial and scalar perspectives are compatible – as opposed to entirely antagonistic – approaches to the study of UK devolution and political change per se. Our central message, then, is to convince the scholar of devolution and constitutional change that political geographical scales and networks of spatial connectivity are mutually constitutive rather than mutually exclusive aspects of contemporary social spatiality. Brief conclusions follow.

Interpreting the geographies of devolution: scales, networks, territories

Writing in *The Times Higher* in 2002, the political scientist, Fred Nash, contended that the body of research deriving from the ESRC's *Devolution and Constitutional Change* was bedevilled by a 'lamentable lack of theoretical and conceptual grounding' (Nash, 2002: 30). Given that this was voiced only two years into the research programme, perhaps it was a rather unforgiving assessment. Nonetheless, we would concur with the view that the *process* of devolution certainly presents researchers from a range of sub-disciplines with an unprecedented opportunity to conceptualize a profound period of transformation; economics, geography, history and politics in the making (Bulmer *et al.*, 2002; Hazell, 2003; *Regional Studies*, 2002; 2005). In the discussion below, we critically assess two theoretical approaches to spatial politics that have begun to punctuate the debate on devolution.

The politics of scalar structuration and state spatiality

A growing number of contributions on UK devolution seem to be interpreting it as a rescaling of the state and a reworking of the geographies of government and governance (MacLeod and Goodwin, 1999). This has been a key message in some analyses of England's Regional Development Agencies (Gibbs and Jonas, 2001; Jones, 2001; Jones and MacLeod, 1999) and the creation of Regional Spatial Strategies. Others, often inspired by the neo-Marxist state theory of Bob Jessop (2002), conceptualize this rescaling of governance as part and parcel of a 'hollowing out' of the national state and a corresponding 'filling in' at other scales such as England's regions, the Northern Ireland Assembly, and the Welsh Assembly and its four Regional Divisions. Indeed several contributors to this volume deploy the concepts of scale and rescaling to illustrate the geographical relocation of government responsibilities and governance capacities.

Nonetheless, it is our contention that to date, many of the scalar-informed analyses of UK devolution have deployed a relatively superficial reading of the concept of rescaling and that a deeper engagement with the now highly enriched theoretical vocabulary on the 'politics of scalar structuration'[3] (Brenner, 2001; 2004) could prove most instructive in examining the UK's unique brand of asymmetrical devolution. A fundamental premise of this approach is that geographical scale is conceptualized as socially constructed rather than ontologically pre-given, and that through this socio-spatial process of structuration, geographic scales 'are themselves implicated in the constitution of social, economic and political processes' (Delaney and Leitner, 1997: 93). In this relational approach (Howitt, 2003), scale is conceived as a 'representational trope', as structured 'relationally within a community of producers and readers who give the practice of scale meaning' (K. Jones, 1998: 27). It also infers that spatial scales – such as those associated with the territorial organization of the state – are not merely the settings of political conflicts but one of their principal 'stakes' (Brenner, 2004). Thus, as cogently argued by Neil Brenner:

> traditional Euclidian, Cartesian and Westphalian notions of geographical scale as a fixed, bounded, self-enclosed and pregiven container are currently being superseded by a highly productive emphasis on process, evolution, dynamism and sociopolitical contestation.
>
> (Brenner, 2001: 603)

In accordance with this foundational principle, Swyngedouw makes a compelling case for an ontologically *process-based* approach to scale that:

> does not in itself assign greater validity to a global or local [or we would add national or nation-state] perspective, but alerts us to a series of sociospatial processes that changes the importance and role of certain geographical scales, re-asserts the importance of others, and sometimes creates entirely new significant scales. Most importantly, however, these scale redefinitions alter and express changes in the geometry of social power by strengthening the power and the control of some while disempowering others.
>
> (Swyngedouw, 1997: 141–2)

Furthermore, Jones (1998: 26) informs us how this process of scalar structuration is performed through a *politics of representation* whereby political agents discursively (re-)present their political struggles across scales; action that, in turn, implicates spatial imaginaries like regions, cities and nation states to be continuously implicated as 'active progenitors', offering an already partitioned geographical 'scaffolding' in and through which such practices and struggles take place (N. Smith, 2003; Brenner, 2001). In our view, this framework unlocks considerable potential for analyzing the geohistory *and* contemporary spatiality of UK devolution; what we might term *state spatiality* (Brenner, 2004). For on one level, it offers scope to examine the crucial role of nationalist and devolutionary political campaigners – as in the Scottish Constitutional Convention and the Campaign for a Welsh Assembly – to discursively present their political struggles through a scalar narrative but also across scales, stretching to London and beyond. Instructive in this regard is Agnew's work on Italy, where political parties have been central players in 'writing the scripts of geographical scale [... and where] ... The boundaries they draw [...] define the geographical scales that channel and limit their political horizons' (Agnew, 1997: 101).

A politics of scalar structuration approach might also enable us to understand how England's regional planning boundaries, which were established during wartime in the 1940s, came to represent 'active progenitors' in shaping the post-1994 map of Regional Government Offices and the post-1999 Regional

Development Agencies (RDAs).[4] And, in turn, through the introduction of particular spatially selective policies and state strategies – *state spatial strategies* (Brenner, 2004) – England's newly revived regional scales are given additional licence to become both the 'objects' of state policy and active 'subjects' in delivering policy (cf. Jones and MacLeod, 2004).[5] Not that any of this should imply some Russian doll-like neatly layered structure of territorial spheres or levels each containing a discrete package of political powers and responsibilities. For as lucidly outlined by Jamie Peck, political strategies and policy endeavours explicitly *tangle and confound scales*, with the result that:

> the scalar location of specific political-economic functions is historically and geographically contingent, not theoretically necessitated. Functions like labor regulation or the policing of financial markets do not naturally reside at any one scale, but are variously institutionalized, defended, attacked, upscaled, and down-scaled in the course of political-economic struggles. Correspondingly, the present scalar location of a given regulatory process is neither natural nor inevitable, but instead *reflects an outcome of past political conflicts and compromises*.
>
> (Peck, 2002: 340, emphasis added)

Deployed in this way, it may be that the theoretical and methodological principles of this politics of scalar structuration perspective can uncover the ways in which contemporary devolution is characterized by a rescaling of policy and planning responsibilities alongside the formation of new and/or revived and/or strengthened state spaces and the extent to which associated strategies to enact *territorial development* are intricately intertwined with the redefinition of state intervention (Brenner, 2004).

Relational regions and networked topologies

In recent years, the merits of a scalar/territorial approach to the understanding of socio-spatiality have been questioned, largely though not exclusively from a coterie of geographers based in England who advocate a more radically *relational approach to space* (see inter alia Allen *et al.*, 1998; Massey, 2004; 2005; Amin, 2002; 2004). The critique has two dimensions. The first is concerned with normative democratic politics and has been most explicitly articulated in a pamphlet entitled *Decentering the Nation: A Radical Approach to Regional Inequality*, where the authors (Amin *et al.*, 2003) disavow the 'spatial grammar' that punctuates the debate and the practice of UK devolution.[6] Indeed they proclaim that by following the well-trodden path of a territorially rooted political discourse and strategy, New Labour's devolution – particularly in relation to England – has done little to disturb the London-centrism that has characterized the business of politics and economics for over the last 100 years.

 In order to confront this entrenched hegemony of London, they advocate replacing the territorial politics of devolution with a 'politics of dispersal'. This envisages different parts of England playing *equal* roles in conducting a more mobile politics, perhaps involving 'national' institutions like Parliament travelling from London to the various 'provinces', although obviously this very process would presumably lead such regions to be de-peripheralized. Amin *et al.* (2003)

judge that these acts of dispersal would instil new spatial imaginaries of the nation – multi-nodal as opposed to deeply centralized – enabling regions to become effective national players whilst also stretching the cognitive maps of regional actors to embrace external connectivity in fostering economic prosperity and social and cultural capital. Amin (2004: 37) has since argued that, in contrast to conventional mappings where devolution politics is territorially 'grounded in an imaginary of the region as a space of intimacy, shared history or shared identity, and community of interest or fate', this *relational spatial grammar* works with the variegated processes of spatial stretching and territorial perforation associated with globalization and a society characterized by transnational flows and networks.

This leads directly onto the second dimension of the critique of a scalar or territorial logic: the possibility of an alternative ontology and conceptual orientation towards relational processes and network forms of organization, that defy a linear distinction between place and space (Amin, 2002). Amin's reasoning for this relates to how:

> In this emerging new order, spatial configurations and spatial boundaries are no longer necessarily or purposively territorial or scalar, since the social, economic, political and cultural inside and outside are constituted through the topologies of actor networks which are becoming increasingly dynamic and varied in spatial constitution. [...] The resulting excess of spatial composition is truly staggering. It includes radiations of telecommunications and transport networks around (and also under and above) the world, which in some places fail to even link up proximate neighbours. [...] It includes well-trodden but not always visible tracks of transnational escape, migration, tourism, business travel, asylum and organized terror which dissect through, and lock, established communities into new circuits of belonging and attachment, resentment and fear. [...] It includes political registers that now far exceed the traditional sites of community, town hall, parliament, state and nation, spilling over into the machinery of virtual public spheres, international organizations, global social movements, diaspora politics, and planetary or cosmopolitan projects.
>
> (Amin, 2004: 33–4)

Viewed through this ontology of relational space, cities, regions and nation states thereby come with no automatic promise of territorial integrity 'since they are made through the spatiality of flow, juxtaposition, porosity, and relational connectivity' (2004: 34). In turn, Amin cautions against fetishizing places as 'communities' that 'lend themselves to territorially defined or spatially constrained political arrangements and choices' (2004: 42; Massey, 2004). Moreover, and without wishing to denigrate any calls for building effective regional voice and representation, he is deeply suspicious of *any* assumption that there is a defined 'manageable' geographical territory to rule over.

This mode of reasoning certainly offers a fundamental challenge to perspectives on the politics of scalar structuration, whose language of 'nested scales and territorial boundaries' is deemed to omit 'much of the topology of economic circulation and network folding' characteristic of contemporary capitalism (Amin, 2002: 395). Moreover, relational thinking provides some alternative avenues for conceptualizing the identity spaces and – drawing on the insights of Peter Taylor (1999) – the emerging 'space-place tensions' of devolution: not least in that if spatial identities are indeed fostered through a mixture of flows and connections across different scales, in the words of Mike Raco, 'then any attempt to "fix" them through policy initiatives will be characterised by over-simplification and an inability to capture their dynamism and ever-changing character' (this volume).

Nations and regions: relational, scalar, networked

So where do these contrasting approaches take us? Is it appropriate to view the process of devolution as a rescaling of state spatiality and territorial restructuring, or as a topology of spatially stretched, variegated flows and territorially perforating trans-regional networks? Or perhaps it is actually quite unhelpful to be posing the question in either/or terms? This would certainly seem to be the view of John Agnew. In the process of introducing his theoretical 'mapping' of politics in modern Italy, Agnew talks of an 'intellectual standoff' between those who perhaps overstate the novelty and impact of networks and those who may remain too committed to the enduring significance of territorial spheres. For Agnew (2002: 2), '[P]art of the problem is the way the debate is posed, as if networks invariably stand in opposition to territories [as if ...] networks are seen as a completely new phenomenon without geographical anchors'.

In an explicit endeavour to transcend this seemingly polarized debate between scalar/territorial and non-scalar/topological perspectives, Harriet Bulkeley (2005) posits two key arguments. The first concerns the false assumption that approaches to the politics of scale – we prefer *politics of scalar structuration* – somehow offer a naïve view of political scales as pre-given, homogeneous and intact. She then adds that such accounts conceptualize the very processes through which such scalar constructions emerge with an emphasis on the fact that they are not neatly bound in territorial terms but take place through various actor networks and spaces of engagement (Cox, 1998; Jones and MacLeod, 2004). A good example is Adams' (1996) analysis of the way telecommunications networks create new linkages across space and scales, helping to link up, or even serve to create new, scales of governance. Bulkeley's second objection to the oppositional positioning of scales and networks concerns the extent to which networks – at least in terms of the objects which are enrolled in networks and their very scope – are themselves scaled (cf. Leitner *et al.*, 2002). It then follows that:

> once the concept of scale is freed from notions of contained and contiguous territories, it is clear that networks have a scalar dimension, both in terms of the ways in which they operate and the ways in which they are framed and configured by other networks/coalitions of actors.
>
> (Bulkeley, 2005)

Thus any conceptualization of the politics of scalar structuration or of rhizomatic or network topologies should recognize that 'scales evolve relationally within tangled hierarchies and dispersed interscalar networks' (Brenner, 2001: 605); and, moreover, that 'geographical scales and networks of spatial connectivity are mutually constitutive rather than mutually exclusive aspects of social spatiality' (2001: 605).

In the remainder of this chapter we suggest that these claims can be demonstrated effectively in the two examples discussed below. In the case of the South West of England, there is powerful evidence of a scalar politics and territorially oriented praxis, where official governmental organizations and oppositional political actors identify contrasting spatial scales – respectively, the South West regional boundary and Cornwall – around and through which to wage their quite explicitly territorial politics of engagement and representation. And yet, we can also interpret that the everyday enactment of these political geographies is being conducted through the trans-scalar movement and spatially variegated actor networks of people, objects (trains, cars), information, ideas and technologies (cf. Mol and Law, 1994). In the second case, that of the 'Northern Way', we encounter an explicit endeavour to create a pan-regional economic strategy for the north of England forged through a multi-nodal inter-urban network: a truly networked space. But again, this is a networked topology that interacts in, through and around the scalar geometries of RDA/Government Office administrative boundaries alongside other scales and territories of government.

We acknowledge that these are two non-conventional mappings of contemporary UK devolution. This makes them interesting for all sorts of reasons but we would also argue that any given political space lends itself towards an analysis of both scalar and network perspectives: the particular emphasis being dependent on the nature of the object being examined and the research questions being posed. Thus, and in stark contrast to those who advocate a network approach whilst simultaneously jettisoning the concept of scale altogether (Marston *et al.*, 2005) or dismissing it as an 'intuitive fiction' (R. Smith, 2003), we contend that it is most fruitful to envision UK devolution as a site, generator and product of *both* scalar politics *and* networked topologies, *and, crucially*, as a relational process in motion.

Convening England's South West: territorial fixes, scalar narratives, contested boundaries

'Programmed spatiality': building the South West region

The Greater South West (GSW) – later termed the South West region – was created by central government during the 1930s following surveys conducted by the Board of Trade and Ministry of Labour on the UK population and economy (Linehan, 2003): a moment of top-down regionalization or what Henri Lefebvre (1976) would have termed state-led 'programmed spatiality' (Jones and MacLeod, 2004). The region has a population of nearly 5 million and suffers relative geographical peripherality, which also translates itself politically and economically. For unlike Scotland and some English regions, the South West has never formed a powerbase for the hegemonic political party of Government. With a GDP at 95 per cent of the UK average, it hosts a mixed economy of resource-based industries alongside port, seaside and tourist activities, and a military high-tech hub enveloping the Bristol city-region (Bridges, 2002).

Prior to the era of RDAs, the South West had a number of economic development agencies operating at different spatial scales, thereby creating fragmented partnerships at the 'standard region' level. Accordingly, the South West RDA (SWRDA) received relatively widespread support for its potential to 'help make the region more competitive and more unified' (DETR, 1997: 64). Like all its

RDA counterparts, the SWRDA confronts this mission tooled with a regional economic strategy that bristles with bravado about competing as a 'world region' (SWRDA, 2000). It is 'shadowed' by the South West Regional Chamber, which comprises representatives from local authorities and social partners, and aims to instil coherence to economic development and spatial planning in the South West. However, in some instances the post-1999 territorial fix has spawned numerous tangled hierarchies and perplexing policy networks which, far from rationalizing the landscape of governance, have intensified its complexity. Indeed the Chamber – now the South West Regional Assembly – has identified 'that the nature of the relationships between the key players within the region and the boundaries between them are at times, unclear' (SWRA, 2002: 15). All of which has left some organizations anxiously groping to define a territorial identity.

Insurgent scalar narratives/networked choreographies: the case of Cornwall

> We note that Cornwall's boundary with Southwest England was fixed in 936 A.D. with only minor changes since, and there is a strong, and very long-lived, sense that Cornwall is a region in its own right.
> (Cornish Constitutional Convention, 2003: 10)

The fragmented region of the South West does include the one territory that has experienced the only recent and notable manifestation of popular regionalism in England (Deacon, 2004). Cornwall is thought to have many characteristics that distinguish it from other counties in the South West and England more generally. It has a distinctive history, its own flag (the Saint Piran) and language, and a case has been put forward to the Council of Europe to obtain national minority status (Deacon, 1999). In recent years, a key stimulus in generating a territorial political narrative has been Cornwall's economic condition: it is seriously lagging, and with a GDP of 65–70 per cent of the British average it is the poorest county in England (Sandford, 2002). Moreover, Cornwall has no major urban metropolitan centre, but rather contains twelve medium-sized towns; and it has relatively meagre networks of communication. When added together, these factors help to explain why Cornwall was granted European Union Objective One status in 2002 (CCC, 2003).

The most active insurgence against New Labour's territorial fix has been led by Mebyon Kernow. Formed in the 1950s, Mebyon Kernow is a grass-roots regional movement, modelled on Breton–Welsh–Celtic lines and combining claims for cultural rights with strategies for economic devolution (Parks and Elcock, 2000). Its activists regularly fought general and local elections with low level success, but throughout the 1990s Mebyon Kernow gained credibility by developing closer alliances with the Liberal Democrats, the hegemonic mainstream party in the South West. Mebyon Kernow's own approach to the RDA model of regionalization is encapsulated in Bernard Deacon's contention that: 'By refusing to debate regionalism the UK government is threatening Cornwall's institutional integrity. It has placed Cornwall in an artificial regional construct – the South West which is very large and culturally incoherent' (Deacon, 1999: 3).

This venture to disturb the territorial shape of southwest Britain was given a major impetus in 2000 with the formation of a Cornish Constitutional Convention (Senedh Kernow) – a cross-party organization supported by

Cornwall's four Liberal Democrat MPs, members of political parties, community and cultural activists – and its campaign for a Cornish Assembly (Deacon *et al.*, 2003). By late 2001, over 50,000 people had signed the petition for a Cornish Assembly and this was taken to the House of Commons for the attention of the Minister for Regions (Mebyon Kernow, 2001). Building on this work, the landmark documents Devolution for One and All (CCC, 2002) and The Case for Cornwall (CCC, 2003) offer a multidimensional blueprint – drawing connections between territory, identity, politics and economics – for a fully-devolved Assembly modelled on the experiences of Wales and the Isles of Scilly and legitimized by Cornwall's 'variable geometry' (2003: 10).

The political strategy of Senedh Kernow also combines a territorial politics of scaling with a networked choreography of place-making. For whilst it is the case that – as briefly alluded to above – most contemporary political endeavours enrol a spatially variegated and trans-scalar topology of actor networks, the very practice of relational networking is brought to life in the CCC's claim that devolution is about 'cutting Cornwall in' to a partnership of the regions of the British Isles, Europe and the world. And also that 'strong relationships will need to be established and maintained with Cornwall's "peer group" of UK regions and nations ... In addition, relationships will need to be renewed with regions and nations along the "Atlantic Arc", and new relationships developed in Europe' (CCC, 2002: 7). Nonetheless, in highlighting all this there is little denying that, as recognized by Bernard Deacon from the Institute of Cornish Studies at the University of Exeter, the primary objectives of Senedh Kernow are being waged through a politics of scalar structuration: 'While being studiously ignored in the Government's 2002 White Paper on the English regions [...], this "inconvenient periphery" provides one of the few explicit examples of a struggle over scale' (Deacon, 2004: 215). This struggle over politics, territory and identity continues, despite the move against formalized regional assemblies (see below).

The Northern Way: enacting a multi-nodal networked region

The territorial boundaries of England's RDAs were established by New Labour during 1997–99. They were largely synonymous with the Government Office regional boundaries, which themselves were mapped onto the administrative boundaries that had been established for a pre-Fordist economy during the conditions of 1940s wartime (Jones and MacLeod, 2004). Since 1999, and in accordance with government priorities, each RDA has worked hard to develop effective regional and sub-regional partnerships and to foster a robust Regional Economic Strategy that will be recovered to dovetail with a Regional Spatial Strategy. However, in recent years, and especially following the rejection (from November 2004) of a directly elected regional assembly in the North East of England, the Government has been encouraging the creation of alternative types of 'region', including three of the four new growth areas announced in February 2003: the M11 corridor (Cambridge to Stansted), Thames Gateway (East London–North Kent), and Milton Keynes–South Midlands (ODPM, 2003) (see the contribution to this volume by Baker and Wong). The fourth newly invented 'region' is 'The Northern Way' (see Deas' discussion in this volume). Prepared by the three northern RDAs – One North East, Yorkshire

Forward, and Northwest – and with the backing of the Office of the Deputy Prime Minister (ODPM),[7] The Northern Way was launched in September 2004 and sees: 'The three regions [...] unit[ing] in a common purpose – to develop the full potential of the North and narrow the £30 billion economic divide with the rest of England' (John Prescott, Foreword in NWSG, 2005b: 3).

The geographical shape of The Northern Way is particularly interesting given the context of this chapter. For a start, the RDA boundaries magically disappear. And this trans-regional porosity is given deeper inflection with those lines that do actually feature: rail and automobile routes and tributaries emphasizing mobility, linkage, networks. But perhaps the most notable signifier concerns the prominence given to eight city regions: Liverpool/Merseyside, Central Lancashire, Manchester, Sheffield, Leeds, Hull and Humber Ports, Tees Valley, and Tyne and Wear. These are presented as relational assets and the 'principal spatial focus' promoting faster economic growth (NWSG, 2004). Substantively, the city-regions correspond to traditional travel to work areas, shopping catchment areas and housing markets. Nonetheless, again the geographical references of the Northern Way discourse appear to be intentionally fuzzy and the spatial ontology relational, as each node is acknowledged to 'cover areas extending well beyond the city centres at their core [and...]. They contain a spectrum of towns, villages and urban fringe areas, and they have mutually inter-dependent relationships with the countryside around them' (NWSG, 2005a). The period between autumn 2004 and late spring 2005 saw the stakeholders in each city-region prepare City Region Development Programmes, which provided:

> for the first time an overview of the economic development potential and requirements of the North's major urban economies. They look at the flow of markets across administrative boundaries and draw out the consequences for the development of policy and investment in a coherent way within *these new geographies*.
>
> (NWSG, 2005b: 9; our emphasis)

The vision of the Northern Way Steering Group (NWSG) is as unambiguous as it is ambitious: 'nothing less than the transformation of the North of England to become an area of exceptional opportunity, combining a world-class economy with a superb quality of life' (NWSG, 2005b: 6). To achieve this, the NWSG proposes three broad types of action:

- Investments that are pan-Northern and add real value by operating across all three regions, such as joint marketing programmes;
- Activities which need to be embedded into mainstream programmes in each region, such as meeting employer skills needs;
- Potential investments for which further evidence must be developed to demonstrate the long-term benefits which will accrue to the North's economy, such as major transport infrastructure.

Indeed, of the ten investment priorities that have been identified thus far, three relate explicitly to transport, euphemistically described as improving 'the North's connectivity'. This involves the preparation of a Northern Airports Priorities Plan designed to 'improve surface access' to key northern airports;

improve access to the North's sea ports; and create a premier transit system in each city-region and stronger linkages between city-regions. Other significant initiatives include: three science cities (Manchester, York and Newcastle); an integrated technology transfer network structure across the North; the creation of up to four world-class research centres and an enhanced programme of Knowledge Transfer Partnerships; the Northern Enterprise in Education Programme (NEEP); a pan-northern Women into Enterprise programme; pan-northern projects in chemicals, food and drink and advanced engineering; environmental technologies, financial and professional services and logistics (NWSG, 2005a). City-region plans are also themselves placing the emphasis on openness, porosity and permeability, with Leeds aiming to 'improve city regional, pan-regional and international connectivity', and Hull and Humber Ports advertising as 'a global gateway'. All in all:

> The Northern Way represents perhaps the most significant economic development collaboration in Europe in the current decade. Therefore, it has required a new way of working of the three RDAs and their partners in the regions, local partners in City Regions and Government Departments.
>
> (NWSG, 2005b: 6)

Even acknowledging its primordial stage of development, the audacious claims being championed on behalf of The Northern Way can be called into question. First, it involves a minute budget: one that is top-sliced from existing RDA funds. Second, the old assertions about trans-regional institutional capacity and networking need to be measured alongside the mundane reality of centrally policed, and territorially defined, targets for mainstream government programmes. In stating this, however, the proliferation of 'newer regionalisms' such as The Northern Way may actually be indicative of how the UK state is seeking to deal in ever more complex ways with the diversity of territorially articulated policy demands: neighbourhood, local, regional, trans-regional, trans-national. At the same time, we can begin to identify how relational processes and trans-regional networked forms of governing are being opened up to permit fresh approaches with which different policy actors can communicate and work together more effectively not simply within their sector but across sectors and across scales. In turn, these emerging debates are helping us move towards a more relational understanding of how policy development proceeds.

Conclusions

In this chapter we have discussed two alternative approaches to the analysis of space, politics and territory. The first concerns a 'politics of scalar structuration'. In recent years this has assumed increasing popularity as an approach with which to capture the relationships between a purported rescaling of policy and planning responsibilities and the transformation of existing, or indeed the creation of new, state spaces, themselves deeply intertwined with the geographical specificities of state power and state intervention. In discussing this, we wish to reiterate how the territorialization of political life is *never* fully accomplished, but remains a precarious and deeply contentious outcome of historically specific state and non-state projects. Consequently, spatial, territorial, and scalar relations are neither pre-given nor naturally necessary 'bounded' features of statehood but are rather deeply *processual* and practical outcomes of strategic initiatives undertaken by a wide range of social forces (Brenner *et al.*, 2003). In the context of the South West of England, the Cornish – in the various guises of Mebyon Kernow and Senedh Kernow – are particularly keen to denaturalize the contemporary territorialization of UK political life, promoting their brand of nationalist regionalism as a processual and practical route through which to confront the perceived contradictions of statist technocratic regionalization.

The second approach to spatial politics has thus far been presented as a counter weight to that of scalar structuration. Deploying this topological approach would envision a radically relational interpretation on devolution and constitutional change emphasizing the networked practices of 'performing' devolution, and would make a case that an emphasis on the notions of territory and scale do violence to the open-ended and actor-centred politics of becoming. This approach is particularly powerful as a way of interpreting how, in the age of globalization, something like economic development can be conceived as a stretched-out process enacted through variegated flows of trans-regional fluids and territorially perforating moments of mobility and exchange, which transcend boundaries, territories and scales. The value of this thinking is clearly evident from our brief description of The Northern Way. Nonetheless, we contend that many everyday *realpolitik* acts of spatial politics – as in the case of a central government classifying a region as a 'problem' or local activists campaigning for devolved government and cultural rights – often distinguish a pre-existing or aspirant spatial scale or territorially articulated space of dependence through which to conduct their actually existing politics of engagement (Cox, 1998). Thus, when the various objectives and strategic priorities defined in the name of The Northern Way are finally tabled, there is every likelihood that RDA and city-regional territorial boundaries and borders will re-emerge as 'active progenitors': with the whole process of 'who getting what' type of investment being waged on territorially demarcated and scalar-defined terms.

Instead of leaving the debate here, with instances and counter examples being offered in defence of each or the other perspective, we make a case for these two approaches to be compatible (cf. Purcell, 2003). For on one level, we consider it both politically naïve and theoretically negligent to ignore the fact that much of the political challenge to devolution prevailing across England and elsewhere is being practised through an avowedly territorial narrative and scalar ontology. However, it would equally be quite absurd to deny that these practices and performances are also often enacted through topologically heterogeneous trans-regional and cross-border networks of 'fluidity' and circulation (Mol and Law, 1994). In short, then, our bottom line is that mobility and fluidity should not be seen as standing in opposition to territories. As Anssi Paasi has remarked:

> There is no doubt that networks do matter, but so do 'geography', boundaries and scales as expressions of social practice, discourse and power. Geography, boundaries and scales are not 'intuitive fictions' and their rejection/acceptance can hardly be 'written away' or erased in our offices but have to be reconceptualized perpetually in order to understand their material/discursive meaning in the transforming world.
>
> (Paasi, 2004: 541–2)

We therefore call for a retaining of territorially oriented readings of political economy and, when and where appropriate, their conjoining with non-territorial or topological approaches (Allen *et al.*, 1998; Amin, 2002). Nonetheless, it is important to underline that in order to proceed in this manner requires us to make analytical distinctions between *territory* (appropriated enacted space), *territoriality* (the sum of relations between subjects, belonging to a collectivity, with the environment), and *territorialization* (the networked and other processes through which these relations are established) (and see Raffestein, 1986; Philo and Södertström, 2004). In making this plea we also invite scholars of relational, topological and network perspectives to refrain from creating a non-discriminatory and crude caricature of all territorial and scalar contributions, whereby all bar none are conveniently disregarded as non-relational. On the contrary, we posit two arguments: 1) as we have outlined above, the growing body of work on the politics of scalar structuration is explicitly relational in its approach to territorial form; and 2) an approach which combines scalar and networked perspectives might also offer some scope to analyze 'topological territories' (Jones, 2004); i.e. the actively performed co-constitution of bounded and unbounded processes of territorialization. In turn, the question becomes how far territorialized forms of political power can be folded or bent without losing their identity – an issue that will no doubt surface as the discourses and practices of The Northern Way unfold. Territory and its governance are, in practice, 'plastic achievements' and future research needs to 'focus attention on the tangle of socio-material agents and frictional alignments in which it is suspended and to recognise that they harbour other possibilities' (Whatmore, 2002: 87). The ongoing rounds of devolution and constitutional change certainly offer the context for stretching these analyses further.

Notes

1 In accordance with most federal systems, UK devolution broadly encompasses three blocks of competences: those devolved from London to the nations and regions, those reserved for Westminster–Whitehall, and those shared between the two (Keating, 2005).
2 Perhaps this is itself an oxymoron given Britain's persistent lack of a written Constitution (Partridge, 1999).
3 Neil Brenner has usefully woven the conceptual vocabulary of Giddens's (1984) structuration theory – 'where structuration connotes a developmental dynamic in which the basic structures of collective social action are continually reproduced, modified and transformed in and through collective social action' (Brenner 2001: 603) – in order to highlight how '[p]rocesses of scalar structuration are constituted and continually reworked through everyday social routines and struggles'.
4 Joe Painter (2002) contends that, in an initiative seemingly inspired by the growth dynamic of the European Union's 'blue banana', the chief task of the RDAs is, quite literally, to *constitute* regional economies.
5 Scalar approaches have less to say about the 'subjective' dimensions of state space. As part of his endeavour to develop a hybrid approach to theorizing devolution, Raco (this volume) provides some suggestive ways in which certain neo-Foucauldian perspectives may be mobilized to help conduct this.
6 The origins of this thinking are not necessarily new and can be traced back to the post-Marxist thinking of Laclau and Mouffe (1985), who explicitly challenged topographical approaches to political action.
7 The Government agreed to match the RDA contribution to create a £100 million Northern Way Growth Fund – to translate the strategy into practical action.

References

Adams, P. (1996) 'Protest and the scale politics of telecommunications', *Political Geography*, 15(5): 419–41.

Agnew, J. (1997) 'The dramaturgy of horizons: geographical scale in the "reconstruction of Italy" by the new Italian political parties, 1992–1995', *Political Geography*, 16(2): 99–121.

Agnew, J. (2002) *Place and Politics in Modern Italy*, Chicago: University of Chicago Press.

Allen, J., Massey, D. and Cochrane, A. (1998) *Rethinking the Region*, London: Routledge.

Amin, A. (2002) 'Spatialities of globalization', *Environment and Planning A*, 34: 385–99.

Amin, A. (2004) 'Regions unbound: towards a new politics of place', *Geografiska Annaler*, 86B: 33–44.

Amin, A., Massey, D. and Thrift, N. (2003) *Decentering the Nation: A Radical Approach to Regional Inequality*, London: Catalyst.

Blair, T. (2000) 'Speech on Britishness', The Labour Party, Milbank, London.

Brenner, N. (2001) 'The limits to scale? Methodological reflections on scalar structuration', *Progress in Human Geography*, 15, 525–548.

Brenner, N. (2004) *New State Spaces*. Oxford University Press.

Brenner, N., Jessop, B., Jones, M. and MacLeod, G. (2003) 'State space in question', in N. Brenner, B. Jessop, M. Jones and G. MacLeod (eds) *State-Space: A Reader*, Oxford: Blackwell.

Bridges, T. (2002) 'The South West', in J. Tomaney and J. Mawson (eds) *England: The State of the Regions*, Bristol: Policy Press: 95–108.

Bulkeley, H. (2005) 'Reconfiguring environmental governance: towards a politics of scales and networks', *Political Geography*, 24: 875–902.

Bulmer, S., Burch, M., Carter, C., Hogwood, P. and Scott, A. (2002) *British Devolution and European Policy-making: Transforming Britain into Multi-level Governance*, London: Palgrave.

Cooke, P. and Clifton, N. (2005) 'Visionary, precautionary and constrained "varieties of devolution" in the economic governance of the devolved UK territories', *Regional Studies*, 39: 437–51.

Cornish Constitutional Convention (2002) *Devolution for One and All: Governance for Cornwall in the 21ˢᵗ Century*, Truro: Senedh Kernow.

Cornish Constitutional Convention (2003) *Your Region, Your Choice: The Case for Cornwall. Cornwall's Response to the Government's Devolution White Paper*, Truro: Senedh Kernow.

Cox, K. (1998) 'Spaces of dependence, spaces of engagement and the politics of scale, or: looking for local politics', *Political Geography*, 17: 1–23.

Deacon, B. (ed.) (1999) *The Cornish and the Council of Europe Framework for the Protection of National Minorities*, York: Joseph Rowntree Reform Trust.

Deacon, B. (2004) 'Under construction: culture and regional formation in south-west England', *European Urban and Regional Studies*, 11: 213–25.

Deacon, B., Cole, D. and Tregidga, G. (2003) *Mebyon Kernow and Cornish Nationalism*, Cardiff: Welsh Academic Press.

Delaney, D. and Leitner, H. (1997) 'The political construction of scale', *Political Geography*, 16(2): 93–7.

DETR (1997) *Building Partnerships for Prosperity: Sustainable Growth, Competitiveness and Employment in the English Regions. Cm 3814*, London: Stationery Office.

Gibbs, D. and Jonas, A. (2001) 'Rescaling and regional governance: the English Regional Development Agencies and the environment', *Environment and Planning C: Government and Policy*, 19(2), 269–80.

Giddens, A. (1984) *The Constitution of Society*, Cambridge: Polity Press.

Giordano, B. and Roller, E. (2004) 'Te para todos? A comparison of the process of devolution in Spain and the UK', *Environment and Planning A*, 36: 2163–81.

Goodwin, M., Jones, M. and Jones, R. (2005) 'Devolution, constitutional change and economic development: explaining and understanding the new institutional geographies of the British state', *Regional Studies*, 39: 421–36.

Hazell, R. (2000) *The State and the Nations 2003: The First Year of Devolution in the United Kingdom*, Exeter: Imprint Academic.

Hazell, R. (2003) *The State of the Nations 2003: The Third Year of Devolution in the United Kingdom*, Exeter: Imprint Academic.

Howitt, R. (2003) 'Scale', in J. Agnew, K. Mitchell and G. Toal (eds) *A Companion to Political Geography*, Oxford: Blackwell: 138–57.

Jeffery, C. (2003) 'An Introduction to the Devolution and Constitutional Change Programme', in C. Jeffery (ed.) *Devolution and Constitutional Change: A Research Programme of the Econonmic and Social Research Council*, Birmingham: ESRC Research Programme, University of Birmingham.

Jeffery, C. and Mawson, J. (2002) 'Introduction: beyond the White Paper on the English Regions', *Regional Studies*, 36(7): 715–20.

Jessop, B. (2002) *The Future of the Capitalist State*, London: Polity.

Jones, K. (1998) 'Scale as epistemology', *Political Geography*, 17(1): 25–8.

Jones, M. (2001) 'The rise of the regional state in economic governance: "Partnerships for prosperity" or new scales of state power', *Environment and Planning A*, 33: 1185–211.

Jones, M. (2004) 'Social justice and the region: grass-roots regional movements and the English question', *Space and Polity*, 8: 157–89.

Jones, M., Jones, R. and Goodwin, M. (2005) 'State modernization, devolution, and economic governance: an introduction and guide to debate', *Regional Studies*, 39: 397–403.

Jones, M. and MacLeod, G. (1999) 'Towards a regional renaissance? Reconfiguring and rescaling England's economic governance', *Transactions of the Institute of British Geographers*, 24: 295–313.

Jones, M. and MacLeod, G. (2004) 'Regional spaces, spaces of regionalism: territory, insurgent politics, and the English question', *Transactions of the Institute of British Geographers*, 29: 433–52.

Keating, M. (2005) *The Government of Scotland: Public Policy Making After Devolution*, Edinburgh: Edinburgh University Press.

Laclau, E. and Mouffe, C. (1985) *Hegemony and Socialist Strategy*, London: Verso.

Lefebvre, H. (1976) *The Survival of Capitalism: Reproduction of the Relations of Production*, London: Allison and Busby.

Leitner, H., Pavlik, C. and Sheppard, E. (2002) 'Networks, governance and the politics of scale: inter-urban networks and the European Union', in A. Herod and M. Wright (eds) *Geographies of Power: Placing Scale*, Oxford: Blackwell: 274–303.

Linehan, D. (2003) 'Regional surveys and the economic geographies of Britain 1930–1939', *Transactions of the Institute of British Geographers*, 28: 96–122.

MacLeod, G. (1998) 'Ideas, spaces and "Sovereigntyscapes": dramatizing Scotland's production of a new institutional fix', *Space and Polity*, 2: 207–33.

MacLeod, G. (2002) 'Identity, hybridity and the institutionalisation of territory: on the geohistory of Celtic devolution', in D.C. Harvey, R. Jones, N. McInroy and C. Milligan (eds) *Celtic Geographies: Old Cultures, New Times*, London: Routledge.

MacLeod, G. and Goodwin, M. (1999) 'Reconstructing an urban and regional political economy: on the state, politics, scale, and explanation', *Political Geography*, 18: 697–730.

Marr, A. (2000) *The Day Britain Died*, London: Profile Books.

Marston, S., Jones, J.P. and Woodward, K. (2005) 'Human geography without scale', *Transactions of the Institute of British Geographers*, 30: 416–32.

Mebyon Kernow (2001) '50,000 to London', *Cornish Nation*, 24: 3–5.

Mol, A. and Law, J. (1994) 'Regions, networks and fluids: anaemia and social topology', *Social Studies of Science*, 24: 641–71.

Morgan, K. (2002) 'The English question: regional perspectives on a fractured nation', *Regional Studies*, 36: 797–810.

Nash, F. (2002) 'Devolution dominoes', *The Times Higher*, 1 February 2: 30.

NWSG (2004) *Moving Forward: The Northern Way*, Newcastle upon Tyne: Northern Way Steering Group.

NWSG (2005a) *Moving Forward: The Northern Way: Action Plan – Progress Report Summary*, Newcastle upon Tyne: Northern Way Steering Group.

NWSG (2005b) *Moving Forward: The Northern Way: Business Plan 2005–2008*, Newcastle upon Tyne: Northern Way Steering Group.

ODPM (2003) *Your Region, Your Say*, London: Office of the Deputy Prime Minister.

Paasi, A. (1986) 'The institutionalization of regions: a theoretical framework for understanding the emergence of regions and the constitution of regional identity', *Fennia*, 16, 105–46.

Painter, J. (2002) 'Governmentality and regional economic strategies', in J. Hillier and E. Rooksby (eds) *Habitus as a Sense of Place*, Aldershot: Ashgate.

Parks, J. and Elcock, H. (2000) 'Why do regions demand autonomy?' *Regional and Federal Studies*, 10: 87–106.

Partridge, S. (1999) *The British Union State: Imperial Hangover or Flexible Citizens' Home*, London: Catalyst.

Paterson, L. (1994) *The Autonomy of Modern Scotland*, Edinburgh: Edinburgh University Press.

Peck, J. (2002) 'Political economies of scale: fast policy, interscalar relations, and neoliberal workfare', *Economic Geography*, 78: 331–60.

Philo, C. and Södertström, O. (2004) 'Social geography: looking for society in its spaces' in G. Benko and U. Strohmayer (eds) *Human Geography for the 21ˢᵗ Century*, London: Arnold.

Purcell, M. (2003) 'Islands of practice and the Brenner/Marston debate: toward a more synthetic human geography', *Progress in Human Geography*, 27: 317–32.

Raffestein, C. (1986) 'Territorialité: concept ou paradigme de la géographie sociale?', *Geographica Helvetica*, 41: 91–96.

Regional Studies (2002) 'Devolution and the English question', C. Jeffrey and J. Mawson (eds), Special Edition 36: 7.

Regional Studies (2002) 'Devolution and economic governance', M. Jones, M. Goodwin and R. Jones (eds), Special Edition 39: 4.

Sandford, M. (2002) 'The Cornish question: devolution in the South-West region', *Report for the South West Constitutional Convention and the Cornish Constitutional Convention*, London: Constitution Unit.

Smith, N. (2003) 'Remaking scale: competition and cooperation in prenational and postnational Europe', in N. Brenner, B. Jessop, M. Jones and G. MacLeod (eds) *State-Space: A Reader*, Oxford: Blackwell.

Smith, R. (2003) 'World city actor-networks', *Progress in Human Geography*, 27: 25–44.

SWRA (2002) 'Who owns the skills & learning agenda?' *South West Regional Assembly Select Committee on Skills & Learning*, Exeter: South West Regional Assembly.

SWRDA (2000) *Regional Strategy for the South West of England 2000–2010*, Exeter: South West of England Regional Development Agency.

Swyngedouw, E. (1997) 'Neither global nor local: "glocalization" and the politics of scale', in K. Cox (ed.) *Spaces of Globalization*, New York: Guilford Press: 137–66.

Taylor, P. (1999) 'Place, space and Macy's: place-space tensions in the political geography of modernities', *Progress in Human Geography*, 23: 7–26.

Taylor, B. and Thomson, K. (eds) (1999) *Scotland and Wales: Nations Again?* Cardiff: University of Wales Press.

Whatmore, S. (2002) *Hybrid Geographies*, London: Sage.

22 Identifying the determinants of the form of government, governance and spatial plan making

Mark Tewdwr-Jones

Introduction

There have been several theoretical developments over the last 25 years surrounding the significance of regional, devolved and local territories within broader processes of globalization, European integration and the changing form of the nation state. Initially, academic interest emanated from economic geography, and focused on reinvigorating the theory of agglomeration (Scott, 1983, 1986), closely followed by analysis of local outcomes of and responses to global processes of restructuring on localities (Cooke, 1989). During the 1990s, theoretical development centred on assessing the 'new regionalism' and the re-scaling of political processes (Jessop, 1990; Keating, 1997; Lovering, 1997), alongside academic debate concerning the prospect and form of institutional capacity building in territories to organise for economic growth and development (Amin and Thrift, 1992, 1995). Theories of 'associative democracy' (Hirst, 1994), the role of 'relational assets' in regional development (Storper, 1997) and of 'institutional thickness' and 'interactive governance' (Amin and Hausner, 1997) are among some of the most influential within the field of urban and regional development and fall within what Lovering (1999) and Jones (2001) identify as a fourth variant of new regionalist theory. Storper's (1997) theory of relational assets, technological innovation and regional development and Hirst's (1994) theory of associative democracy see a collective communicative rationality or normatively regulated action overcoming distortion in social action and interaction. The work of Amin (1996), Amin and Thrift (1992, 1995), Amin and Hausner (1997), and Phelps and Tewdwr-Jones (2000, 2004) by contrast, identify regional capacity building with notions of diversity in social action in the institutions underpinning local development.

Within the planning discipline, too, there have been several related theoretical developments, each attempting to understand the role of spatial strategy making and spatial governance in emerging forms of sub-national scales that possess development, infrastructure and spatial coordination political objectives. These include notions of: 'collaborative planning' (Healey, 1997) and spatial strategy and governance (Vigar *et al.*, 2000), that emphasizes aspects of spatial coordination and integration within regional and local partnerships; the planning polity (Tewdwr-Jones, 2002), where planning is viewed as a function of vested interests within the political arena of both government and governance; and regional sustainable strategy making (Haughton and Counsell, 2004), that identifies the relationships, structures of and outputs from the variety of regional actors and networks in achieving spatial policy making.

At the same time, events in the practical world have also been developing at a pace. The governmental and institutional structures within which spatial plan making and the planning system reside, are also being reformed, with a push towards the 'rescaling' of governance (MacLeod and Goodwin, 1999), the development of regional government (Tomaney, 2000, 2002), the promotion of governance and partnership (Raco, 2000), the modernization of local government (Morphet, 2004), and an enhanced European dimension to spatial strategy making (Healey, 1996; Tewdwr-Jones and Williams, 2001; Jensen and Richardson, 2004). All of these developments have impacted upon the style of government, policy making and territorial capacity across and between states, and influenced the form and content of academic analysis too.

Whilst there are important differences in the origin and emphasis of each of these formulations, all have – to a greater or lesser extent – concerned the building of regional and local institutional capacity within aspects of spatial governance. They have attracted the interest of both the geography and planning disciplines and, as such, there are some important similarities that may be drawn out between them (Vigar *et al.*, 2000: 43–6), even if protagonists of each of the analyses have, it seemed, preferred for the most part not to draw out such relationships. Most notably, the formulations draw on and discuss a particular type of social action and interaction, drawing on Habermasian, Giddensian and inter- and intra-governmental concepts, concerning interaction, participation, policy implementation and capacity building, both below the level of the nation state, and between the nation state and sub-national entities.

Recently, aspects of the relationship between spatial policy making and government and governance at the regional and local scales have been subjects worthy of analysis (see, for example, Vigar *et al.*, 2000; Tewdwr-Jones, 2002). Such changes, or themes, include a *re-scaling* of planning (Allmendinger and Tewdwr-Jones, 2000), the *integration* and *submergence* of development plans within other strategies and plans (Murdoch and Tewdwr-Jones, 1999), and the continued *marketization* of development plans and planning instruments (Allmendinger, 2003). In line with the British Labour government's push towards enhanced regionalism within England (Amin *et al.*, 2003), there has been a great deal of academic analysis and speculation concerning the various changes to spatial plan making and regional government in England (Counsell and Haughton, 2003), and in respect of devolution to Scotland, Northern Ireland and Wales (Allmendinger, 2001; Berry *et al.*, 2001; Jones *et al.*, 2004). But, whereas the focus to date has been at the regional and strategic scale, the implications of these changes at the local level within the UK have been discussed only briefly (Stoker, 2004). Similarly, analyses of the trajectories of change, how they relate to past approaches and debates, how and whether they constitute aspects of the new regionalism or new localism, and what their relationships might be to existing forms of government, and their impacts on planning, have also been absent. Without such an analysis, the full implications and significance of such changes to spatial strategy making, to government and governance, within an increasingly fragmented nation state, are lost.

The changing nature of governance and the implications of change on government

The contested space of regional and local governance and the changing nature of spatial plan making cannot be fully understood or isolated from wider political, economic and social transformations. The changing nature of the state and the relational view of policy formation at different scales has been the focus of much recent debate (see, for example, Brenner, 2000; Jones, 2001; Jessop, 2000; Amin, 1999). While much of the debate has provided a contested though broadly agreed context for understanding what has been termed the 'new regionalism', the local-ized implications (MacLeod, 2001) and resultant 'scalar' flux (Jones, 2001) provide an appropriate framework for analysing and understanding sub-regional state restructuring and the changing nature of governance and, in turn, spatial plan making.

Jessop (2000) identifies the implications of globalizing capital as the driving force behind contemporary shifts in the nature and role of the state and shifting scales of power:

> What globalisation involves ... is the creation and/or restructuring of scale as a social relation and as a site of social relations. This is evident in the con-tinuing (if often transformed) significance of smaller scales ... as substantive sites of real economic activities; in economic activities oriented to the artic-ulation of other scales into the global – such as glocalisation, 'glurbanisation', international localization, and so forth; and in new social movements based on localism ...
>
> (Jessop, 2000: 341)

The implications of such drivers of change in UK governance include the rise of new scales of policy intervention as well as moves to better coordinate, regulate or integrate new policy scales with existing institutions and processes of coordi-nation (Jones, 2001) – that is, within existing forms of government. Thus, the state is involved in a constant reworking of powers and responsibilities between scales and across policy sectors. At the same time, these contemporary reworkings of state power also represent a jostling between historically layered models of spa-tial power and new forms of activities (Agnew, 1999). Spatial plan making within transforming modes of territorial capacity development provides a good illustra-tion of this jostling of power and activity.

Issues such as coordination, integration and better regulation have been key themes of New Labour's approach to the modernization of governance since 1997. In the view of some, this emphasis amounts to a 'neo-corporatism' (Allmendinger and Tewdwr-Jones, 2000; Allmendinger, 2003) with its emphasis upon 'win–win–win' solutions or the 'triple bottom-line' approach (Counsell *et al.*, 2003), that seek, for example, to avoid trade-offs between economic growth and environmental protection and include a broad range of voices and inputs into policy processes and outcomes. Key to achieving this approach is greater pol-icy integration across and between public and private bodies and a renewed emphasis on partnerships (Newman, 2001).

These driving forces of state restructuring, multi-level governance and central government policy lead to a heady mix of policy initiative across government to which planning, and spatial plan making especially, has been directly and indirectly subjected. Most significantly, legislative bills relating to planning have recently progressed through both the Westminster and Scottish Parliaments. But the reforms to planning have been set within much broader reforms to regional and local government, including new provisions to change the structures, objectives and internal management of local government that have completely transformed not only the practice but also the image of local government (Newman, 2001; Stoker, 2004). Before considering aspects of the jostling of power relations between government and governance, and the agents of spatial policy making, it is appropriate firstly to outline how plan making and development planning activities themselves have undergone restructuring over the last eight years.

The recent focus on devolution in UK planning is not limited to, for example, a potential shift from one form of spatial planning to another. Planning is a contributor to and a reflection of a more fundamental reform of territorial management that aims to, inter alia, improve integration of different forms of spatial development activity, not least economic development. So, at one level, devolution and its implications for spatial planning must be analysed in respect of other aspects of New Labour's regional project, particularly the government's concern with business competitiveness. At another level, the current reforms which privilege regional scale policy interventions will inevitably require changes in the divisions of powers and responsibilities at local and national levels. In other words, devolution and decentralisation involve a major rescaling of both planning and spatial development, which is unfolding rapidly and unevenly across Britain.

The objective of this spatial transformation is to widen the trajectory of planning, or spatial strategy making, in the modernization and governance agendas at both the regional level and the local level within the UK. Although this is resulting in new participatory processes and greater contestation on the form and trajectory of regional and local spatial plan making, it is nevertheless occurring within a planning legislative framework that remains firmly rooted within the agenda of and context provided by the central and local government state and within professional planning duties. Here, the tension surrounding the form of spatial plan making occurs at the level of the state between the broader form of and relationship between governance and government. In essence, and the key question here is, when new styles of plan making occur within broader participatory governance, what happens to older styles of government working and professional practice, both of which remain *in situ* alongside new governance models but have been used to very different forms of working?

State fluidity and spatialization narratives

As the contributors to this book have pointed out, the current forms of government and institutional restructuring, and the emergence of new governance and partnership forms at formal administrative government levels and within the spaces between, is occurring in existing formalized geographical territories and transcending state and spatial boundaries. These processes are juxtaposed next to an intense bout of the ongoing modernization of government, the broadening out and overhaul of planning as an integrating and regulatory tool, and the state's rather sudden conversion to attempt to address the meaning of and activity within places, is providing a vast canvass for academic interpretation, political analysis, and professional introspection.

This book is an attempt to make sense of the changes, and to take stock of the shifting sands of spatial strategy making within and between formal levels of existing and emerging government and territories. Keating provides a thorough account for the historical development of the UK state and distinguishes between types of states, with or without political force, pointing to key variables that determine a state's existence, including the institutional identity, political entrepreneurship, and social and political mobilization. As Goodwin *et al.* discuss, a vital role for academic assessment during these changes is to uncover the political, social and cultural dynamics of institutions and also to consider not only what and how institutions are changing, but also the space between institutions. Only in this way can we develop, in the words of Jensen and Richardson, a 'progressive sense of place' and identify spaces of resistance. This suggests ongoing uncertainties with the form of government–governance and with the state at various levels, a process that Healey refers to as 'jerky', but also with the ability of spatial planning to embrace integration and become more than, to use Healey's description, 'a legalized conflict resolution process'. For academic perspectives, the state is more than anonymous institutions or bland strategies; the state, according to Goodwin *et al.* is a peopled organization, and spatial planning is well placed to consider the nature of places and how they function. Planning has, for far too long, been concerned with official spaces and fixated with (perhaps archaic) established boundaries but, echoing the work of both Massey (2005) and Sandercock (2004), this tends to miss most of what's going on in places, the stories they contain, the interactions between citizens, and with an associated failure to concentrate on space–place tensions.

As the case study illustrations depict from different territories around the UK, the last eight years or so have witnessed increasing attempts of existing and new forms of government to adopt new broader and participatory forms of spatial visioning that transcend, overlap or even replace established boundaries and silo mentalities. It is, as Deas rightly observes in his analysis of the North West region of England, a faltering process of governance that is as much about overcoming barriers, perceptions, and traditional institutional cultures, as it is about creating new sub-national agendas that relate to distinctive problems and distinctive places. This fluidity and duality in purpose suggests the existence of multiple identities within territories that need to be recognized and legitimated in the first instance. But for such identities and territories to form a bedrock upon which successful spatial strategy making can occur, there needs to be a distinction made between the forms of identities, strategy ownership and place-meanings, that may lead to alternative ways of conceptualizing problems. Deas refers to these processes as making a distinction between institutional and popular identities, a theme developed by Counsell and Haughton in their discussion of Yorkshire and Humberside. Alongside a necessity to provide coherence and policy direction within the individual plans of regions and sub-regions, the central state has taken on the mantle of an enforcing role towards integration between diverse, overlapping and out-of-sync strategies and stakeholders.

Harris and Hooper also refer to this tendency in the discussion of the Wales Spatial Plan that offers much in its style of planning, but since the strategy also performs the role of corporate document for the Welsh Assembly Government, its actual function rests somewhere between the political and the functional, where the role of spatial planning is to perform nothing more than a policy integration tool. With this evidence, one can question whether this style of spatial strategy making and planning is starting to address the real space–place tensions that spatial planning as a concept was meant to address. It is a point reflected upon by Ellis and Neill in the Northern Ireland context, where spatial planning is not yet being recognized or approved of by all those fragmented stakeholders who are responsible for delivery and implementation on the ground. They also question whether planners themselves are actually willing to address some fundamental issues associated with space–place tensions, including an ability to address contentious claims on territory and increasing diverse cultural identities of places. Is there not a danger that spatial planning and strategy making are becoming a symbolic but rather hollow style of language?

This suggests that spatial planning and strategy making could adopt multiple personas and be interpreted and utilized in different ways between different territories. Allmendinger, in his discussion of Scotland, for example, suggests that devolution and the embracing of spatial planning there has not materialized into anything distinctive, despite the obvious possibilities legally, politically, and socially, for a different type of practice to emerge north of the border. He suggests that the reasons for this lie in, what he calls, 'the drivers of divergence and convergence' that fill in the space once occupied by previous forms and exercises of the state. These may relate to the governing political power, the existence of national party politics and ideologies, a monolithic civil service, or even global economic forces outwith the territory.

In London, as Hall discusses, where a new metropolitan-wide spatial strategy is already in place, we see slightly different forces at work. Here, there is political commitment existing from a new tier of government, the mayor, to integrate various strategies and governance actors, with distinctive policies and recognition to ground the strategy in place–space concerns. But this optimism is, to some extent, handicapped by the lack of authority given to that political office by the traditional tiers of government at national and local level, and by an attempt to push economic competitiveness and social inclusion alongside a geographical balancing act for growth. In Goode and Munton's discussion of the attempts to formalize sustainable development agendas within London, similar concerns are raised, with devolved powers only able to be realized when they are acceptable to existing tiers of government or agencies of governance. Perhaps this is a hallmark of new forms of governance that require negotiation and agreement in order to stand any chance of acceptance and implementation. But we also know that the landscape of institutional government, and the mix of formal government and *ad hoc* governance structures, is one of an inequitable distribution of power, where some agencies can rely on the state for legitimacy, political support and resources and are more securely fixed than others.

Goode and Munton's discussion indirectly suggests that there may be four quite distinct stages in the bedding down of state restructuring and modernization as they are brought to bear on spatial strategy making. These comprise: first, a concerted move to institutionally restructure, that may take a considerable period of time and effort; second, there is the requirement for these new institutions and processes of governing to formulate their own strategies and vision statements, and to seek consultation, participation and approval for their contents; third, there is a necessity for the institutions and agencies to perform a joining-up or integrating role, across different strategies that may already exist, to ensure they are compatible and can be taken into account by other actors in the formulation of their own strategies; and fourth, there is the last and possibly elusive stage, to actually generate a visionary content to a strategy that meets place–space requirements but which may only be possible to achieve on a redrafting or revision of the document once the first three stages have been progressed and completed.

These stages of spatial strategy making exclude the various stages of bargaining, contention, implementation and monitoring that would be essential once a strategy was fit for purpose. And they may be compounded by the degree to which the strategies are able to stand alone or rest on certain agencies and the state for their survival and respect. These are themes taken up by Baker and Wong in their discussion of the regional housing growth problem in England where the enthusiasm for a spatial framework for housing allocation and development has been tempered by the rigidity and fixed boundaries of existing ways of working. Associated with this has been an inability on the part of central government to provide strategic policy without central direction, and thereby allow regional and local distinctiveness to emerge within the new processes. Vigar, too, in his narrative of regional transport planning in the North East of England talks of the problems caused by the panoply of strategies causing bureaucratic complexity, and problems being resolved by playing to the lowest common denominator to placate different vested interests, that in turn devalues the political drive and vision in spatial strategy making. Since strategists spend an inordinate amount of time striving for consistency and compatibility, and want to ensure progress in strategy formulation over time, the resulting strategies themselves may be overtly timid since they have avoided tackling problems boldly.

Cooke and Clifton comment upon this timidity and failure to realize spatial strategy distinctiveness in their discussion of the economic policies of Wales when compared to the other devolved Celtic countries since 1999. They point to the speed with which newly devolved governing processes initiate institutional restructuring and the wholesale effort concentrated on restructuring and reorganizing the departments and agencies of government rather than on unique policy directions. This relates to Allmendinger's point, previously referred to, about the actual degree of autonomy within new devolved and decentralized processes of the state to enact distinctive policy and bring about change. But, equally, it suggests that there may be failure in either the ability or determination on the part of politicians and strategy makers at the sub-national level, or that the new processes of government are 'over-institutionalised'. In order to be seen to be bringing about change and demonstrate their value, they will be prone to only enacting change that is firmly within the terms of possibility permitted by the central state in its design, monitoring, funding and working relationship with sub-government tiers.

The ability of devolved and decentralized institutions to demonstrate their worthiness, is also commented upon by Mac Ginty in his critical deconstruction of public attitudes toward the Northern Ireland Assembly. The form of devolution there was somewhat different to the rest of the UK but there are certain similarities in perceptions toward the new governing bodies, including a suspicion as to their value and role, and justification for the number of elected representatives within territorial boundaries while the central state discusses attempts to forward participatory, rather than representative, governance. He comments that the new forms of institutional processes are set within, what he refers to as, 'decidedly old notions of territory'. It has given rise to a form of re-territorialization, in that each of the Celtic nations have tended to look over their shoulders at what is happening institutionally and in terms of policy development, but whether this could be described as a search for unique policy solutions, or rather a blatant lack of territorial vision, is a debatable point.

Lloyd and Peel take up this latter point in their discussion of city-regionalism, and refer to what they call a 'new mercantilism of spatiality', where ideas of spatial strategy making and the use of spatial language and labels, are traded across territories, places and governments. In the case of city-regionalism, this is leading to the celebration of diversity and identity within metropolitan areas but is nevertheless firmly articulated through national spatial contexts and international flows and competitiveness.

The final section of the book deals with changing relationships between existing forms and tiers of government and modernization and decentralization processes, and a search for identity within strategy making. Musson *et al.*, in their study of decentralization, come to the conclusion that the new forms of English regionalization are merely serving to reinforce the centre, and that even where new strategies are emerging, they are not matched by the establishment of delivery mechanisms on the ground. This means that either a strategy can remain wishful, or else floats above formal processes of the state; if it lands within the formal institutions of government, it is prone to the traditional and fixed ways of implementation, the very type of process it was perhaps meant not to be; or else it relates to an issue of such strategic importance, that the state parachutes in new *ad hoc* processes to ensure delivery, for example, the development of the Thames Gateway (with the Prime Minister chairing a cabinet committee) or the Olympics 2012 development.

At the local level, Morphet highlights the notion of 'new localism' as another dimension of state restructuring and modernization that has been pursued by Labour since 2002. The concept of devolving power and resources away from central control and towards local democratic institutions and agents and local consumers and communities is one that provides both a mechanism for further pursuing the obsession with standards and value while countering the notion that Labour is centralizing. However, there are also strong practical justifications for this move as a way of addressing problems associated with managing and integrating the plethora of initiatives among a range of actors and institutions. According to some one of the reasons why successive Thatcherite administrations failed to implement a wide variety of initiatives aimed at local government was the excessive central control that failed to recognize or chose to ignore the discretion available to local managers and professionals in interpreting and implementing policy (see, for example, Marsh and Rhodes, 1992; Allmendinger and Thomas, 1998). Better involving local level groups and organizations can be an effective way of ensuring that policy is integrated and implemented.

The final three chapters of the book return to some of the more conceptual issues discussed at the outset. Raco, in his discussion of identity, devolution and governance, suggests that there is now a hybridism of differing ideas and practices emerging in different parts of the UK. He maintains that the state is fixed with the necessity to draw boundaries and set parameters, or limits, to governing, which is convenient and assists in attempting to embed new processes within particular scales. This also benefits a concern to fix identities for the purpose of governing and to seek a consensus to enable the electorate to interact with their political institutions, but it does little to resolve place–space tensions.

Similarly, MacLeod and Jones draw out differences in conceptualizing spaces and territories, and how an overt concentration on the politics of scalar structuration potentially misses the attributes and substance of places and their significance. They argue for a topological approach to find a conceptual focus alongside debates in rescaling order to capture, or at least appreciate, the looseness of exchanges, the way people occupy and use space, so as to transcend rigid and archaic boundaries, and various territories. It is an attractive proposition that many analysts are now considering, and one that ties in neatly to some of the discussion elsewhere in the book, relating to the need to consider the 'spaces between' institutions and indirectly to some of the frustrations and difficulties associated with the otherwise-optimistic new forms of spatial strategy making.

As the book has attempted to make sense of the restructuring, modernizing, decentralizing, and institutional fluidity of government and governance within these small islands, it is an interesting conclusion to suggest that we may need to start concentrating on the 'otherness' and 'outside', beyond the established parameters of existing fields of study, existing boundaries of the state and its scales of policy making, and the forms of spatial plan making as they are only now starting to emerge in various territories.

Conclusions

This chapter has sought to debate the new and emerging relationships between government and governance, between spatial strategy making and identity, and between fixed and fluid territories. Academic interest to date has been focused on what happens at the local, sub-regional and regional territorial scales when either the nation state permits decentralization and devolution to lower tiers, or when those tiers develop their own capacities – institutionally, territorially and politically – to claim responsibility for policy making. What seems to be lacking in these debates is an attempt to analyse two issues: first, what happens to traditional or pre-existing styles of government when those new governance processes and tiers are created; and second, what happens to the spaces outside government and governance – how do they create, produce and perform, and who inhabits them? This is where the existing theory is currently starting to emerge. Within the spatial planning field in particular, questions over the role of the nation state and its relationship to the decentralized tiers or informal spaces remain extremely pertinent fields of enquiry, since so much of planning as a social and economic activity remains 'directed and steered' by central government departments in the so-called national interest. A concern with continued economic growth and competition, the need for further housebuilding, the provision of new or replacement transport infrastructure, the protection of the best landscapes, and investment in information technology, are just some of the issues that the UK government regard as (its) priority policy areas. The difficulty the UK government possesses, however, is how to fulfil its election manifesto commitments to establish decentralized governance tiers while at the same time retaining some interest, or even control, over core policy areas that the nation state needs to concern itself with.

The creation and promotion of formal and informal institutions of governance at the regional, sub-regional and local level since 1997 has led to the welcomed reduction in the role of the nation state in determining the fortunes of sub-national territories. But such decentralization and devolution processes, structurally and institutionally, have been overlain by simultaneous policy-making transfers from the local and regional to the centre. This twin process of institutional decentralization and policy transfer has created a heady mix of different styles of government and governance across the UK territory that remain fluid and difficult to comprehend. The confusing array of styles of government and governance is resulting in continued jostling of power between existing forms of government and the newly emerging forms of governance. This power-tension does not necessarily occur between different agencies and different governmental officials; it is often the case that the same governmental personnel have the duty to deliver formal governmental requirements at the same time as mediating or participating as an actor within governance networks and partnerships. The confusion is also occurring, therefore, within agencies in addition to between them.

As many of the actors within governance arrangements are beginning to discover, the promotion of spatial planning and participatory governance within each scale is a goal worth pursuing, but it still requires a form of strategic direction to set some sort of political vision if agreement by lowest common denominator is to be avoided. That mediation and direction role seems to be being referred to traditional forms of government and professionalism, the very institutions that were characteristic of pre-governance styles of working. What governance within spatial strategy making is creating is greater discretion and autonomy within particular scales and across particular territories. When differences occur in reaching consensus between participants, actors – somewhat regrettably – look to more established and more legally certain agents to resolve their difficulties. Those agents are, predominantly, agencies of the nation state. Central government agencies may resent the desire for their involvement in processes that they have transferred to sub-national tiers. But equally, such involvement may serve a useful purpose in planning in order to provide a perspective reflective of the national interest with regards to the economy, transport and housing. Such practices may become more formalized over time as the nation state finds it increasingly difficult to let go of policy-making areas on subject areas that it has vested interests in.

What these difficulties suggest is the need to resolve the dilemma of ensuring enhanced 'self-determination' for sub-national territories alongside the need for strategic direction and political vision that extends not only within regions and localities, but also within and between Europe and the UK government. The institutional structures at the regional and local level, promoted and legitimized by the nation state for the most part, are leading to more bottom-up policy making and enhanced democratic participation in governance. But they may do so at the possible expense of more sub-national political vision and place-focused narratives that are developing unique responses appropriate for particular territories. Within this complex framework of the future government and governance processes within the UK, those officials responsible for its continued operation may attempt to make sense of the upheaval by becoming more concerned within inter- and intra-governmental power jostling, and attempts to stabilize, fix or control fluid governmental processes. If that is allowed to continue, the nation state will increasingly bear a heavy hand in designing (and, by implication, controlling) sub-national institutions, legitimizing and formalizing governance practices, and policing and monitoring emerging processes and strategies. The temptation may be too great for central government to withstand: by unleashing the governance beast, government will feel increasingly inclined to attempt to control it. Government does not cease to exist merely because new governance processes are created; government transforms too, shifts discretion from different parts of the process to others, and repositions itself within the institutional landscape. Such practices may seem inevitable in wider global restructuring and modernization processes, but these may invariably produce greater standardization of processes, structures and policies. Ironically, these are the regimes that were the very features of government that governance seemed to be designed to be an antithesis to.

References

Agnew, J. (1999) 'Mapping political power beyond state boundaries: territory, identity, and movement in world politics', *Millennium*, 28: 499–521.

Allmendinger, P. (2001) 'The head and the heart: national identity and urban planning in a devolved Scotland', *International Planning Studies*, 6(1): 33–54.

Allmendinger, P. (2003), 'From New Right to New Left in UK planning', *Urban Policy and Research*, 21(1): 57–79.

Allmendinger, P. and Tewdwr-Jones, M. (2000) 'New Labour, new planning? The trajectory of planning in post-New Right Britain', *Urban Studies*, 37(8): 1379–1402.

Allmendinger, P. and Thomas, H. (eds.) (1998) *Urban Planning and the British New Right*, London: Routledge.

Amin, A. (1996) 'Beyond associative governance', *New Political Economy*, 1(3) 309–33.

Amin, A. (1999) 'An institutionalist perspective on regional economic development', *International Journal of Urban and Regional Research*, 23: 365–78.

Amin, A. and Hausner, J. (eds.) (1997) 'Interactive governance and social complexity', in *Beyond Market and Hierarchy: Interactive Governance and Social Complexity*, Cheltenham: Edward Elgar: 1–31.

Amin, A., Massey, D. and Thrift, N. (2003) *Decentering the Nation: A Radical Approach to Regional Inequality*, London: Catalyst.

Amin, A. and Thrift, N. (1992) 'Neo-marshallian nodes in global networks', *International Journal of Urban and Regional Research*, 16: 571–87.

Amin, A. and Thrift, N. (1995) 'Globalisation, institutional "thickness" and the local economy', in P. Healey, S. Cameron, S. Davoudi, S. Graham and A. Mandani-Pour (eds) *Managing Cities: The New Urban Context*, Chichester: John Wiley: 91–108.

Berry, J., Brown, L. and McGreal, S. (2001) 'The planning system in Northern Ireland post-devolution', *European Planning Studies*, 9(6): 781–91.

Brenner, N. (2000) 'The urban question as a scale question: reflections on Henri Lefebvre, urban theory and the politics of scale', *International Journal of Urban and Regional Research*, 24: 361–78.

Cooke, P. (ed.) (1989) *Localities*, London: Unwin Hyman.

Counsell, D. and Haughton, G. (2003) 'Regional planning tensions: planning for economic growth and sustainable development in two contrasting English regions', *Environment and Planning C: Government and Policy*, 21(2): 225–39.

Counsell, D. Haughton, G. Allmendinger, P. and Vigar, G. (2003) 'From Land Use Plans to Spatial Development Strategies: New Directions in Strategic Planning in the UK', *Town and Country Plannng*, January: 15–19.

Haughton, G. and Counsell, D. (2004) *Regions, Spatial Strategies and Sustainable Development*, London: Routledge.

Healey, P. (1996) 'Consensus-building across difficult divides: new approaches to collaborative strategy making', *Planning Practice & Research*, 11(2): 207–16.

Healey, P. (1997) *Collaborative Planning*, Basingstoke: Macmillan.

Hirst, P. (1994) *Associative Democracy*, Cambridge: Polity Press.

Jensen, O.B. and Richardson, T. (2004) *Making European Space: Mobility, Power and Territorial Identity*, London: Routledge.

Jessop, B. (1990) *State Theory: Putting Capitalist States in Their Place*, Cambridge: Polity Press.

Jessop, B. (2000) 'The crisis of the national spatio-temporal fix and the tendential ecological dominance of globalizing capitalism', *International Journal of Urban and Regional Research*, 24(2): 323–59.

Jones, M. (2001) 'The rise of the regional state in economic governance: "partnerships for prosperity" or new scales of state power?', *Environment and Planning A*, 33: 1185–211.

Jones, R.A., Goodwin, M., Jones, M. and Simpson, G. (2004) 'Devolution, state personnel, and the production of new territories of governance in the United Kingdom', *Environment and Planning A*, 36(1): 89–109.

Keating, M. (1997) 'The invention of regions: political restructuring and territorial government in western Europe', *Environment and Planning C, Government and Policy*, 15: 383–98.

Lovering, J, (1997), 'Misleading and misreading Wales: the new regionalism', Papers in Planning Research 166, Department of City and Regional Planning, Cardiff University.

Lovering, J. (1999) 'Theory led by policy? The inadequacies of the "new regionalism" in economic geography illustrated from the case of Wales', *International Journal of Urban and Regional Research*, 23(2): 379–95.

MacLeod, G. (2001) 'New regionalism reconsidered: globalization and the remaking of political economic space', *International Journal of Urban and Regional Research*, 25(4): 804–29.

MacLeod G. and Goodwin M. (1999) 'Space, scale and state strategy: rethinking urban and regional governance', *Progress in Human Geography*, 23(4): 503–27.

Marsh, D. and Rhodes, S. (eds.) (1992) *Policy Networks in British Government*, Oxford: Oxford University Press.

Massey, D. (2005) *For Space*, London: Sage.

Morphet, J. (2004) 'Integration in planning: a scoping paper for the RTPI', Paper given to the Planning and Development Research Conference, University of Aberdeen.

Murdoch, J. and Tewdwr-Jones, M. (1999) 'Planning and the English regions: conflict and convergence amongst the institutions of regional governance', *Environment and Planning C: Government and Policy*, 17(6): 715–29.

Newman, J. (2001) *Modernising Governance: New Labour, Policy and Society*, London: Sage.

Phelps, N.A. and Tewdwr-Jones, M. (2000) 'Scratching the surface of collaborative and associative governance: identifying the diversity of social action in institutional capacity building', *Environment and Planning A*, 32(1): 111–30.

Phelps, N.A. and Tewdwr-Jones, M. (2004) 'Institutions, collaborative governance and the diversity of social action', in A. Wood and D. Valler (eds.) *Governing Local and Regional Economies: Institutions, Politics and Economic Development*, Aldershot, UK: Ashgate: 93–120.

Raco, M. (2000) 'Assessing community participation in local economic development: lessons for the new urban policy', *Political Geography*, 19: 573–99.

Sandercock, L. (2004) *Cosmopolis II: Mongrel Cities*, London: Continuum.

Scott, A.J. (1983) 'Industrial organisation and the logic of intra-metropolitan location, I: theoretical considerations', *Economic Geography*, 59: 233–50.

Scott, A.J. (1986) 'Industrial organisation and location: division of labour, the firm and spatial process', *Economic Geography*, 62: 215–31.

Stoker, G. (2004) *Transforming Local Governance: From Thatcher to New Labour*, Basingstoke: Palgrave.

Storper, M. (1997) *The Regional World: Territorial Development in a Global Economy*, London: Guilford.

Tewdwr-Jones, M. (2002) *The Planning Polity: Planning, Government and the Policy Process*, London: Routledge.

Tewdwr-Jones, M. and Williams, R.H. (2001) *The European Dimension of British Planning*, London: Spon Press.

Tomaney, J. (2000) 'End of the Empire State? New Labour and devolution in the United Kingdom', *International Journal of Urban and Regional Research*, 24(3): 677–90.

Tomaney, J. (2002) 'The evolution of regionalism in England', *Regional Studies*, 36(7): 721–31.

Vigar, G., Healey, P., Hull, A. and Davoudi, S. (2000) *Planning, Governance and Spatial Strategy in Britain: An Institutionalist Analysis*, London: Macmillan.

Index